METHODS IN MOLECULAR BIOLOGY

Series Editor
John M. Walker
School of Life and Medical Sciences
University of Hertfordshire
Hatfield, Hertfordshire, AL10 9AB, UK

For further volumes:
http://www.springer.com/series/7651

Protein-Carbohydrate Interactions

Methods and Protocols

Edited by

D. Wade Abbott

Agriculture and Agri-Food Canada, Lethbridge, AB, Canada

Alicia Lammerts van Bueren

Biotechnology Institute, University of Groningen, Groningen, The Netherlands

Humana Press

Editors
D. Wade Abbott
Agriculture and Agri-Food Canada
Lethbridge, AB, Canada

Alicia Lammerts van Bueren
Biotechnology Institute
University of Groningen
Groningen, The Netherlands

ISSN 1064-3745 ISSN 1940-6029 (electronic)
Methods in Molecular Biology
ISBN 978-1-4939-6898-5 ISBN 978-1-4939-6899-2 (eBook)
DOI 10.1007/978-1-4939-6899-2

Library of Congress Control Number: 2017933839

Printed on acid-free paper

This Humana Press imprint is published by Springer Nature
The registered company is Springer Science+Business Media LLC
The registered company address is: 233 Spring Street, New York, NY 10013, U.S.A.

Preface

Protein-carbohydrate interactions are involved in diverse processes required for life, including the microbial degradation of plant biomass and marine polysaccharides and human health and nutrition. Understanding and predicting how carbohydrates are recognized and modified by carbohydrate-active enzymes (i.e., CAZymes) therefore is an important area of basic research that spans multiple disciplines and holds vast promise for informing future innovations in renewable resource utilization and medicine. Since the turn of the millennia, the field of protein-carbohydrate interactions has been transformed by high-throughput and ultrasensitive instrumentation, which has enabled us to study complex carbohydrate utilization systems at the levels of metagenomes, metatranscriptomes, and metaproteomes. This increase in technology has opened new doors for CAZyme discovery and application. Here within we will provide a wide-ranging resource for studying protein-carbohydrate interactions that extends from traditional biochemical methods to state-of-the-art techniques, both of which will continue to propel the field forward in the coming years. In particular, this volume will focus on four different research themes.

Part I describes methods for screening and quantifying CAZyme activity. These chapters will survey each class of CAZyme, including glycoside hydrolases (Chap. 1, Copper-Bicinchoninic Acid; Chap. 2, High-Performance Anion-Exchange Chromatography; and Chap. 3, 3,5-Dinitrosalicylic Acid Assays), polysaccharide lyases (Chap. 4), carbohydrate esterases (Chap. 5), glycosyltransferases (Chap. 6), and lytic polysaccharide monooxygenases (Chap. 7). In addition, a method for investigating carbohydrate depolymerization by cellulosomes, which can contain multiple enzyme classes and activities (Chap. 8), is provided.

Part II contains methods for investigating the interactions between proteins and carbohydrate ligands. These techniques include affinity gel electrophoresis of catalytic modules (Chap. 9), microscale thermophoresis (Chap. 10), and NMR spectroscopy (Chap. 11). The final chapter in this section describes current methods for detecting the biomechanical activity of expansions (Chap. 12), a class of proteins involved in the loosening of plant cell wall networks.

Part III discusses methods for the visualization of carbohydrates and protein-carbohydrate complexes. These chapters include a novel bioinspired plant cell wall assembly for measuring protein interactions by fluorescence (Chap. 13), using carbohydrate-binding modules as probes within plant cell walls (Chap. 14), and investigating the subcellular localization of CAZymes within Gram-negative bacteria (Chap. 15). These are followed by three different methods for investigating carbohydrate structure. First is a method for using Fourier transform mid-infrared spectroscopy to characterize the composition of plant cell walls (Chap. 16); this is followed by methods for studying fluorescent glycans by electrophoresis (Chap. 17) and capillary electrophoresis (Chap. 18).

Finally, Part IV focuses on structural and "omics" approaches for studying systems of CAZymes. First, a "dissect and build" approach for determining multimodular CAZyme structure involving combinatorial small-angle X-ray scattering and X-ray crystallography is described (Chap. 19) followed by methods describing the development of "omics" techniques to identifying novel CAZyme systems using metagenomics (Chap. 20), transcriptomics (Chap. 21), and proteomics (Chapter 22) approaches.

We anticipate that this collection of methods for studying carbohydrate modification and protein-carbohydrate interactions will be a valuable resource to the glycomics research community. As the field continues to advance, methods included within this volume will have utility for illuminating the biology of glycomics, driving biotechnological innovation, and developing solutions for human health and for sustainable resources within the emerging green economy.

Lethbridge, AB, Canada D. Wade Abbott
Groningen, The Netherlands Alicia Lammerts van Bueren

Contents

Contributors

D. Wade Abbott • *Functional Genomics of Complex Carbohydrate Utilization, Lethbridge Research and Development Centre, Agriculture and Agri-Food Canada, Lethbridge, AB, Canada; Department of Chemistry and Biochemistry, University of Lethbridge, Lethbridge, AB, Canada*

Jane Wittrup Agger • *Center for BioProcess Engineering, Department of Chemical and Biochemical Engineering, Technical University of Denmark, Lyngby, Denmark*

Gregory Arnal • *Michael Smith Laboratories, University of British Columbia, Vancouver, BC, Canada*

Magnus Ø. Arntzen • *Department of Chemistry, Biotechnology and Food Science, Norwegian University of Life Sciences, Ås Akershus, Norway*

Lior Artzi • *Faculty of Biochemistry, Department of Biomolecular Sciences, The Weizmann Institute of Science, Rehovot, Israel*

Jathavan Asohan • *Michael Smith Laboratories, University of British Columbia, Vancouver, BC, Canada*

Mohamed A. Attia • *Michael Smith Laboratories, University of British Columbia, Vancouver, BC, Canada; Department of Chemistry, University of British Columbia, Vancouver, BC, Canada*

Ajay Badhan • *Lethbridge Research and Development Centre, Agricultures and Agri-Food Canada, Lethbridge, AB, Canada*

Louise Badruna • *LISBP, Université de Toulouse, CNRS, INRA, INSA, Toulouse, France*

Edward A. Bayer • *Faculty of Biochemistry, Department of Biomolecular Sciences, The Weizmann Institute of Science, Rehovot, Israel*

Alisdair B. Boraston • *Department of Biochemistry and Microbiology, University of Victoria, Victoria, BC, USA*

Andreas Bremges • *Computational Biology of Infection Research, Helmholtz Centre for Infection Research, Braunschweig, Germany; German Center for Infection Research (DZIF), Braunschweig, Germany*

Harry Brumer • *Michael Smith Laboratories, University of British Columbia, Vancouver, BC, Canada; Department of Chemistry, University of British Columbia, Vancouver, BC, Canada; Department of Biochemistry and Molecular Biology, University of British Columbia, Vancouver, BC, Canada*

Vincent Burlat • *Laboratoire de Recherche en Sciences Végétales, UMR 5546 UPS/CNRS, Castanet-Tolosan, France*

Darrell Cockburn • *Department of Microbiology and Immunology, University of Michigan, Ann Arbor, MI, USA*

Daniel J. Cosgrove • *Department of Biology, Pennsylvania State University, University Park, PA, USA*

Mirjam Czjzek • *UPMC Univ. Paris 06, CNRS, UMR 8227, Integrative Biology of Marine Models, Sorbonne Universite, Roscoff, Bretagne, France*

Hayden J. Danyluk • *Simon Fraser University, Department of Molecular Biology and Biochemistry, Burnaby, BC, Canada*

Claire Dumon • *LISBP, Université de Toulouse, CNRS, INRA, INSA, Toulouse, France*
Daniel M. Durachko • *Department of Biology, Pennsylvania State University, University Park, PA, USA*
Vincent G.H. Eijsink • *Department of Chemistry, Biotechnology and Food Science, Norwegian University of Life Sciences, Ås Akershus, Norway*
Elizabeth Ficko-Blean • *UPMC Univ. Paris 06, CNRS, UMR 8227, Integrative Biology of Marine Models, Sorbonne Universite, Roscoff, Bretagne, France*
Carlos M.G.A. Fontes • *CIISA – Faculdade de Medicina Veterinária, Universidade de Lisboa, Lisbon, Portugal*
Robert J. Forster • *Lethbridge Research and Development Centre, Agriculture and Agri-Food Canada, Lethbridge, AB, Canada*
Julie M. Grondin • *Lethbridge Research Center, Agriculture and Agri-Food Canada, Lethbridge, AB, Canada*
Robert J. Gruninger • *Lethbridge Research and Developmental Centre, Agriculture and Agri-Food Canada, Lethbridge, AB, Canada*
Nathan K. Hepler • *Department of Biology, Pennsylvania State University, University Park, PA, USA*
Joanne K. Hobbs • *Department of Biochemistry and Microbiology, University of Victoria, Victoria, BC, Canada*
G. Douglas Inglis • *Department of Chemistry and Biochemistry, University of Lethbridge, Lethbridge, AB, Canada*
Darryl R. Jones • *Functional Genomics of Complex Carbohydrate Utilization, Lethbridge Research and Development Centre, Agriculture and Agri-Food Canada, Lethbridge, AB, Canada*
Benoit J. Kunath • *Department of Chemistry, Biotechnology and Food Science, Norwegian University of Life Sciences, Ås, Norway*
David N. Langelaan • *Department of Biochemistry and Molecular Biology, Dalhousie University, Halifax, NS, Canada*
Galina Mai-Gisondi • *Department of Bioproducts and Biosystems, Aalto University, Espoo, Aalto, Finland*
Emma R. Master • *Department of Chemical Engineering and Applied Chemistry, University of Toronto, Toronto, ON, Canada*
Tim A. McAllister • *Lethbridge Research and Development Centre, Agriculture and Agri-Food Canada, Lethbridge, AB, Canada*
Nicholas McGregor • *Michael Smith Laboratories, University of British Columbia, Vancouver, BC, Canada; Department of Chemistry, University of British Columbia, Vancouver, BC, Canada*
Alice C. McHardy • *Computational Biology of Infection Research, Helmholtz Centre for Infection Research, Braunschweig, Germany*
Lauren S. McKee • *Division of Glycoscience, School of Biotechnology, KTH, Royal Institute of Technology, AlbaNova University Centre, Stockholm, Sweden*
Richard McLean • *Functional Genomics of Complex Carbohydrate Utilization, Lethbridge Research and Development Centre, Agriculture and Agri-Food Canada, Lethbridge, AB, Canada; Department of Chemistry and Biochemistry, University of Lethbridge, Lethbridge, AB, Canada*
Cédric Y. Montanier • *LISBP, Université de Toulouse, CNRS, INRA, INSA, Toulouse, France*

SARAH MORAÏS • *Faculty of Biochemistry, Department of Biomolecular Sciences, The Weizmann Institute of Science, Rehovot, Israel*
STEVEN C. MOSIMANN • *Department of Chemistry and Biochemistry, University of Lethbridge, Lethbridge, AB, Canada*
MARIA NGO • *Department of Chemistry and Biochemistry, Wilfrid Laurier University, Waterloo, ON, Canada*
GABRIEL PAËS • *FARE laboratory, INRA, University of Reims Champagne-Ardenne, Reims, France*
PHILLIP P. POPE • *Department of Chemistry, Biotechnology and Food Science, Norwegian University of Life Sciences, Ås, Norway*
IAN REID • *Centre for Structural and Functional Genomics, Concordia University, Montreal, QC, Canada*
MÉLISSA ROBB • *Department of Biochemistry and Microbiology, University of Victoria, Victoria, BC, Canada*
LEONA K. SHUM • *Department of Chemistry, The University of British Columbia, Okanagan, Kelowna, BC, Canada*
STEVEN P. SMITH • *Department of Biomedical and Molecular Sciences, Queen's University, Kingston, ON, Canada*
JOHANNA STERN • *Faculty of Biochemistry, Department of Biomolecular Sciences, The Weizmann Institute of Science, Rehovot, Israel*
MICHAEL D.L. SUITS • *Department of Chemistry and Biochemistry, Wilfrid Laurier University, Waterloo, ON, Canada*
BIRTE SVENSSON • *Enzyme and Protein Chemistry, Department of Systems Biology, Technical University of Denmark, Kongens Lyngby, Denmark*
ADRIAN TSANG • *Centre for Structural and Functional Genomics, Concordia University, Montreal, QC, Canada*
RICHARD R.E. UWIERA • *Department of Agricultural, Food and Nutritional Sciences, University of Alberta, Edmonton, AB, Canada*
GUSTAV VAAJE-KOLSTAD • *Department of Chemistry, Biotechnology and Food Science, Norwegian University of Life Sciences, Ås Akershus, Norway*
EDWARD R. WAGNER • *Department of Biology, Pennsylvania State University, University Park, PA, USA*
YUXI WANG • *Lethbridge Research and Development Centre, Agriculture and Agri-Food Canada, Lethbridge, AB, Canada*
AARON WEIMANN • *Computational Biology of Infection Research, Helmholtz Centre for Infection Research, Braunschweig, Germany*
BJØRGE WESTERENG • *Department of Chemistry, Biotechnology and Food Science, Norwegian University of Life Sciences, Ås Akershus, Norway*
CASPER WILKENS • *Department of Chemical and Biochemical Engineering, Technical University of Denmark, Kongens Lyngby, Denmark*
HAIYANG WU • *LISBP, Université de Toulouse, CNRS, INRA, INSA, Toulouse, France*
WESLEY F. ZANDBERG • *Department of Chemistry, The University of British Columbia, Okanagan, Kelowna, BC, Canada; Department of Chemistry, Kelowna, BC, Canada*

Part I

Analysis of Carbohydrate-Active Enzyme Activity

Chapter 1

A Low-Volume, Parallel Copper-Bicinchoninic Acid (BCA) Assay for Glycoside Hydrolases

Gregory Arnal, Mohamed A. Attia, Jathavan Asohan, and Harry Brumer

Abstract

The quantitation of liberated reducing sugars by the copper-bicinchoninic acid (BCA) assay provides a highly sensitive method for the measurement of glycoside hydrolase (GH) activity, particularly on soluble polysaccharide substrates. Here, we describe a straightforward method adapted to low-volume polymerase chain reaction (PCR) tubes which enables the rapid, parallel determination of GH kinetics in applications ranging from initial activity screening and assay optimization, to precise Michaelis–Menten analysis.

Key words Glycoside hydrolase (GH), Glycosidase, Carbohydrate-active enzymes (CAZymes), Copper-bicinchoninic acid (BCA), Polysaccharide, Enzymology, Reducing sugar

1 Introduction

The cleavage of a glycosidic bond in an oligo- or polysaccharide by a glycoside hydrolase (glycosidase, EC 3.2.1) results in the generation of a new hemiacetal chain end, the aldehyde form of which can be oxidized with the concomitant reduction of metal ions to lower oxidation states. This reaction is the basis for the classic silver-based Tollen's and copper-based Fehling's qualitative tests for "reducing sugars" [1]. In particular, these free "reducing ends" readily convert cupric (Cu^{2+}) to cuprous (Cu^{1+}) ions, the latter of which can be bis-chelated by disodium 2,2′-bicinchoninic acid (BCA) to yield a stable, intense purple complex (λ_{max} 562 nm). The formation of the Cu^{1+}-BCA complex was first used for carbohydrate detection following chromatography [2], an application that in fact predates its widespread use for protein quantitation [3]. Subsequent development of the copper-BCA assay for carbohydrates included quantitative analysis of reducing monosaccharides and oligosaccharides with up to nanomole sensitivity [4–7], application to glycoside hydrolase (GH) activity assays [8], and miniaturization in a micro-well plate format [9–11]. In these applications, the unique accuracy and sensitivity of the BCA assay *vis-à-vis* other

D. Wade Abbott and Alicia Lammerts van Bueren (eds.), *Protein-Carbohydrate Interactions: Methods and Protocols*, Methods in Molecular Biology, vol. 1588, DOI 10.1007/978-1-4939-6899-2_1, © Springer Science+Business Media LLC 2017

reducing-end methods (e.g., Nelson-Somogyi, dinitrosalicylate) has been underscored [7, 8].

In this chapter, we describe the further adaptation of a miniaturized, parallel BCA glycoside hydrolase assay to standard polymerase chain reaction (PCR) tubes. This enables the use of widely available PCR thermocyclers/thermoblocks for temperature control for both enzymatic reactions and color development steps. Additionally, the use of capped PCR tubes in strip format facilitates sample handling. Furthermore, the intrinsic tight seal of PCR tubes (which are designed for high-temperature use), together with heated thermocycler/thermoblock lids, significantly minimizes errors due to sample evaporation from small volumes, thereby improving assay precision. This method is directly applicable for most standard biochemistry and molecular biology laboratories, and does not inherently require advanced robotics for implementation.

2 Materials

All solutions should be prepared using ultrapure water with a resistivity of 18 MΩ-cm at 25 °C.

2.1 BCA Reagent

1. Analytical balance.
2. Magnetic stirrer and stir bars.
3. 100 mL graduated cylinder.
4. Solution A (54.28 g/L Na_2CO_3, 24.20 g/L $NaHCO_3$, and 1.94 g/L disodium 2,2′-bicinchoninic acid). Weigh 2.714 g (25.6 mmol) of Na_2CO_3, 1.210 g (14.4 mmol) of $NaHCO_3$, and 0.097 g (0.25 mmol) of bicinchoninic acid disodium salt hydrate. Dissolve by stirring in ca. 25 mL of ultrapure water in a graduated cylinder and dilute to a final volume of 50 mL. This solution is stable for 1 month when stored at room temperature in the dark.
5. Solution B (1.25 g/L $CuSO_4{:}5H_2O$, 1.26 g/L L-serine). Weigh 0.062 g (0.25 mmol) of $CuSO_4.5H_2O$ and 0.063 g (0.60 mmol) of L-Serine. Dissolve by stirring in ca. 25 mL of ultrapure water in a graduated cylinder and dilute to a final volume of 50 mL. This solution is stable for 1 month when stored at 4 °C in the dark.
6. The BCA working reagent is prepared daily by mixing equal volumes of Solution A and Solution B (1:1 ratio), which results in a light blue solution.

2.2 Carbohydrate Solutions

1. Analytical balance.
2. Vortex.

3. Beakers or Erlenmeyer flasks.
4. 1.7 mL plastic microcentrifuge tubes.
5. 15 mL plastic centrifuge tube.
6. In a 15 mL conical tube, prepare a fresh 10 mM D-glucose solution by weighing 18.01 mg of D-glucose and dissolving in 10 mL of ultrapure water (*see* **Note 1**).
7. Prepare carbohydrate substrate stock solutions at 5 mg/mL, or the highest practical concentration, by dissolving in ultrapure water in a beaker or flask (*see* **Notes 2** and **3**).

2.3 Buffer and Enzyme Solutions

1. Prepare all buffers according to standard protocols [12] at ten times (10×) the final concentration desired in the assay.
2. Enzyme stock solutions and suitable dilutions should be prepared following best practices for protein handling [13].

2.4 Apparatus for Performing Assays

1. Strips of eight thin-walled PCR tubes (200 μL volume).
2. Two multichannel reagent reservoirs (*see* **Note 4**).
3. 1–10, 10–100, and 100–1000 μL micropipettes and disposable plastic tips.
4. 1–10 and 10–300 μL multichannel pipettes.
5. PCR thermocycler or thermoblock for PCR tubes, capable of accurately maintaining a wide range of temperatures (4–100 °C). An Eppendorf Thermomixer C equipped with the ThermoTop is especially recommended; PCR tubes are completely encompassed by the corresponding block, which guarantees a homogeneous temperature distribution in 200 μL reactions, while the heated ThermoTop efficiently prevents assay solution condensation on tube lids.
6. Benchtop centrifuge equipped with a PCR tube rotor.
7. Flat-bottom polystyrene microplate suitable for absorbance measurements.
8. Microplate reader suitable for A_{562} measurements.

3 Methods

Figure 1 provides a schematic overview of an assay designed for a thermocycler/thermoblock with a 6 × 8 heating block, which includes enzyme assays in triplicate, blank assays in duplicate, and one series of reducing-sugar standards. The number of replicates of each may be expanded depending upon available thermal equipment. Optimization of enzyme assays to generate accurate, publication-quality data is necessarily an iterative procedure involving determination of the optimal substrate(s), enzyme concentration, buffer and

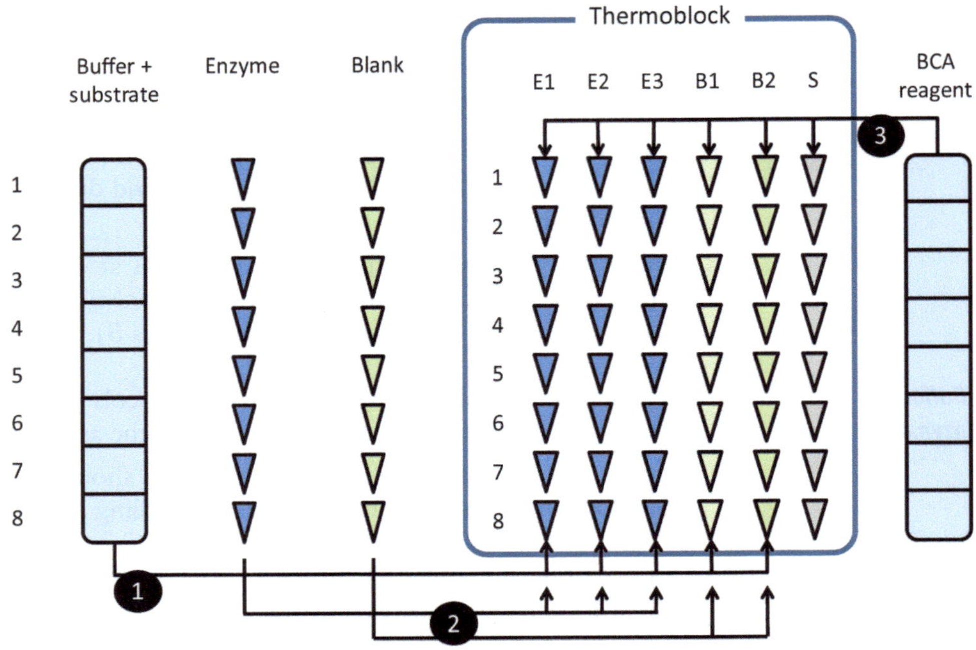

Fig. 1 Assay scheme. A 6 × 8 tube layout is shown that corresponds to common PCR thermocycler/thermoblock format, in which up to eight different conditions can be tested in parallel. A solution of substrate in the assay buffer and the BCA working reagent are distributed into multichannel reservoirs; enzyme in buffer and the corresponding buffer blanks are distributed into 8-PCR tube strips. These different solutions are subsequently distributed using multichannel pipettes in the following steps. *Step 1*: Substrate (columns E1, E2, E3, B1, B2) and carbohydrate standard (column S) solutions in the assay buffer are distributed and equilibrated at the assay temperature in a PCR thermocycler/thermoblock. *Step 2*: Assays are initiated by adding the enzyme solutions to the tubes in columns E1, E2, and E3; blank (buffer only) solutions are added to the columns B1 and B2. *Step 3*: Reactions are stopped by the addition of BCA working reagent and placement on ice, prior to readjusting the block temperature to 80 °C for color development

pH value, and temperature [14]. As such, we provide here a general assay protocol, in which individual parameters may be varied (Fig. 2).

3.1 Thermally Equilibrate Thermocycler/Thermoblock

1. Set the thermocycler/thermoblock temperature to the desired assay temperature (*see* **Note 5**).

3.2 Preparation of the BCA Reagent Reservoir

1. Prepare the multichannel reservoir by distributing at least 700 μL of freshly prepared BCA reagent in the wells 1–8 (*see* **Note 4**).
2. Keep the reservoir on ice until use.

3.3 Preparation of Carbohydrate Standard Solutions

Standard solutions are freshly prepared by dilution to yield a series of 8 carbohydrate (e.g., D-glucose) standard solutions of 0–125 μM (*see* **Note 1**).

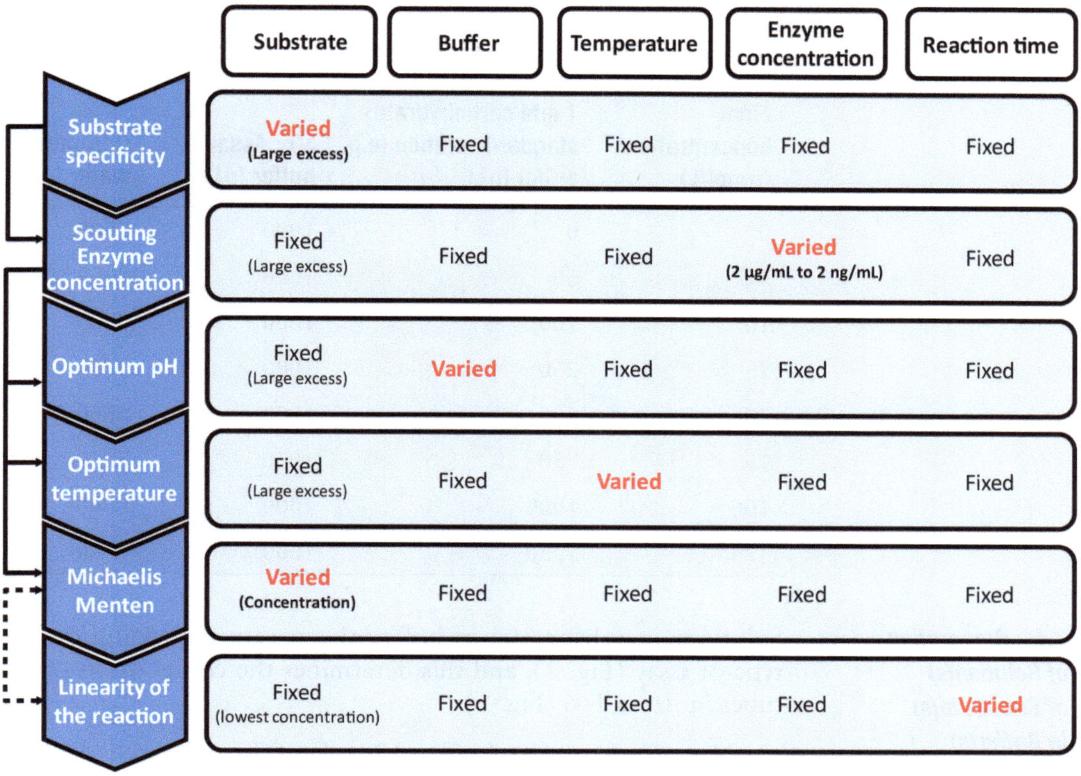

Fig. 2 Examples of various glycoside hydrolase (GH) operational assays that can be performed using the method described here, with the fixed and varied parameters indicated for each. Kinetic analysis of previously uncharacterised enzymes will involve iteration of these assays to determine optimal assay parameters. Typically, GH characterization begins with substrate screening to determine enzyme specificity. Consequently, enzyme concentration is scouted to assure that the reaction is linear over a convenient assay time (e.g., 10–30 min), and that the amount of reducing ends released falls within the standard curve (0–125 µM). Optimal pH and temperature can then be determined using a suitable enzyme concentration (e.g., one that gives A_{562} of ca. 0.8). Finally, Michaelis–Menten analysis can be performed by varying substrate concentration (e.g., 0.1 K_M to 10 K_M), taking special care to ensure that assay linearity is maintained at the lowest substrate concentration [14]

1. Dilute the carbohydrate standard stock solution (10 mM) by a factor of 10 by adding 9 mL of ultrapure water 1 mL of stock solution to, resulting in a 1 mM carbohydrate solution.
2. Dilute the 1 mM carbohydrate standard solution in the assay buffer as indicated in Table 1 to obtain 8 different glucose solutions at different concentrations (0–125 µM) (*see* **Note 6**).
3. In an 8-PCR-tube-strip (Fig. 1, "S"), distribute 100 µL of each solution into individual tubes.
4. Keep the standards on ice until use.

Table 1
Preparation of carbohydrate (e.g., D-glucose) standard solutions

Final concentration (μmol/L)	1 mM carbohydrate standard solution (e.g., D-Glc) (μL)	10× Assay buffer (μL)	Ultrapure water (μL)
0	0	1000	9000
5	50	1000	8950
10	100	1000	8900
25	250	1000	8750
50	500	1000	8500
75	750	1000	8250
100	1000	1000	8000
125	1250	1000	7750

3.4 Preparation of Solution(s) of Substrate(s) in Buffer(s)

Solutions of substrate(s) in buffer(s) can vary depending on the type of assay (Fig. 2), and this determines the composition of the tubes in Rows 1–8 (Fig. 1).

1. Mix 60 μL of 10× buffer, 120 μL of 5× polysaccharide (*see* **Note 7**), and 360 μL of ultrapure water in Row 1 of the multichannel reagent reservoir (*see* **Note 4**). This 540 μL total volume is sufficient for 6 assays (3 enzyme assay replicates and 2 blanks, with additional volume to allow accurate pipetting).
2. Repeat the step above for the remaining wells of the multichannel reservoir to set the conditions of the seven other assays, varying maximally one solution parameter (Fig. 2, e.g. substrate or buffer concentration, or type).

3.5 Preparation of Enzyme and Blank Solutions

The preparation of enzyme and blank solutions can vary depending on the type of assay (Fig. 2).

1. Prepare a working enzyme solution in Row 1 of the "Enzyme" PCR-tube strip (Fig. 1) by dilution of the enzyme stock solution in 1× assay buffer (e.g., 50 mM) containing 0.1 mg/mL of BSA (*see* **Note 8**) to a final volume of at least 40 μL (sufficient volume for three replicates plus additional volume to allow accurate pipetting). This working enzyme solution will be further diluted an additional factor of 10 in the final assay solution (vide infra).
2. Repeat the step above for the remainder of the tubes in "Enzyme" strip, unless performing an enzyme concentration scouting experiment by serial dilution (Fig. 2).
3. Prepare the blanks in a PCR-tube strip (*see* Fig. 1, "Blank") by filling each tube with at least 30 μL of 1× assay buffer (e.g., 50 mM) containing 0.1 mg/mL of BSA (*see* **Note 9**).

3.6 Performing the Assay

1. In a thermocycler/thermoblock, organize five empty strips of eight PCR tubes (corresponding to columns E1, E2, E3, B1, and B2 as indicated in Fig. 1) and include the strip of standard solutions "S" that was prepared in Subheading 3.3.
2. Using a multichannel pipette, transfer 90 μL of the substrate in buffer solution(s) into Strips E1, E2, E3, B1, and B2.
3. Equilibrate the reaction mixtures at the set assay temperature in the thermocycler/thermoblock for 10 min.
4. Ca. 20 s before starting the reaction, open the thermocycler/thermoblock lid and *carefully* open the E1 PCR tube lids.
5. Start the reactions in E1 by adding 10 μL of the enzyme solution(s) from the "Enzyme" PCR-tube strip (Fig. 1) using a multichannel pipette. Mix thoroughly by reciprocal pipetting (three times; be consistent in mixing method and time). Close the lids of PCR tubes and thermocycler/thermoblock.
6. Maintaining a strict 1-min time interval between enzyme additions, repeat **step 5** for Strips E2 and E3.
7. Again maintaining a strict 1-min time interval between additions, add the "Blank" solution from the corresponding PCR-tube strip (Fig. 1) to Strips B1 and B2.
8. Ca. 20 s before the end of the desired assay duration (e.g., 10 min since the addition of enzyme to Strip E1; *see* **Note 10**) prepare to stop the reactions in Strip E1 by opening the thermocycler/thermoblock lid and the PCR tube lids (carefully).
9. Precisely at the end of the assay duration, stop the reactions in Strip E1 by adding 100 μL of BCA reagent with a multichannel pipette and mix by reciprocal pipetting 2 times (be consistent in time). Place the reactions on ice to prevent immediate color development (*see* **Note 11**). Close the lid of the thermocycler/thermoblock.
10. Maintaining strict 1-min time intervals, repeat **step 9** for Strips E2, E3, B1, B2, and S.
11. Develop the color of the copper-BCA complex solution as described below.

3.7 Copper-BCA Color Development

1. Equilibrate the thermocycler/thermoblock at 80 °C.
2. Incubate Strips E1, E2, E3, B1, B2, and S1 (Fig. 1) simultaneously at 80 °C for 20 min.
3. Cool all tubes on ice (*see* **Note 12**).
4. Briefly centrifuge the tubes and transfer 190 μL of the solution to a flat-bottom polystyrene microplate using a multichannel pipette.
5. Measure A_{562} using a microplate reader that has been zeroed versus 190 μL ultrapure water.

3.8 Calculations

3.8.1 Standard Curve

1. Subtract the A_{562} value of the blank (0 μM carbohydrate) from all the A_{562} values of the carbohydrate standard solutions.
2. Plot the background-corrected absorbance values versus carbohydrate concentration.
3. Use linear regression to obtain the best fit line through all points. The intercept of the best-fit line should be zero (or negligible) and hence the slope (m_{std}) will be used to directly determine the concentrations of reducing ends generated in the enzyme assays (vide infra) (Fig. 3a).

3.8.2 Calculation of Specific Activity

1. Calculate the average A_{562} values of the enzyme replicates in Strips E1, E2, and E3; $A_{562(\text{enzyme})}$.
2. Calculate the average A_{562} values of the blanks in both series, Strips B1 and B2; $A_{562(\text{blank})}$.
3. Subtract the averaged enzyme and blank A_{562} values and divide by the slope of carbohydrate standard curve (*see* Subheading 3.8.1, **step 3**) to obtain the concentration of reducing ends in each assay in μM (*see* **Note 13**).

$$[reducing\ ends](\mu M) = \frac{A_{562(enzyme)} - A_{562(blank)}}{m_{std}\left(A_{562} \cdot \mu M^{-1}\right)}$$

4. Convert concentration values into μmole values by multiplying by the assay volume, 10^{-4} L (≡ 100 μL).

$$reducing\ ends(\mu mol) = [reducing\ ends](\mu M) \times 10^{-4}\,\text{L}$$

5. Divide the amount of product formed (μmol) by the assay time to yield activity values (μmol/min). If care was taken to ensure

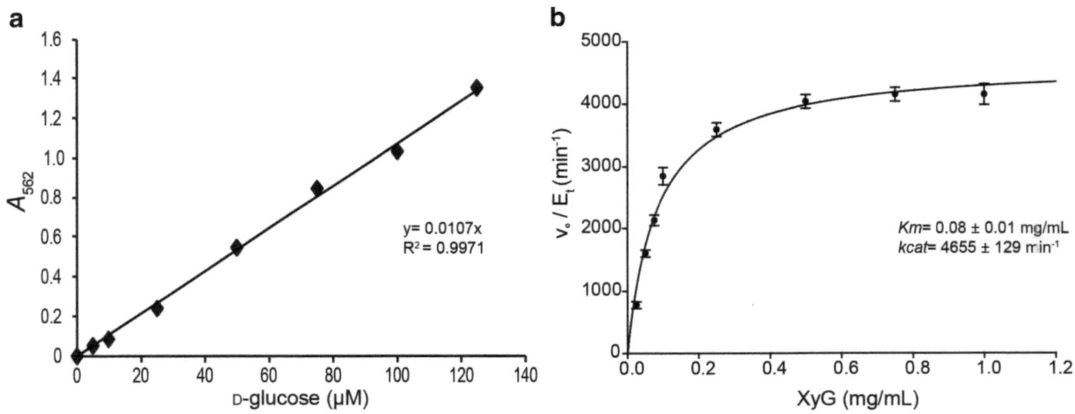

Fig. 3 (**a**) Example background-corrected, best-fit line obtained from 0 to 125 μM D-glucose standard solutions in 50 mM sodium phosphate buffer. (**b**) Example Michaelis–Menten kinetic analysis of a bacterial GH74 enzyme on tamarind seed xyloglucan (XyG; data from [15])

that the assay was performed under initial-rate conditions, this value is equivalent to v_o.

$$activity(\mu mol / \min) = \frac{reducing\ ends(\mu mol)}{t(\min)}$$

6. Calculate the mass of enzyme (mg) in each assay. The enzyme stock concentration (mg/mL) should be divided by a factor that accounts for any intermediate dilutions to reach the working enzyme solution (*see* Subheading 3.5). The resulting value should be then multiplied by the volume of working enzyme solution used in the assay solution, 0.010 mL (≡ 10 μL).

$$enzyme\ mass(mg) = \frac{[E]_{stock}(mg / mL)}{working\ dilution\ factor} \times 0.010 mL$$

7. Specific activity values in units of μmol/min/mg are calculated by dividing the activity values (μmol/min, **step 5**) by the enzyme mass (mg, **step 6**).

$$specific\ activity(\mu mol / \min / mg) = \frac{activity(\mu mol / \min)}{enzyme\ mass(mg)}.$$

3.8.3 Michaelis–Menten Kinetic Analysis

For Michaelis–Menten kinetic analysis, it is necessary to plot the ratio of initial velocity over total enzyme concentration in the assay, $v_0/[E]_t$ (units of reciprocal time), versus a range of initial substrate concentrations, [S].

1. Calculate the initial velocity of the enzyme-catalyzed reaction by dividing the micromolar concentration of reducing-ends formed (calculated according to Subheading 3.8.2, **step 3**) by the assay time.

$$v_o(\mu M / \min) = \frac{[reducing\ ends](\mu M)}{t(\min)}$$

2. Calculate $[E]_t$ in μM according to the equation below, accounting for any intermediate dilutions to reach the working enzyme solution (*see* Subheading 3.5), subsequent dilution of the working enzyme solution (10 μL) in the total assay volume (100 μL), and the enzyme molar mass, M_r. (molecular weight) in units of g/mol (*see* **Note 14**).

$$[E]_t(\mu M) = \frac{[E]_{stock}(g \cdot L^{-1})}{working\ dilution\ factor} \cdot \frac{10 \mu L}{100 \mu L} \cdot \frac{1 mol}{M_{r(enz)} g} \cdot \frac{10^6 \mu mol}{1 mol}$$

3. $v_0/[E]_t$ values in units of min^{-1} are then obtained by simply dividing the initial velocity value (μM/min) by the enzyme concentration (μM). k_{cat} and K_m values can be obtained by

fitting the Michaelis–Menten equation to a plot of $v_0/[E]_t$ (min^{-1}) versus [S] (*see* Fig. 3b and **Note 15**).

$$\frac{v_o}{[E]_t}\left(\min^{-1}\right)=\frac{v_o\left(\mu M/\min\right)}{enzyme\left(\mu M\right)}$$

4 Notes

1. Glucose has been widely used in establishing copper-BCA standard curves, regardless of the polysaccharide tested, due to its low cost and wide general availability. Several studies have demonstrated that different monosaccharides and oligosaccharides exhibit minor [8, 16] or major [5, 7, 11] deviations from the response observed for glucose, which should be considered when establishing the standard curve for a particular enzyme assay. Glucose standard curves are linear up to 125 μM in a wide range of buffers that we have tested in our lab.
2. Solubilization of high molecular mass polysaccharides can be challenging, often requiring portion-wise addition, heating, and vigorous stirring.
3. Low molecular weight polysaccharides or oligosaccharides may exhibit unacceptably high background levels in the assay, due to intrinsic reducing ends. Reduction of saccharides to alditols with sodium borohydride [17] can be used to resolve this problem.
4. Can be substituted by 1.7 mL microfuge tubes.
5. Temperature gradient thermocyclers/thermoblocks can be effectively used for temperature-activity profile analysis (cf. Fig. 2).
6. Large-volume, single-step dilutions are used to minimize systematic and accumulated pipetting errors. Weighing solutions on an analytical balance can further improve dilution accuracy and reduce assay error.
7. For screening and assay optimization, polysaccharide concentrations of 0.5–1.0 mg/mL (or higher) typically ensure that less than 1% of the substrate is hydrolyzed at absorbance values falling within the linear range of the standard curve. As such, enzyme assays typically may be assumed to be linear with respect to time and thus under initial-rate conditions.
8. Addition of BSA limits nonspecific protein adsorption onto plastic surfaces and thus loss of activity. Although proteins can react with the copper-BCA reagent [3], thus potentially interfering with reducing-sugar quantitation, the BSA concentration in the final assay provides a negligible contribution to

color development, which is nonetheless compensated by the blank solutions.

9. Although proteins can react with the copper-BCA reagent [3], thus potentially interfering with reducing-sugar quantitation, typical concentrations of purified enzymes in the assay (<2 μg/mL) do not significantly contribute to A_{562}. However, when using crude enzyme preparations or other cases involving high enzyme (protein) concentration, the "Blank" PCR-tube strip should be prepared to contain an inactivated enzyme control.
10. It is recommended to use extended assays (1 h to overnight) when screening enzyme activity toward different substrates, to capture low-level activities.
11. These solutions are stable on ice up to 1 h.
12. The BCA assay is not a true end-point method; therefore, color continues to develop at 80 °C, but is greatly slowed at 4 °C. It is important to respect incubation times and to use a series of standard solutions incubated in parallel with test assays to obtain reliable and reproducible results.
13. It is necessary to calculate propagation of error (standard deviation or standard error) for this and for all further calculations, *see* [18].
14. The ExPASy ProtParam tool [19] can be used to accurately determine the molar mass of enzymes of known amino acid sequence: http://web.expasy.org/protparam/.
15. If the data does not appear to outline the classic rectangular hyperbola described by the Michaelis–Menten equation, the analysis should be repeated by increasing or decreasing the range of substrate concentrations as appropriate to bracket the apparent K_M value from ca. 0.1 K_M to 10 K_M [14]. If saturation of the enzyme with substrate cannot be achieved, the ratio of k_{cat}/K_M can be obtained from the slope of a linear fit to the data at [S] < K_M.

Acknowledgments

Funding from the Natural Sciences and Engineering Research Council of Canada (NSERC) via the Strategic Partnership Grants for Networks (for the Industrial Biocatalysis Network) and Discovery Grant programs is gratefully acknowledged. Equipment infrastructure was funded by the Canada Foundation for Innovation and the British Columbia Knowledge Development Fund. We thank Sean McDonald (Brumer group, UBC) for comments on an early version of this chapter.

References

1. Sinnott M (2013) Carbohydrate chemistry and biochemistry: structure and mechanism, 2nd edn. RSC Publishing, London
2. Mopper K, Gindler EM (1973) New noncorrosive dye reagent for automatic sugar chromatography. Anal Biochem 56:440–442
3. Smith PK, Krohn RI, Hermanson GT, Mallia AK, Gartner FH, Provenzano MD, Fujimoto EK, Goeke NM, Olson BJ, Klenk DC (1985) Measurement of protein using bicinchoninic acid. Anal Biochem 150:76–85
4. McFeeters RF (1980) A manual method for reducing sugar determinations with 2,2'-bicinchoninate reagent. Anal Biochem 103: 302–306
5. Waffenschmidt S, Jaenicke L (1987) Assay of reducing sugars in the nanomole range with 2,2'-bicinchoninate. Anal Biochem 165: 337–340
6. Doner LW, Irwin PL (1992) Assay of reducing end-groups in oligosaccharide homologs with 2,2'-bicinchoninate. Anal Biochem 202:50–53
7. McIntyre AP, Mukerjea R, Robyt JF (2013) Reducing values: dinitrosalicylate gives over-oxidation and invalid results whereas copper bicinchoninate gives no over-oxidation and valid results. Carbohydr Res 380:118–123
8. Garcia E, Johnston D, Whitaker JR, Shoemaker SP (1993) Assessment of endo-1,4-beta-D-glucanase activity by a rapid colorimetric assay using disodium 2,2'-bicinchoninate. J Food Biochem 17:135–145
9. Fox JD, Robyt JF (1991) Miniaturization of 3 carbohydrate analyses using a microsample plate reader. Anal Biochem 195:93–96
10. Kenealy WR, Jeffries TW (2003) Rapid 2,2'-bicinchoninic-based xylanase assay compatible with high throughput screening. Biotechnol Lett 25:1619–1623
11. Meeuwsen PJA, Vincken JP, Beldman G, Voragen AGJ (2000) A universal assay for screening expression libraries for carbohydrases. J Biosci Bioeng 89:107–109
12. Stoll VS, Blanchard JS (1990) Buffers: principles and practice. In: Deutscher MP (ed) Methods in enzymology: guide to protein purification. Academic Press, Cambridge, MA, pp 24–38.
13. Deutscher MP (1990) Maintaining protein stability. In: Deutscher MP (ed) Methods in enzymology: guide to protein purification. Academic Press, Cambridge, MA, pp 83–89
14. Cornish-Bowden A (2012) Practical aspects of kinetics. In: Fundamentals of enzyme kinetics, 4th edn. Wiley-Blackwell, Hoboken, NJ, pp 85–106
15. Attia M, Stepper J, Davies GJ, Brumer H (2016) Functional and structural characterization of a potent GH74 endo-xyloglucanase from the soil saprophyte *Cellvibrio japonicus* unravels the first step of xyloglucan degradation. FEBS J 283:1701–1719
16. Kongruang S, Han MJ, Breton CIG, Penner MH (2004) Quantitative analysis of cellulose-reducing ends. Appl Biochem Biotechnol 113:213–231
17. Abdelakher M, Hamilton JK, Smith F (1951) The reduction of sugars with sodium borohydride. J Am Chem Soc 73:4691–4692
18. Skoog DA, Holler FJ Crouch SR (2006) Appendix I. In: Douglas A et al (eds) Principles of instrumental analysis, 6th edn. Brooks Cole, Pacific Grove, CA.
19. Gasteiger E, Hoogland C, Gattiker A, Duvaud S, Wilkins MR, Appel RD, Bairoch A (2005) Protein identification and analysis tools on the ExPASy server. In: Walker JM (ed) The proteomics protocols handbook. Humana Press, New York City, NY, pp 571–607

Chapter 2

Quantitative Kinetic Characterization of Glycoside Hydrolases Using High-Performance Anion-Exchange Chromatography (HPAEC)

Nicholas McGregor, Gregory Arnal, and Harry Brumer

Abstract

High-performance anion-exchange chromatography coupled to pulsed amperometric detection (HPAEC-PAD) is a powerful analytical technique enabling the high-resolution separation and sensitive quantification of oligosaccharides. Here, we describe a general method for the determination of glycoside hydrolase kinetics that harnesses the intrinsic power of HPAEC-PAD to simultaneously monitor the release of multiple products under conditions of low substrate conversion. Thus, the ability to track product release under initial-rate conditions with substrate concentrations as low as 5 μM enables the determination of Michaelis–Menten kinetics for glycosidase activities, including hydrolysis and transglycosylation. This technique may also be readily extended to other carbohydrate-active enzymes (CAZymes), including polysaccharide lyases, and glycosyl transferases.

Key words Carbohydrate-active enzyme, Kinetics, Hydrolysis, Transglycosylation, HPLC, HPAEC-PAD, Oligosaccharide

1 Introduction

The assembly and deconstruction of the great variety of complex carbohydrates found in nature is facilitated by a corresponding diversity of carbohydrate-active enzymes (CAZymes), which have been classified into hundreds of protein sequence-based families [1]. Of these, the glycoside hydrolases (GHs) and polysaccharide lyases (PLs) mediate glycan breakdown to component monosaccharides, and are therefore of broad fundamental and applied importance [2–5]. Detailed substrate-specificity, kinetics, and product analyses are central to CAZyme discovery and characterization, and moreover form the bedrock of advanced protein structure-function analyses and refined bioinformatic predictions [6–9]. These data, in turn, underpin biocatalyst engineering and applications.

In harness with classical quantitative assays (*see* the chapters by McKee (Chapter 3) and Arnal et al. (Chapter 1) in this volume),

D. Wade Abbott and Alicia Lammerts van Bueren (eds.), *Protein-Carbohydrate Interactions: Methods and Protocols*, Methods in Molecular Biology, vol. 1588, DOI 10.1007/978-1-4939-6899-2_2, © Springer Science+Business Media LLC 2017

high-performance anion exchange chromatography coupled to pulsed amperometric detection (HPAEC-PAD, [10–12]) is a powerful tool for CAZyme functional analysis. HPAEC-PAD is widely used for CAZyme product characterization in end-point, or limit-digest, analyses (e.g., *see* refs. 13–15 and references therein). Our group has subsequently adapted the sensitivity of HPAEC-PAD for initial-rate enzyme kinetic analysis of glycoside hydrolases and transglycosylases under conditions of low substrate conversion. Significantly, this technique allows the simultaneous quantitation of all enzyme products and determination of Michaelis–Menten kinetic parameters for native (non-derivatized) saccharides [13, 16–20]. We detail here a general method for the initial-rate analysis of glycosidases, which can be readily adapted to a wide range of HPAEC-PAD chromatographic media.

2 Materials

All solutions must be prepared from ultrapure water having a resistivity of 18 MΩ·cm at 25 °C and should be filtered using a suitable 0.22 μm nominal pore size filter before use. All reagents should be obtained at analytical grade or other suitable high-purity grade. Prepare and store all reagents under ambient conditions unless otherwise indicated. All waste should be appropriately collected and disposed of according to local regulations. We recommend using dedicated glassware for eluent preparation to avoid contamination and maintain chromatographic reproducibility. As described below, oligosaccharide analyses were performed using a Dionex ICS-5000 with a 4-channel gradient pump, AS-AP autosampler, and ED electrochemical cell equipped with a 1 mm gold electrode and Ag/AgCl reference electrode (*see* **Note 1**). Oligosaccharides were separated on a 3 × 250 mm CarboPac PA200 column with a 3 × 50 mm CarboPac PA200 guard column at a flow rate of 0.5 mL/min. Data acquisition and analysis was performed using Chromeleon 7 chromatography software. The methods below may be adapted to other comparable HPAEC-PAD systems, including various columns appropriate to specific mono- and oligosaccharides (reviewed in [12]).

2.1 HPAEC-PAD

1. Eluent A, 1 M sodium acetate solution: Carefully weigh 82.03 g of anhydrous sodium acetate (BioUltra grade, Sigma cat. No. 71183; or comparable) into a 1 L beaker. Add ca. 500 mL of water and mix to dissolve. Quantitatively transfer the solution to a 1 L volumetric flask, add water to 1 L and invert to mix. Vacuum-filter the solution and transfer to a clean and dry plastic HPLC reservoir bottle (*see* **Note 2**).
2. Eluent B, 1 M sodium hydroxide: To ca. 500 mL of water in a 1 L graduated cylinder or volumetric flask, add 52.8 mL of 50% w/w carbonate-free sodium hydroxide solution (IC grade, Sigma, cat. no. 72064, or comparable). Add water to

1 L and mix well. Vacuum-filter the solution and transfer to a clean and dry plastic HPLC reservoir bottle (*see* **Note 3**).

3. Eluent C, ultrapure water: Vacuum-filter and transfer to a clean and dry plastic HPLC reservoir bottle.
4. 2 mL glass vials with pre-slit soft caps and microvolume vial inserts (maximum volume of 200 μL, or comparable microvolume vials).

2.2 Enzymes and Oligosaccharides

1. A set of 2–20, 10–100, and 100–1000 μL pipettes (or comparable nominal volumes) and disposable plastic tips.
2. Thin-walled PCR tubes (200 μL volume).
3. Two PCR thermocyclers or heat blocks with heated lids and 200 μL PCR tube blocks.
4. Enzyme stock solutions should be prepared in a suitable assay buffer [21], including any necessary additives or cofactors to maintain activity (*see* **Note 4**), and handled using best practices [22]. Although the methods below assume that the optimal conditions (e.g., pH, temperature, buffer, cofactors) are known through independent experiments, these may also be determined by HPAEC-PAD assay prior to kinetic characterization.
5. Appropriate oligosaccharide substrates and products of known molecular composition should be dissolved in ultrapure water (or a suitable buffer solution of minimal concentration (*see* **Note 5**)). Stocks should in general be made at the highest practical carbohydrate concentration (e.g., 10 mM for oligosaccharides); volumes should be chosen to minimize errors due to weighing milligram amounts, in light of carbohydrate availability and cost.

3 Methods

3.1 Gradient Design

The success of initial-rate kinetics using HPAEC-PAD is dependent on efficient separation of all enzyme products at <10 % (ideally, <1 %) conversion from a correspondingly large amount of oligosaccharide substrate. Broad substrate peaks that disturb the baseline are a particular concern, even when perfectly Gaussian in the absence of peak fronting and tailing. At low substrate conversion, separation of products from minor impurities present in the starting material must also be considered. Given the potential diversity of carbohydrate analytes and HPAEC media, it is not possible to provide a single-gradient solution here. Generally, however, the gradient should be as short as possible to maximize sample throughput while at the same time providing suitable analyte separation. The program should also include a sufficiently long re-equilibration step to maximize retention time reproducibility.

Table 1
Example gradients for HPAEC-PAD separations

Gradient suitable for	Time (min)	1 M NaOH (%)	1 M NaOAc (%)	Refs.
Xyloglucan oligosaccharides	0	10	6	[17] *see also* [16]
	5	10	6	
	17	10	30	
	17.1	50	50	
	18	10	6	
	22	10	6	
Cello-oligosaccharides	0	10	2	[20]
	4	10	2	
	12	10	20	
	12.1	50	50	
	13	10	2	
	17	10	2	
Malto-oligosaccharides	0	10	5	[19]
	4	10	5	
	8	10	25	
	8.1	50	50	
	9	10	5	
	13	10	5	

Some exemplar gradients employed in our lab for enzyme kinetics with linear and branched oligosaccharide substrates are given in Table 1.

3.2 Determining PAD Response for Saccharide Products

1. From individual monosaccharide and oligosaccharide stock solutions, prepare six 40 μL solutions of 0.5, 1.0, 5.0, 10, 30, and 50 μM in microvolume HPLC vials by dilution (*see* **Note 6**). Independent replicates may be included as necessary to ensure statistical accuracy and precision.
2. Using the programmable autosampler, sequentially inject 10 μL of each solution (*see* **Note 6**) into the HPAEC-PAD system and run the gradient program.
3. For each chromatogram, integrate the product peaks (*see* **Note 7**) and plot integration (nC*min) against concentration.
4. Apply linear regression analysis to determine the linear working range of the assay and calculate the slope (nC*min/μM) for each product standard (*see* **Note 8**).

3.3 Scouting Enzyme Concentration

Optimization of enzyme assays to generate accurate, publication-quality data is necessarily an iterative procedure in which substrate concentration, enzyme concentration, buffer, pH value, and temperature are key parameters. Prior to performing detailed

Michaelis–Menten kinetic analysis, it is advisable first to estimate a suitable enzyme concentration for assaying each substrate of interest.

1. Prepare a series of seven fivefold serial dilutions of the enzyme stock in 20 μL of the assay buffer in 200 μL PCR tubes (*see* **Note 9**). This results in the following broad range of dilutions suitable for initial scouting: 1:5, 1:25, 1:125, 1:625, 1:3125, 1:15,625, 1:78,125. A more limited range of dilutions may be appropriate if prior information on enzyme specific activity exists.
2. From the substrate stock solution (*see* Subheading 2.2, **item 5**), prepare 0.5 mL of working solution in the assay buffer at a concentration roughly half of the anticipated K_M value of the substrate. A working solution of 0.2 mM is often a useful starting point for many oligosaccharides (this concentration is typically below the K_M value).
3. Initiate the enzymatic reaction by adding 20 μL of substrate working solution to the first 20 μL solution of diluted enzyme. Gently mix by reciprocal pipetting or gently tapping the PCR tube; consistently mixing individual samples is essential for assay reproducibility.
4. Immediately transfer the tube to a PCR thermocycler preheated to the temperature optimum of the enzyme.
5. Repeat **steps 3** and **4** for each enzyme dilution, spacing the additions precisely by 30 s or 1 min.
6. Include a blank reaction consisting of 20 μL of the working substrate solution mixed with 20 μL of buffer.
7. Precisely 1 h after the addition of substrate to each tube, stop the reaction by heating to 95 °C for 10 min (*see* **Note 10**).
8. Transfer the samples to microvolume HPLC vials and inject 10 μL of each sample into the HPAEC-PAD system using the programmable autosampler (*see* **Note 6**).
9. For each chromatogram, integrate the product peak(s) and subtract the integration of any peak observed at the same elution time in the blank sample.
10. Convert integration values to product concentration(s), [P], using the linear calibration(s) (*see* Subheading 3.2).
11. For each product in each sample, plot [P] against the enzyme concentration in the assay ($[E]_t$). In this stopped assay, enzyme concentrations that result in <10 % conversion of the original substrate concentration (as determined by the sum of all products) can be reasonably assumed to give linear, initial-rate kinetics suitable for subsequent, detailed Michaelis–Menten analysis. $[E]_t$ values producing <1 % substrate conversion are preferred [23].

3.4 Determining Kinetic Michaelis–Menten Parameters for an Oligosaccharide Substrate

1. Based on results from Subheading 3.3, prepare 700 μL of a diluted working enzyme solution from the enzyme stock solution (*see* **Note 9**).
2. In PCR tubes, prepare 80 μL each of eight dilutions of the oligosaccharide substrate, spanning a concentration range of ca. 0.2–20 times the anticipated K_M value (e.g., 0.2, 0.4, 0.8, 1.5, 3.0, 5.0, 10, and 20 × K_M). These solutions will be subsequently diluted 1:2 in the assay to give substrate concentration values, [S], spanning 0.1–10 K_M [23].
3. Equilibrate these solutions in the thermocycler at the desired assay temperature.
4. Initiate the enzyme reaction in the first tube by adding 80 μL of enzyme solution to 80 μL of substrate solution. Gently mix by reciprocal pipetting or gently tapping the PCR tube; consistently mixing individual samples is essential for assay reproducibility. Place the tube back in the thermocycler.
5. Repeat **steps 3** and **4** for each substrate concentration, spacing the additions precisely by 30 s or 1 min. Randomization of the order of enzyme addition to the substrate tubes is recommended to minimize systematic error resulting from any time-dependent enzyme inactivation in the working enzyme solution.
6. Precisely 15, 30, 45, and 60 min after the addition of enzyme to each tube, sample 40 μL into individual PCR tubes and stop the reaction by heating to 95 °C for 10 min (*see* **Notes 10** and **11**).
7. Transfer the samples to microvolume HPLC vials and inject 10 μL of each sample into the HPAEC-PAD system using the programmable autosampler (*see* **Note 6**).
8. For each chromatogram, integrate the product peak(s) (*see* Fig. 1a).
9. Convert integration values to product concentration(s), [P], in μM using the linear calibration(s) (*see* Subheading 3.2).
10. For each initial substrate concentration in the assay, plot [P] against time for each product (*see* **Note 12**). Visually inspect the data for linearity and use linear regression to determine the initial enzyme-catalyzed reaction rate, v_o, in units of μM/min from the slope of the linear range (*see* Fig. 1b and **Note 13**).
11. Starting from the stock enzyme concentration in g/L (= mg/mL) calculate $[E]_t$ in μM according to the equation below. Account for any intermediate dilutions to reach the working enzyme solution (*see* Subheading 3.3), subsequent dilution of the working enzyme solution (80 μL) in the total assay volume (160 μL), and the enzyme molecular weight molar mass, M_r (i.e., molecular weight) in units of g/mol.

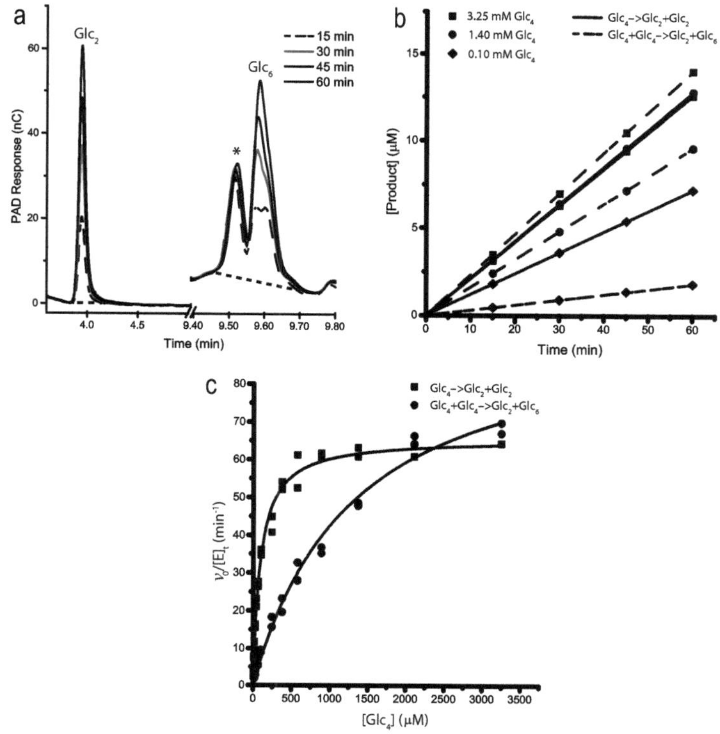

Fig. 1 Example data collection and processing. Panel (**a**) shows HPAEC-PAD traces for the activity of an endo-glucanase on a single initial concentration of cellotetraose (β1,4-linked Glc_4), which produces two molecules of cellobiose (Glc_2) by hydrolysis ($Glc_4 \rightarrow Glc_2 + Glc_2$) and cellobiose ($Glc_2$) plus cellohexaose ($Glc_6$) by disproportionation/transglycosylation ($Glc_4 + Glc_4 \rightarrow Glc_2 + Glc_6$) in competing catalytic reactions. Peaks were integrated using a consistent baseline (*dotted line*) obtained by adjusting the parameters of the Cobra algorithm within Chromeleon 7. Panel (**b**) shows the evolution of Glc_2 (*solid lines*) and Glc_6 (*dashed lines*) concentrations over the assay period as determined from the detector response calibration curves for both products and the stoichiometry of both reactions ($Glc_4 \rightarrow 2Glc_2$ and $2Glc_4 \rightarrow Glc_2 + Glc_6$). A select subset of the data is shown, corresponding to hydrolysis and transglycosylation reactions above and below the apparent K_M values for both (*n.b.* the subtraction of a substrate blank lacking enzyme was applied to these data for each substrate concentration). Panel (**c**) shows the plot of $v_o/[E]_t$ versus [S] for the full dataset including all initial substrate concentrations. The Michaelis–Menten equation was fit to extract apparent k_{cat} and K_M values for both the hydrolysis and transglycosylation reactions

$$[E]_t\,(\mu mol / L) = \frac{[E]_{stock}\,(g/L)}{\text{working dilution factor}} \cdot \frac{80\,\mu L}{160\,\mu L} \cdot \frac{1\,mol}{M_{r(enz)}\,g} \cdot \frac{10^6\,\mu mol}{1\,mol}$$

12. Calculate $v_o/[E]_t$ (min^{-1}) for each product by dividing v_o (μM/min) by $[E]_t$ (μM) and plot versus [S].

13. Calculate k_{cat} and K_M values for each product-forming reaction by fitting the Michaelis–Menten equation to the data (*see* Fig. 1c and **Note 14**).

4 Notes

1. It is important to polish the working electrode regularly (every 1–2 months) and replace the reference electrode regularly (every 6–12 months) to maximize reproducibility, dynamic range, and sensitivity of the electrochemical detector. The waveform described by Rocklin et al. cycled at 2 Hz will minimize wear of the working electrode [10].
2. For highly anionic carbohydrates, sodium nitrate is a good alternative eluent providing stronger elution and lower viscosity than many alternatives [24].
3. To obtain more consistent retention times, carbonate contamination due to absorption of atmospheric CO_2 by aqueous sodium hydroxide should be avoided. Work efficiently and consistently when preparing the sodium hydroxide solution (do not let stand exposed to the atmosphere or in glass vessels; the latter to avoid silicate dissolution). 50% NaOH stock solution bottles should be disturbed as little as possible and the solution should be drawn from the middle of the bottle to avoid withdrawing any precipitated carbonate. Once diluted, maintaining the sodium hydroxide solution under CO_2-free atmosphere will improve retention time reproducibility; equipment manufacturers can include a sparging apparatus for this purpose. Alternatively, solutions may be made fresh and replaced weekly to maximize reproducibility.
4. Enzyme solutions should be prepared in such a way so as to minimize carbohydrate contamination. Enzymes obtained from culture medium should be thoroughly purified. Note that commercial enzymes obtained as freeze-dried powders may contain nonreducing carbohydrates (e.g., sucrose, trehalose, raffinose) as stabilizing/bulking agents, which may limit working concentration ranges in the absence of further purification (e.g., dialysis). Polyethyleneglycol and detergents (especially anionic detergents) should be avoided; these can bind irreversibly to the column and reduce chromatographic performance. In our experience, most buffers work well, but the addition of hydroxylated compounds (e.g., tris, glycerol) will generate a large peak at the beginning of the chromatogram, thereby obscuring weakly retained carbohydrates. The presence of proteins, peptides, or amino acids can also produce small, unretained peaks. High concentrations of certain anions such as halides and mineral salts can cause unpredictable

elution times. In general, it is best to use the lowest practical concentration of buffer compounds and other additives.

5. Substrates should be prepared at high purity with the same considerations as enzyme solutions. We have found that most research-grade purified oligosaccharides (e.g., from Sigma Chemical Company, St. Louis, USA; Megazyme International, Wickow, Ireland; or Carbosynth, Compton, UK) are directly suitable for use in HPAEC-PAD following dissolution and filtration. If solubility is an issue, substrates may first be dissolved in DMSO and diluted with ultrapure water to a concentration that does not cause enzyme inhibition; DMSO does not interfere with chromatographic detection.
6. Sample volumes are chosen to enable 10 μL partial-loop injections, which require 30 μL plus a sufficient excess for accurate autosampling.
7. Chromeleon 7 offers the ability to automatically process large numbers of chromatograms to quickly extract peak integrations; the built-in Cobra algorithm can automatically identify and integrate peaks, including defining baselines. Cobra peak-detection parameters should be adjusted to maximize the consistency of baseline prediction (*see* Fig. 1a). The resulting integration values can be easily transferred into external graphing programs for analysis.
8. With the HPAEC-PAD system described, saccharide concentrations of 0.5–50 μM typically give responses within the linear range of the detector. If deviation from linearity is observed at high concentrations, electrode polishing may be required.
9. Low protein binding (e.g., siliconized) microfuge tube or the addition of bovine serum albumin at a concentration of 0.1 mg/mL should be used to prevent loss of enzyme activity due to adsorption on plastic surfaces.
10. Alternatively, and especially for thermostable enzymes, stop the reaction by adding 10 μL of 0.5 M aqueous NaOH to give a final concentration of 0.1 M. To minimize base-catalyzed hydrolysis, maintain at 4 °C and do not store for extended periods.
11. For substrate concentrations greater than 1 mM, a dilution step is highly recommended. High concentrations of substrate can negatively impact chromatographic performance and a large substrate peak can interfere with peak detection and integration. Quantitatively dilute the reaction to 1 mM initial substrate concentration immediately following enzyme inactivation.
12. Measuring change in concentration over time abrogates the need to subtract a blank.
13. Deviations from linearity may indicate a loss of enzyme activity during the assay, substrate depletion, or abnormally strong

enzyme inhibition by reaction products. Verify that the sum of all product concentrations at 60 min is <10% of the initial substrate concentration. Reduce $[E]_t$ and repeat the complete Michaelis–Menten analysis. Consider reducing the assay temperature or exploring the use of additives that may improve enzyme stability.

14. If the data does not appear to outline the classic rectangular hyperbola described by the Michaelis–Menten equation, the analysis should be repeated by increasing or decreasing the range of substrate concentrations as appropriate to bracket the apparent K_M value (*see* Subheading 3.4, **step 2**) [23]. If saturation of the enzyme with substrate cannot be achieved, the ratio of k_{cat}/K_M can be obtained from the slope of a linear fit to the data at $[S] < K_M$.

Acknowledgments

Funding from the Natural Sciences and Engineering Research Council of Canada (NSERC) via the Strategic Partnership Grants for Networks (for the Industrial Biocatalysis Network) and Discovery Grant programs is gratefully acknowledged. Equipment infrastructure was funded by the Canada Foundation for Innovation and the British Columbia Knowledge Development Fund. We thank Kazune Tamura (Brumer group, UBC) for comments on an early version of this chapter.

References

1. Lombard V, Ramulu HG, Drula E, Coutinho PM, Henrissat B (2014) The carbohydrate-active enzymes database (CAZy) in 2013. Nucleic Acids Res 42:D490–D495
2. Fushinobu S, Alves VD, Coutinho PM (2013) Multiple rewards from a treasure trove of novel glycoside hydrolase and polysaccharide lyase structures: new folds, mechanistic details, and evolutionary relationships. Curr Opin Struct Biol 23:652–659
3. Hemsworth GR, Déjean G, Davies GJ, Brumer H (2016) Learning from microbial strategies for polysaccharide degradation. Biochem Soc Trans 44:94–108
4. Sweeney MD, Xu F (2012) Biomass converting enzymes as industrial biocatalysts for fuels and chemicals: recent developments. Catalysts 2:244–263
5. Henrissat B, Sulzenbacher G, Bourne Y (2008) Glycosyltransferases, glycoside hydrolases: surprise, surprise! Curr Opin Struct Biol 18:527–533
6. Lombard V, Bernard T, Rancurel C, Brumer H, Coutinho PM, Henrissat B (2010) A hierarchical classification of polysaccharide lyases for glycogenomics. Biochem J 432:437–444
7. Mewis K, Lenfant N, Lombard V, Henrissat B (2016) Dividing the large glycoside hydrolase family 43 into subfamilies: a motivation for detailed enzyme characterization. Appl Environ Microbiol 82(6):1686–1692
8. Aspeborg H, Coutinho PM, Wang Y, Brumer H, Henrissat B (2012) Evolution, substrate specificity and subfamily classification of glycoside hydrolase family 5 (GH5). BMC Evol Biol 12:186
9. Stam MR, Danchin EGJ, Rancurel C, Coutinho PM, Henrissat B (2006) Dividing the large glycoside hydrolase family 13 into subfamilies: towards improved functional annotations of α-amylase-related proteins. Protein Eng Des Sel 19:555–562
10. Rocklin RD, Clarke AP, Weitzhandler M (1998) Improved long-term reproducibility for pulsed amperometric detection of carbohydrates via a new quadruple-potential waveform. Anal Chem 70:1496–1501

11. Rocklin RD, Pohl CA (1983) Determination of carbohydrates by anion exchange chromatography with pulsed amperometric detection. J Liq Chromatogr 6:1577–1590
12. Corradini C, Cavazza A, Bignardi C, Corradini C, Cavazza A and Bignardi C (2012) High-performance anion-exchange chromatography coupled with pulsed electrochemical detection as a powerful tool to evaluate carbohydrates of food interest: principles and applications. Int J Carbohydr Chem 2012: e487564
13. Eklöf JM, Ruda MC, Brumer H (2012) Distinguishing xyloglucanase activity in endo-β(1→4)glucanases. Methods Enzymol 510:97–120
14. van Munster JM, Sanders P, ten Kate GA, Dijkhuizen L, van der Maarel MJEC (2015) Kinetic characterization of *Aspergillus niger* chitinase CfcI using a HPAEC-PAD method for native chitin oligosaccharides. Carbohydr Res 407:73–78
15. Rothenhöfer M, Grundmann M, Bernhardt G, Matysik F-M, Buschauer A (2015) High performance anion exchange chromatography with pulsed amperometric detection (HPAEC-PAD) for the sensitive determination of hyaluronan oligosaccharides. J Chromatogr B 988:106–115
16. Baumann MJ, Eklof JM, Michel G, Kallas AM, Teeri TT, Czjzek M et al (2007) Structural evidence for the evolution of Xyloglucanase activity from Xyloglucan endo-transglycosylases: biological implications for cell wall metabolism. Plant Cell 19:1947–1963
17. Larsbrink J, Izumi A, Ibatullin Farid M, Nakhai A, Gilbert Harry J, Davies Gideon J et al (2011) Structural and enzymatic characterization of a glycoside hydrolase family 31 α-xylosidase from *Cellvibrio japonicus* involved in xyloglucan saccharification. Biochem J 436: 567–580
18. Larsbrink J, Thompson AJ, Lundqvist M, Gardner JG, Davies GJ, Brumer H (2014) A complex gene locus enables xyloglucan utilization in the model saprophyte *Cellvibrio japonicus*. Mol Microbiol 94(2):418–433
19. Larsbrink J, Izumi A, Hemsworth GR, Davies GJ, Brumer H (2012) Structural enzymology of *Cellvibrio japonicus* Agd31B protein reveals α-Transglucosylase activity in glycoside hydrolase family 31. J Biol Chem 287: 43288–43299
20. McGregor N, Morar M, Fenger TH, Stogios P, Lenfant N, Yin V et al (2016) Structure-function analysis of a mixed-linkage β-glucanase/xyloglucanase from the key ruminal Bacteroidetes *Prevotella bryantii* B_14. J Biol Chem 291:1175–1197
21. Stoll VS, Blanchard JS (2009) Chapter 6 buffers: principles and practice. In: Deutscher MP, Burgess RR (eds) Methods enzymol. Academic Press, Cambridge, MA, pp 43–56
22. Deutscher MP (1990) Maintaining protein stability. In: Deutscher MP (ed) Guide to protein purification. Academic Press, Cambridge, MA, pp 83–89.
23. Cornish-Bowden A (2012) Practical aspects of kinetics. In: Fundamentals of enzyme kinetics. Portland Press Ltd, London, pp 85–106
24. Zhang Y, Inoue Y, Inoue S, Lee YC (1997) Separation of oligo/polymers of 5-*N*-acetylneuraminic Acid, 5-*N*-glycolylneuraminic acid, and 2-keto-3-deoxy-d-glycero-d-galactononic acid by high-performance anion-exchange chromatography with pulsed amperometric detector. Anal Biochem 250: 245–251

Chapter 3

Measuring Enzyme Kinetics of Glycoside Hydrolases Using the 3,5-Dinitrosalicylic Acid Assay

Lauren S. McKee

Abstract

Use of the 3,5-dinitrosalicylic acid reagent allows the simple and rapid quantification of reducing sugars. The method can be used for analysis of biological samples or in the characterization of enzyme reactions. Presented here is an application of the method in measuring the kinetics of a glycoside hydrolase reaction, including the optimization of the DNSA reagent, and the production of a standard curve of absorbance and sugar concentration.

Key words Reducing sugars, Enzyme kinetics, Glycoside hydrolase, UV–visible spectrophotometry, 3,5-dinitrosalicylic acid

1 Introduction

The 3,5-dinitrosalicylic acid (DNSA) assay utilizes the inherent chemical reactivity of DNSA to detect and quantify reducing sugars. The method was first developed with some variations in the 1920s and 1940s [1, 2] in publications which partly established the chemistry of the reaction and the impact of assay conditions on the accuracy of measurements. The method was formalized in 1959 by GL Miller [3]. This seminal paper established the reaction parameters and described an optimized reagent composition. The article had been cited more than 19,000 times by January 2017, underlining the huge utility of the assay. The method is simple and rapid to perform, and robust enough to be useful for analysis of many kinds of samples. Citing papers include thousands of studies on glycoside hydrolase activity, among them descriptions of purified recombinant proteins and natural mixtures of enzymes, and analyses of plant biomass degradation by chemical and enzymatic hydrolysis. Presented here is an approach to use the method for analysis of glycoside hydrolase reaction kinetics (*see* **Note 1**).

The chemistry of the reaction is quite simple, and is shown in Fig. 1. The aldehyde or ketone functional group of a sugar is

D. Wade Abbott and Alicia Lammerts van Bueren (eds.), *Protein-Carbohydrate Interactions: Methods and Protocols*, Methods in Molecular Biology, vol. 1588, DOI 10.1007/978-1-4939-6899-2_3, © Springer Science+Business Media LLC 2017

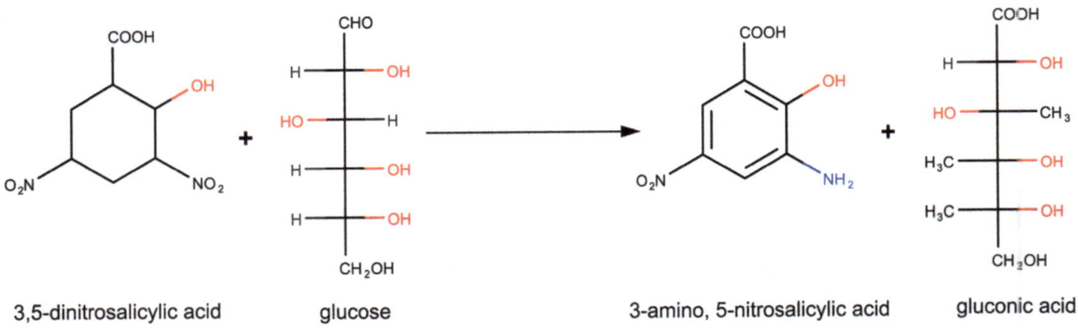

Fig. 1 The chemistry exploited by the DNSA assay. The aldehyde (e.g., in glucose) or ketone (e.g., in fructose) group is oxidized, and 3,5-dinitrosalicylic acid (DNSA) is oxidized to 3-amino, 5-nitrosalicylic acid (ANSA), resulting in a color change in the solution

oxidized to a carboxyl group, and in the process the yellow DNSA compound is reduced to 3-amino, 5-nitrosalicylic acid (ANSA), which has a reddish-brown color and can be detected by measuring absorbance at 575 nm. This simple reaction scheme implies 1:1 stoichiometry of sugar oxidation and DNSA reduction, but there are confounding factors. There may be side reactions, which is why different monosaccharides will lead to different color intensities, and other factors such as the presence of oxygen can also lead to inaccuracies in measurement. As a result, although this method is simple and highly flexible in its application, care must be taken to optimize the precise composition of the DNSA working reagent for each experiment, and to run multiple control experiments every time.

The major component of the working analytical reagent is DNSA dissolved in water. The protocol involves adding an equal volume of the working reagent to a sample, heating to 95–100 °C for up to 20 min to develop the color change reaction, then reading absorbance at 575 nm with a UV–visible spectrophotometer. Additional components in the reagent serve to increase the accuracy and reliability of measurements.

The inclusion of phenol in the working reagent leads to an approximate five-fold increase in color change: the maximum color intensity for any given reducing sugar concentration is achieved with 0.2% phenol, so its inclusion is always advised. Lower concentrations of phenol can lead to a loss of linearity in the color change reaction [2].

Additional accuracy of measurements derives from the inclusion of sodium sulphite (Na_2SO_3) in the working reagent. This compound serves to scavenge oxygen, preventing a loss of sugar due to destruction by oxidation. Early versions of the method flushed samples with nitrogen gas to achieve the same effect [3]. The inclusion of Na_2SO_3 has been shown to reduce glucose destruction by 70% [2]. It is therefore recommended that sulphite is included in the working reagent to improve accuracy, especially when measuring very low concentrations, as it leads to a noticeable

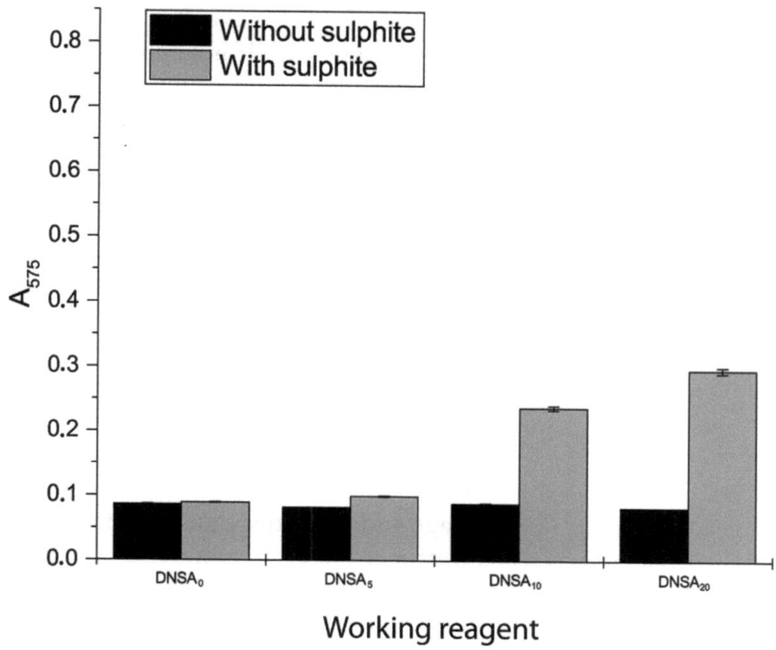

Fig. 2 The effect of the inclusion of Na_2SO_3 in the working reagent. Absorbance values are increased as less glucose is lost due to oxidative destruction. For these samples, working reagent containing 0, 5, 10, or 20 μL glucose per 20 mL of DNSA solution (referred to respectively as $DNSA_0$, $DNSA_5$, $DNSA_{10}$, and $DNSA_{20}$) was added to an equal volume of water prior to color development

increase in baseline absorbance, particularly when using a working reagent with a higher glucose content (Fig. 2).

The most significant variable in the preparation of the working reagent is the concentration of glucose included in the solution. While the DNSA reagent allows for precise measurement of reducing sugar concentration, it does so over a limited linear range of concentrations. It is therefore necessary to include glucose in the reagent in order to raise the overall reducing sugar concentration in analytical samples to be within this range. Standard curves with variable amounts of glucose in the working reagent are presented later in this chapter to show how increasing the amount of glucose extends the linearity of the assay's response to cover lower sample concentrations. However, the increased absorbance values resulting from high glucose contents will also saturate the reagent in samples with high reducing sugar concentrations. It is therefore important that an appropriate amount of glucose be included in the working reagent, to allow for accurate measurements over the whole range of reducing sugar concentrations in the samples under analysis.

Many users add Rochelle salt (sodium potassium tartrate) to samples after heating but before cooling and measuring absorbance, to stabilize the color change. The inclusion of Rochelle salt in the working reagent itself is a mistake, as it can interrupt the oxygen scavenging by sodium sulphite, thereby contributing to a loss of glucose [2]. However, the addition of Rochelle salt to a sample immediately after color development does stabilize the color change, leading to more repeatable measurements [2, 4], although a significant color change is required for this to be useful.

2 Materials

All solutions should be prepared using ultrapure water and analytical grade reagents. Solutions should be stored at room temperature, and will be stable for around 1 month, unless specified otherwise. Take heed of local regulations concerning chemical storage and waste disposal. The dry DNSA compound is an irritant and proper personal protective equipment should be worn when handling it. The DNSA solution in water has a pH of around 12.5, which also requires proper handling and disposal. Phenol should be dispensed and added to the reagent in a fume hood. Additionally, the DNSA solution stains surfaces, equipment, textiles and skin, so care must be taken when dispensing the liquid.

2.1 Reagents

1. DNSA stock reagent: 1% (w/v) DNSA, 0.2% (v/v) phenol, 0.05% (w/v) Na_2SO_3, and 1% NaOH. Weigh 5 g DNSA and 0.25 g Na_2SO_3 into a glass beaker. Add 400 mL water and stir using a magnetic stirrer. Add 0.1 mL phenol, and make up to 500 mL with water. Store at room temperature.
2. 20% (w/v) stock of glucose: Weigh 100 g of glucose into a beaker and add 500 mL water. To improve the long-term storage of this solution, it should be autoclaved or filter sterilized. The solution can then be stored at room temperature.
3. *Optional.* 40% (w/v) solution of potassium sodium tartrate (Rochelle salt): weigh 40 g of Rochelle salt into a beaker and add 100 mL water. Stir using a magnetic stirrer. Store at room temperature.
4. DNSA working reagent: take 20 mL of the stock reagent (*see* **step 1**), and add glucose from the stock prepared in **step 2**. Typically you will add 10–20 μL of glucose solution to the DNSA stock to make the working reagent, but this should be optimized for each new experiment to ensure that the linear range of response covers the reducing sugar concentrations in the samples under analysis. The working reagent is not suitable for long-term storage, as the presence of glucose will cause the color change to develop over time, leading to inaccurate sample analysis. A small volume of working reagent is

therefore produced when needed, and stored at room temperature for no more than 2 days.

2.2 Samples

1. For the standard curve, prepare solutions of glucose (or the most appropriate monosaccharide for your analyses, *see* **Note 2**) over a range of concentrations in the same buffer as will be used to generate the experimental samples. It is recommended that a first-time user of the method begins with 20 samples between 0 and 1 g/L of monosaccharide, to become familiar with the linear range of the response.

2.3 Equipment

Samples should be prepared in heat-stable 1.5 mL eppendorf tubes. A boiling water bath or high temperature heat-block is required for incubation of samples to induce the color change. A UV–visible spectrophotometer is required for measuring sample absorbance. The examples included in this chapter were analyzed using a Cary 50 spectrophotometer (Varian) and the Simple Reads software. Disposable plastic cuvettes can be utilized in the spectrophotometer. Plotting software is required to generate curves and to analyze kinetic data.

3 Methods

All steps can be performed at room temperature, unless specified otherwise. The example presented is the hydrolysis of tamarind xyloglucan by a GH5 enzyme [5]. We therefore present a standard curve of glucose. You should use the most appropriate monosaccharide for your standard curve. For instance, if you will examine the hydrolysis of xylan, make a standard curve of xylose.

3.1 Initial Scoping

First, perform some enzyme reactions to scope the likely range of reducing sugar concentration in the samples (*see* **Note 3**).

1. Prepare a DNSA working reagent by adding 10 μL of glucose (20% stock) to 20 mL DNSA stock. This is working reagent $DNSA_{10}$.
2. Set up enzyme reactions in a final volume of 200–500 μL. The reactions should contain buffer, substrate, and enzyme, all at known concentrations. For this example, a reaction volume of 250 μL containing 50 mM sodium citrate buffer, an enzyme concentration of 4.8 μM, and a substrate concentration of 0.5 g/L are utilized.
3. Incubate the reactions at an appropriate temperature for 1 h, or overnight if the activity is low, and then stop the reaction by adding an equal volume of DNSA working reagent.

4. Develop the color of the samples by placing the tubes in boiling water or in a heat-block set to 95 °C, and incubating for 20 min. Maximum color change may be achieved after a shorter time, and this can be optimized as an experiment proceeds.
5. *Optional step.* Add 300 μL of Rochelle salt to stabilize the color change (*see* **Note 4**).
6. Place the tubes on ice briefly (up to 5 min) to cool to room temperature, to prevent fluctuations in the absorbance reading.
7. Centrifuge the tubes for 1 min at 10,000 × *g* to pellet insoluble material, including precipitated denatured protein, which could interfere with the absorbance measurements.
8. Transfer the colored solutions to disposable plastic cuvettes. Read absorbance at A_{575}. The typical expected range of color change is depicted in Fig. 3. At very high sugar concentrations, the color change will be saturated, and samples will appear black. Highly concentrated samples must be diluted *before* the addition of the DNSA reagent: diluting after the reagent is added and color developed will not give an accurate measurement of sugar concentration in the sample.
   (a) If the absorbance of the samples falls outside of the range of 0.1–1.0, fresh working reagent should be prepared containing more or less glucose (i.e., prepare working reagent $DNSA_5$ to decrease the color change, or $DNSA_{15}$ to increase the color change).
   (b) The absorbance values shown in these examples are obtained without "blanking" the spectrophotometer against a water sample. The "autozero" function on the spectrophotometer can be used with a water sample before you start, but this must be consistently performed for the generation of the standard curve and the analysis of analytical samples.

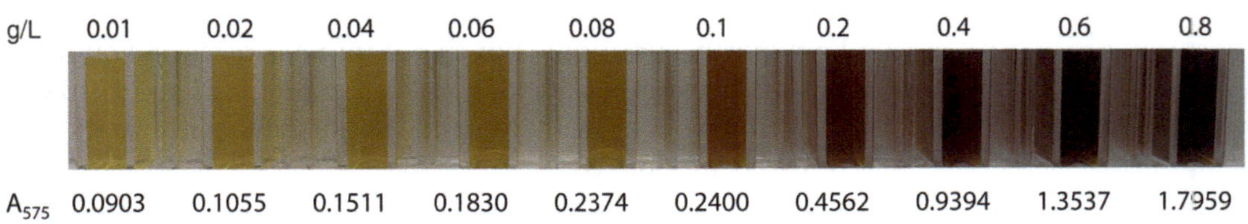

Fig. 3 Typical color change expected for the DNSA reaction. These samples were prepared using $DNSA_{10}$, a DNSA working reagent containing 10 μL glucose (from a 20% stock) per 20 mL DNSA stock reagent. Samples represent a serial dilution of glucose, from 0.01 to 0.8 g/L, from left to right

3.2 Generate a Standard Curve

Prepare a standard curve of monosaccharide concentration vs absorbance at 575 nm, preparing the sugar solutions in the same buffer conditions as the analytical samples. The example shown is a standard curve for glucose in 50 mM sodium citrate buffer.

1. Prepare serial dilutions of the monosaccharide from 0 to 1 g/L. Prepare 200–500 μL of each concentration, in triplicate.
2. Add an equal volume of the appropriate DNSA working reagent, selected in the initial scoping process (*see* **Note 5**).
3. Develop the color of the samples by placing the tubes in boiling water or a heat-block set to 95 °C, and incubating for 20 min.
4. *Optional step.* Add 300 μL of Rochelle salt to stabilize the color change.
5. Place the tubes on ice briefly (up to 5 min) to cool to room temperature, to prevent fluctuations in the absorbance reading.
6. Centrifuge the tubes for 1 min at 10,000 × *g* to pellet any insoluble material which could interfere with the absorbance measurements.
7. Transfer the colored solutions to disposable plastic cuvettes. Read absorbance at 575 nm.
8. Use a plotting software to produce a scatter plot of absorbance vs sugar concentration, and determine the equation of the straight line: this will be used to calculate the reducing sugar concentration in analytical samples using the absorbance measurements.

Figure 4 shows standard curves for the same range of samples using different amounts of glucose in the DNSA working reagent. With increasing amounts of glucose included, the linear range of the assay is expanded to cover lower sample concentrations, and the absorbance values increase, toward the limits of detection in a typical spectrophotometer. Absorbance values should be within 0.1 and 1.0 to maintain accuracy.

3.3 Analysis of Enzyme Kinetics

1. Set up enzyme reactions in a final volume of 200–500 μL. The reactions should contain buffer, substrate, and enzyme, all at known concentrations. In the example given here, we utilized 50 mM sodium citrate buffer, an enzyme concentration of 4.8 μM, and substrate concentrations from 0 to 8 g/L in a final reaction volume of 250 μL.
   (a) For each substrate concentration, a reaction should be allowed to proceed for several different lengths of time, to allow a time-course of the reaction to be plotted to make sure the reaction is being measured within the linear range of activity (Fig. 5).

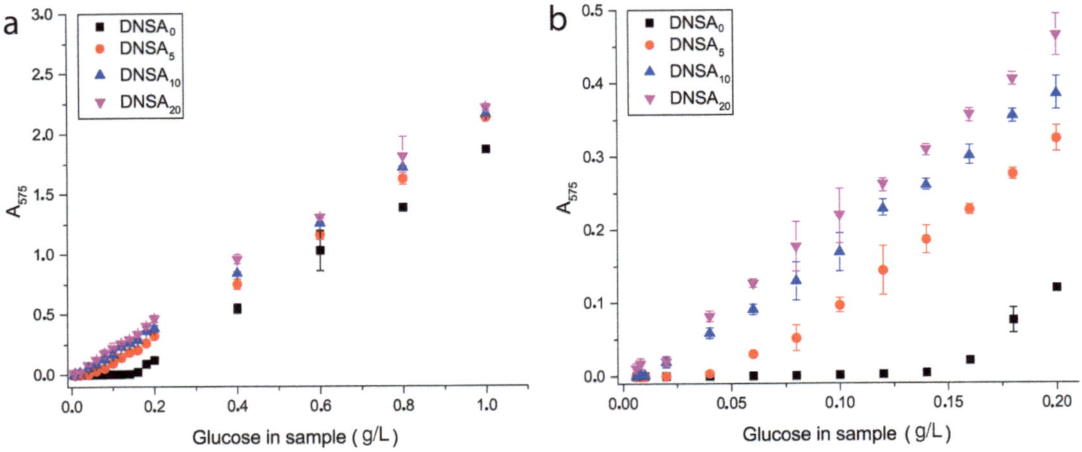

Fig. 4 Standard curve of glucose prepared using four different DNSA working reagents. $DNSA_0$: working reagent with no additional glucose. $DNSA_5$: 5 μL glucose (20%) stock per 20 mL DNSA stock. $DNSA_{10}$: 10 μL glucose (20%) stock per 20 mL DNSA stock. $DNSA_{20}$: 20 μL glucose (20%) stock per 20 mL DNSA stock. (**a**) Full range of absorbance of the samples. (**b**) A closer look at the lower limits of linearity for each reagent

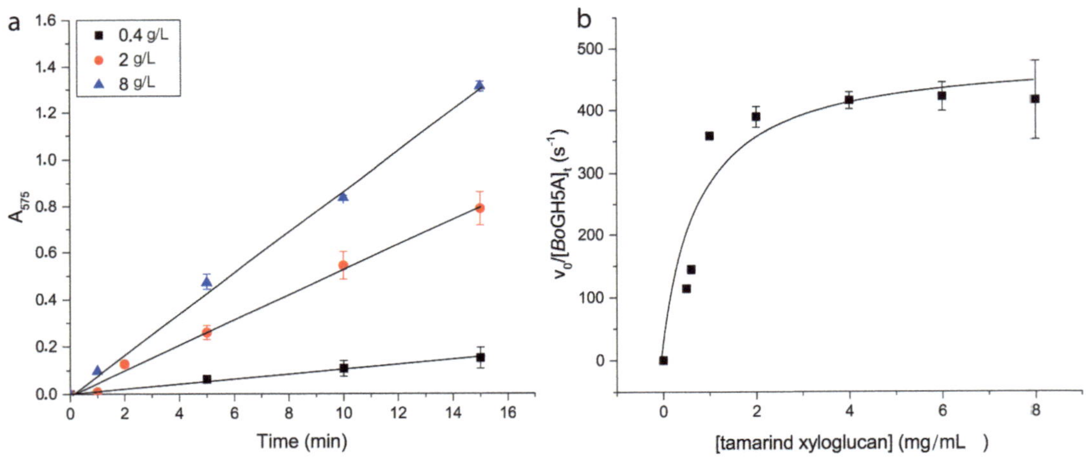

Fig. 5 Example kinetic analysis using the DNSA method, analyzing *endo* hydrolysis of xyloglucan by *Bo*GH5A [5]. (**a**) Time-course of the reaction at different substrate concentrations (0.4, 2, and 8 g/L xyloglucan). (**b**) Full Michaelis–Menten plot of the data. K_{cat}: 435.3 ± 25.6 s^{-1}. K_m: 0.82 ± 0.17 mg/mL. k_{cat}/K_m: 534.0 s^{-1} mg^{-1} mL [5]

2. For each time-course, take the slope of the line (ΔA_{575} min^{-1}) and convert these values to the concentration of glucose released per minute of the assay using the equation from your standard curve. It will be useful to then convert this to molarity of glucose (M glc min^{-1}). Divide these values by enzyme concentration (M) to obtain the rate constant (min^{-1}) for the

reaction at each substrate concentration. These can then be converted to s^{-1} by dividing by 60. Plot these values against substrate concentration, to generate a full kinetic plot for the reaction (Fig. 5).

(a) Use a plotting software to perform nonlinear regression on the data and determine Michaelis–Menten kinetics of the reaction.

4 Notes

1. The DNSA method is particularly useful for analysis of *endo* type hydrolysis of polysaccharide substrates. This is because the initial reducing sugar concentration in the substrate is low due to the length of the carbohydrate chains, and because *endo* chain cleavage generates a lot of new reducing ends. An *exo* type reaction, such as polysaccharide debranching, will generate comparatively fewer reducing ends, so additional glucose in the working reagent may be required. An oligosaccharide substrate will give a more intense background signal than a polysaccharide substrate at the same concentration (g/L) as there is a greater number of reducing ends per equivalent mass of substrate, so assays of oligosaccharide cleavage will require less glucose in the working reagent.
2. When performing an experiment, a great deal of preliminary scoping is required to establish a reagent composition which will give reliable, reproducible measurements over the range of reducing sugar concentrations under analysis. It is important that the standard curve of absorbance vs reducing sugar concentration be performed using the same reagent conditions as the experiment. This initial scoping phase can be laborious, but is preferable to having to begin again with a different reagent composition mid-way through an experiment.
3. The most appropriate monosaccharide to use for your standard curve will typically be the major component of your polysaccharide substrate (i.e., use a standard curve of xylose for measuring hydrolysis of xylan, a standard curve of mannose for measuring hydrolysis of mannan, and so on). Different monosaccharides give different responses to DNSA treatment.
4. The addition of Rochelle salt is part of the inherited protocol in a lot of laboratories, but the decision to include this step varies, as the benefit is not always apparent. The addition of the Rochelle salt should help to stabilize the color change if a significant coloration has been achieved and if added at the correct moment, but can instead dilute a sample and cause loss of color at lower color intensities. It is recommended that a first-time user of the DNSA method

produces standard curves under a range of conditions, both with and without the use of the Rochelle salt stabilizer to become sufficiently familiar with its effects, before true experimental work begins.

5. Ideally, the working reagent composition should be reported in full in any eventual publications arising from the work to allow for future replication and comparison of results. This is rarely seen in practice, but is a helpful detail to include in reports.

Acknowledgments

The author thanks the support of the Knut and Alice Wallenberg Foundation through the Wallenberg Wood Science Centre, Sweden, as well as the KTH Carbohydrate Materials Consortium (CarboMat) supported by the Swedish Research Council FORMAS.

References

1. Sumner J, Graham VA (1921) Dinitrosalicylic acid: a reagent for the estimation of sugar in normal and diabetic urine. J Biol Chem 47:5–9
2. Miller GL (1959) The use of dinitrosalicylic acid for the determination of reducing sugar. Anal Chem 31:426–428
3. Sumner J, Sisler E (1944) A simple method for blood sugar. Arch Biochem 4:333–336
4. Sumner J, Noback CV (1924) The estimation of sugar in diabetic urine, using dinitrosalicylic acid. J Biol Chem 62:287–290
5. Larsbrink J, Rogers TE, Hemsworth GR, McKee LS, Tauzin AS, Spadiut O, Klinter S, Pudlo NA, Urs K, Koropatkin NM, Creagh AL, Haynes CA, Kelly AG, Cederholm SN, Davies GJ, Martens EC, Brumer H (2014) A discrete genetic locus confers xyloglucan metabolism in select human gut Bacteroidetes. Nature 506:498–502

Chapter 4

An Improved Kinetic Assay for the Characterization of Metal-Dependent Pectate Lyases

Darryl R. Jones, Richard McLean, and D. Wade Abbott

Abstract

Pectate lyases are a subset of polysaccharide lyases (PLs) that specifically utilize a metal dependent β-elimination mechanism to cleave glyosidic bonds in homogalacturonan (HG; α-D-1,4-galacturonic acid). Most commonly, PLs harness calcium for catalysis; however, some PL families (e.g., PL2 and PL22) display preferences for transitional metals. Deploying alternative metals during β-elimination is correlated with signature coordination pocket chemistry, and is reflective of the evolution, functional specialization, and cellular location of PL activity. Here we describe an optimized method for the analysis of metal-dependent polysaccharide lyases (PLs). We use an endolytic PL2 from *Yersinia enterocolitica* (*Ye*PL2A) as example to demonstrate how altering the catalytic metal within the reaction can modulate PL kinetics.

Key words Enzyme assay, Kinetics, Polysaccharide lyase (PL), Pectin, Pectate, Metal-dependent, Uronic acid

1 Introduction

Polysaccharide lyases (PLs) are a class of carbohydrate active enzymes (CAZymes) that catalyze the cleavage of uronic acid containing polysaccharides by β-elimination [1]. One of the most common terrestrial uronic acid containing polysaccharides is pectin, which is a heterogeneous matrix composed of three main components homogalacturonan (HG), rhamnogalacturonan-I (RG-I), and rhamnogalacturonan-II (RG-II) [2, 3]. HG is an unbranched homopolysaccharide, the primary pectic component of plant cell walls, and the substrate of pectate lyases [3, 4]. The lytic cleavage of polysaccharides by PLs results in the generation of a new reducing end and a hexenuronic acid moiety (HexA) at the non-reducing end of the leaving group. The formation of the unsaturated HexA provides a unique opportunity to directly monitor product formation using ultraviolet–visible spectroscopy at 232 nm without the need for synthetic substrates.

D. Wade Abbott and Alicia Lammerts van Bueren (eds.), *Protein-Carbohydrate Interactions: Methods and Protocols*, Methods in Molecular Biology, vol. 1588, DOI 10.1007/978-1-4939-6899-2_4, © Springer Science+Business Media LLC 2017

Despite a variety of determined protein folds [4, 5], the PLs demonstrate a high level of structural convergence within their active sites that is likely the result of limitations introduced by the chemistry of the substrate [6, 7]. β-Elimination catalyzed by PLs proceeds through three main steps: (1) neutralization of the charge on uronate group, (2) abstraction of the C-5 proton , charge delocalization, and oxyanion formation, and (3) protonation of the glycosidic oxygen and formation of the unsaturation between C4 and C5 [1, 8, 9]. Neutralization of the uronate most commonly occurs through a bivalent cation, and can be assisted by basic or amidic amino acids [1, 8, 10]. In most metal-dependent PLs, calcium is harnessed to neutralize the uronate group and stabilize the oxyanion intermediate. This is likely because calcium is present at a high concentration (10 μM–10 mM) in the plant cell wall [5, 11]. Interestingly, some PL families (PL2 and PL22) have shown a preference for divalent cations other than calcium, and recent work has shown that these PLs may display greater plasticity in metal selectivity than previously thought [7, 10, 12–14].

The metal cofactor is an integral part of the enzyme active site. Removing the catalytic metal with a strong chelating agent followed by supplementation is the most commonly reported method for investigating metal selectivity in PLs [7, 10, 12]. This approach, however, can introduce high concentration ion microenvironments, which can be detrimental to protein stability. Often these effects result in enzyme precipitation and prevent accurate measures of activity. To avoid these confounding effects, a more conservative approach can be used for exchange of metals within the active site. Recently, this method has been used successfully to report the effects of different metals on the kinetics of *Ye*PL2A, a member of the PL2 family from *Y. enterocolitica* [14]. Here we describe this optimized approach for the exchange of metal cofactors and quantitative analysis of metal selectivity in PLs (Fig. 1).

2 Materials

Prepare all solutions using ultrapure water and analytical grade reagents. Extreme care must be taken to ensure no contamination of solutions with any unwanted ions as trace contamination of optimal ions will be bound preferentially over the ion being tested. All glassware should be rinsed with copious amounts of ultrapure water prior to use. Prepare and store all reagents at room temperature (unless indicated otherwise).

2.1 Dialysis

1. 10× EDTA Buffer: 20 mM EDTA (ethylenedinitrilotetraacetic acid), 200 mM CAPSO (3-(cyclohexylamino)-2-hydroxy-1-propanesulfonic acid), pH 9.0. Add 47.5 g of CAPSO and 5.8 g of EDTA to a graduated cylinder, top to 900 mL with water, and stir until dissolved. Adjust pH to 9.0 with HCl and make up to 1 L with water. Transfer to a glass bottle (*see* **Note 1**).

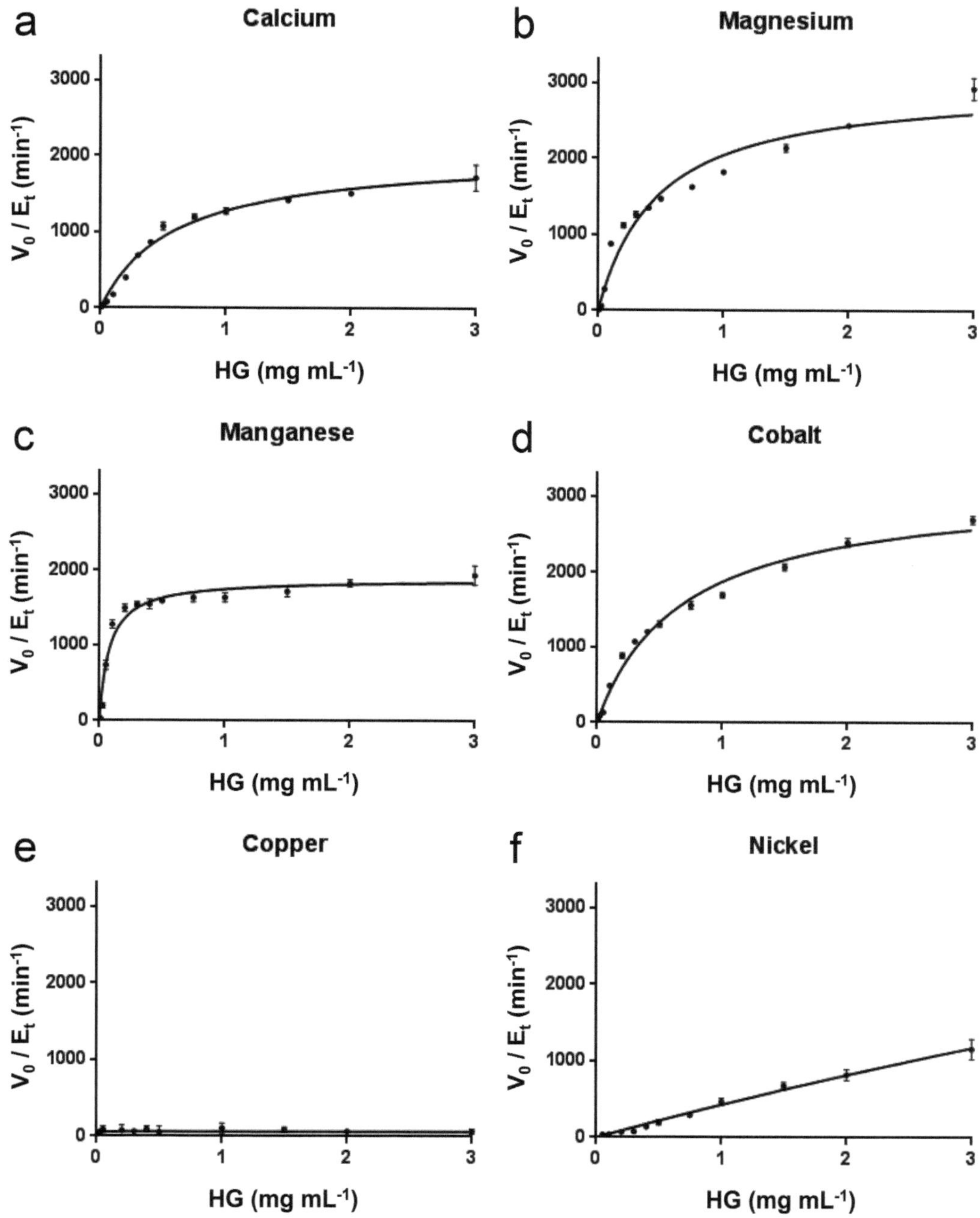

Fig. 1 Michaelis–Menten kinetic curves for PL2A from *Y. enterocolitica* (*Ye*PL2A) on HG after dialysis exchange of divalent metal cations: (**a**) calcium, (**b**) magnesium, (**c**) manganese, (**d**) cobalt, (**e**) copper, and (**f**) nickel. Unsaturated product formation was monitored by ultraviolet–visible spectroscopy at 232 nm. Error bars represent standard deviation over three replicates

2. 10× CAPSO buffer: 0.5 M CAPSO, pH 9.0. Weigh out 118.7 g CAPSO and prepare a 1 L solution as in the previous step.
3. Dialysis membrane with 6–8 kDa molecular weight cutoff, 25.5 mm diameter.

4. 20 L carboy of water cooled to 4 °C.
5. Calcium solution: 100 mM $CaCl_2$. Weigh 0.74 g of calcium chloride dihydrate ($CaCl_2·2H_2O$). Dissolve in 25 mL water in a 50 mL graduated cylinder or volumetric flask and make up to 50 mL with water. Transfer to a 50 mL conical tube.
6. Magnesium Solution: 100 mM $MgCl_2$. Weigh 1.02 g of magnesium chloride hexahydrate ($MgCl_2·6H_2O$). Dissolve in 25 mL water in a 50 mL graduated cylinder or volumetric flask and make up to 50 mL with water. Transfer to a 50 mL conical tube.
7. Manganese solution: 100 mM $MnCl_2$. Weigh 0.99 g of manganese (II) chloride tetrahydrate ($MnCl_2·4H_2O$). Dissolve in 25 mL water in a 50 mL graduated cylinder or volumetric flask and make up to 50 mL with water. Transfer to a 50 mL conical tube.
8. Cobalt solution: 100 mM $CoCl_2$. Weigh 1.19 g of cobalt (II) chloride hexahydrate ($CoCl_2·6H_2O$). Dissolve in 25 mL water in a 50 mL graduated cylinder or volumetric flask and make up to 50 mL with water. Transfer to a 50 mL conical tube.
9. Nickel solution: 100 mM $NiCl_2$. Weigh 1.19 g of nickel (II) chloride hexahydrate ($NiCl_2·6H_2O$). Dissolve in 25 mL water in a 50 mL graduated cylinder or volumetric flask and make up to 50 mL with water. Transfer to a 50 mL conical tube.
10. Copper solution: 100 mM $CuCl_2$. Weigh 0.85 g of copper (II) chloride dihydrate ($CuCl_2·2H_2O$). Dissolve in 25 mL water in a 50 mL graduated cylinder or volumetric flask and make up to 50 mL with water. Transfer to a 50 mL conical tube.

2.2 PL Assay

1. Homogalacturonan (HG) stock: 20 mg mL^{-1} HG. Dissolve 200 mg HG in 10 mL of water. Store in a 15 mL conical tube.
2. Quartz cuvettes, with 1 cm path length (*see* **Note 2**).
3. Spectrophotometer capable of continuous monitoring and recording at 232 nm.

3 Methods

3.1 Dialysis

Carry out all procedures at 4 °C unless specified otherwise.

1. Prepare 1× EDTA Buffer: 2 mM EDTA, 20 mM CAPSO pH 9.0. In a 5 L bucket mix 400 mL 10× EDTA Buffer with 3.6 L of the water prechilled to 4 °C. Add a stir bar and incubate at 4 °C on a stir plate on medium setting.
2. Soak an appropriate length (approximately 12 cm) of dialysis tubing in buffer from the previous step for 2–5 min, seal one end, fill with up to 20 mL enzyme, seal the other end, and place in bucket from **step 1** for at least 2 h (*see* **Note 3**). Keep an aliquot of enzyme from this step ("UNTREATED" positive control).

3. Repeat **step 1**, and transfer dialysis tubing to the fresh bucket of 1× EDTA Buffer for a minimum of 2 h.
4. Prepare 1× CAPSO Buffer: 20 mM CAPSO pH 9.0. In a 5 L bucket mix 400 mL 10× CAPSO Buffer with 3.6 L of water prechilled to 4 °C. Transfer dialysis tubing from **step 3**, add a stir bar, and incubate for 2 h on a stir plate on medium setting.
5. Repeat **step 4**, and transfer dialysis tubing to the fresh bucket of 1× CAPSO Buffer for a minimum of 2 h. Keep an aliquot of enzyme from this step ("EDTA TREATED" ablation control).
6. Prepare metal dialysis buffers, label six 1 L beakers 1–6. Add 50 mL 10× CAPSO buffer and 445 mL chilled water to each beaker. Add 5 mL of the calcium solution to beaker one, 5 mL of the magnesium solution to beaker two, 5 mL of the manganese solution to beaker three, 5 mL of the cobalt solution to beaker four, 5 mL of the nickel solution to beaker five, and 5 mL of the copper solution to beaker six. Stir each beaker continuously (*see* **Note 4**).
7. Cut six lengths of dialysis tubing (approximately 5 cm in length), and soak each in water for 3–5 min. Seal one end of each, and divide the content of the dialysis tubing from **step 5** equally among the six lengths of tubing, fill each tube with no more than 2.5 mL of enzyme solution and seal the other end (*see* **Note 3**). Place one enzyme filled length of dialysis tubing in each beaker from **step 6** and incubate for at least 2 h.
8. Repeat **step 6** and transfer the dialysis tube from **step 7** into the corresponding beakers of fresh buffer, incubate with stirring for at least 2 h.
9. Transfer the dialysis bag contents from beakers 1–6 to tubes labeled "CALCIUM", "MAGNESIUM," "MANGANESE," "COBALT," "NICKEL," and "COPPER" respectively (*see* **Note 5**).

3.2 PL Assay

1. Turn on spectrophotometer and allow UV lamp to warm up.
2. Prepare reaction blank at room temperature; 1 mg mL^{-1} HG, 50 mM CAPSO pH 9.0. Add 50 μL HG stock, 100 μL 10× CAPSO stock to a 1 mL quartz cuvette, and make up to 1 mL with water. Cover cuvette and invert 3–5 times to mix (*see* **Note 6**).
3. Zero spectrophotometer absorbance at 232 nm with the blank sample.
4. Prepare digestion reaction in a clean quartz cuvette as per **step 2**, adding "UNTREATED" enzyme at a total enzyme concentration ($[E]_t$) of 1 μM (*see* **Note 7**).
5. Immediately place digest reaction in spectrophotometer and record absorbance at 232 nm for 3–10 min at room temperature or until absorbance stabilizes (*see* **Note 8**).

6. Repeat **steps 4** and **5** with "EDTA TREATED" enzyme sample (*see* **Note 9**).
7. Repeat **steps 2–5** with "CALCIUM," "MAGNESIUM," "MANGANESE," "COBALT," "NICKEL," and "COPPER" dialyzed enzyme samples with the addition of 1 mM of the respective metal, or 10 μL of the 100 mM metal stock solution in a 1 mL reaction. (*see* **Note 10**).
8. For each digestion reaction, graph the absorbance at 232 nm as a function of time (minutes) and calculate the slope of the plot in units A_{232} min^{-1}.
9. With an extinction coefficient of 5200 M^{-1} cm^{-1} for unsaturated products, the calculated slopes from **step 8**, and a path length of 1 cm calculate the initial reaction velocities, V_0, for each condition using the Beer–Lambert law ($A = \varepsilon\ c\ l$) where A is absorbance, ε is the extinction coefficient, c is the concentration in M, and l is the optical path length in cm (*see* **Notes 11** and **12**).
10. To determine Michaelis–Menten kinetics, initial reaction velocities need to be measured over a range of substrate concentrations. **Steps 1–9** should be repeated at 3.0, 2.5, 2.0, 1.5, 0.8, 0.5, 0.4, 0.3, 0.2, and 0.1 mg mL^{-1} HG.
11. For each enzyme and cation condition, plot the initial velocity, V_0, over final enzyme concentration, $[E]_t$, as a function of substrate concentration, [S]. To determine kinetic parameters from these data, a Michaelis–Menten curve should be fit to the $V_0/[E]_t$, versus [S] plot according to the equation:

$$V_0 = V_{max}([S]/([S]+K_M)$$

where V_{max} is maximal velocity, K_M is the substrate concentration at which half of maximal velocity is achieved (*see* **Note 13**).

4 Notes

1. The β-elimination reaction requires an alkaline pH. If the pH optimum of your enzyme is known the CAPSO buffers can be adjusted to this pH. When comparing multiple enzymes a suitable pH should be selected, where all enzymes are similarly active.
2. Quartz cuvettes are required for analysis at 232 nm.
3. Dialysis buffer to enzyme should be at a minimum 200:1 ratio to ensure complete removal of ions. If there is more than 20 mL total enzyme solution it is recommended to concentrate the enzyme sample by ultrafiltration. If concentration is not an option it is also possible to scale up the amount of dialysis buffer used, split your protein into smaller batches, or increase the number of dialysis exchanges performed.

4. Divalent metal cations selected here are based on reported preferences of PL families 2 and 22, where the metal binding pocket of the enzyme is composed of nitrogen ligands with an optimal geometry for the binding of transitional metals [7, 10, 12–14]. Additional bivalent ions may be screened if the metal binding pocket of the test enzyme does not contain the signature histidine residues associated with transitional metal preference [10, 15].
5. If there is any precipitate in these samples, centrifuge at 10,000 × g for 10 min at 4 °C, then transfer the supernatant to a fresh tube. Quantify protein concentration by measuring absorption at 280 nm and using an appropriate extinction coefficient, which can be accurately predicted from enzyme primary sequence by ProtParam (http://www.expasy.org/tools/protparam.html) [16]. An aliquot of the last dialysis buffer can be used as a blank.
6. If using a smaller size cuvette or a quartz 96-well plate, scale down the digestion to an appropriate volume, be sure to note the path length of the cuvette.
7. Enzyme concentrations resulting in measurable initial velocities will vary greatly depending on the enzyme under investigation as well as the metal co-factor. It is recommended to start with 1 μM final enzyme, and optimize enzyme concentration (1 nM → 1 μM final) such that linear rates can be observed in **step 5**. If the reaction proceeds too rapidly to observe a linear increase in absorbance, the digestions can be prepared at lower enzyme concentrations starting at 500 nM and decreasing to as little as 1 nM. When selecting a final enzyme concentration it is crucial that the same concentration is used for all digestions.
8. Recording time will vary depending on the activity of the enzyme and reaction conditions. It is essential to generate a linear rate for each reaction.
9. Monitor this digestion reaction for the same amount of time as the "UNTREATED" positive control reaction. Increased absorbance should not be detected as EDTA treatment should remove metal cations and ablate activity. If activity is seen repeat method 3.1 Dialysis, increasing the EDTA Buffer concentration in **steps 1–3** from 2 mM EDTA, 20 mM CAPSO pH 9.0 to 5 mM EDTA, 20 mM CAPSO pH 9.0, and performing **step 3** twice.
10. Addition of some ions can cause the gelation and aggregation of HG. Blanks should be made immediately before zeroing the spectrophotometer.
11. If a cuvette with a path length other than 1 cm was used, use the appropriate path length in the initial velocity calculation.

12. Assuming a path length of 1 cm, the calculated slope from **step 8** can be divided by the extinction coefficient to yield the product formation rate in M min^{-1}.
13. Michaelis–Menten curves should be fit by data analysis software such as GraphPad Prism to ensure accurate curve fit. It is imperative to check for goodness of fit as lyases are prone to substrate inhibition and different inhibition curves may better fit the data.

References

1. Lombard V, Bernard T, Rancurel C et al (2010) A hierarchical classification of polysaccharide lyases for glycogenomics. Biochem J 432:437–444. doi:10.1042/BJ20101185
2. Ridley BL, O'Neill MA, Mohnen D (2001) Pectins: structure, biosynthesis, and oligogalacturonide-related signaling. Phytochemistry 57:929–967. doi:10.1016/S0031-9422(01)00113-3
3. Atmodjo MA, Hao Z, Mohnen D (2013) Evolving views of pectin biosynthesis. Annu Rev Plant Biol 64:747–779. doi:10.1146/annurev-arplant-042811-105534
4. Garron ML, Cygler M (2010) Structural and mechanistic classification of uronic acid-containing polysaccharide lyases. Glycobiology 20:1547–1573. doi:10.1093/glycob/cwq122
5. Garron ML, Cygler M (2014) Uronic polysaccharide degrading enzymes. Curr Opin Struct Biol28:87–95. doi:10.1016/j.sbi.2014.07.012
6. Charnock SJ, Brown IE, Turkenburg JP et al (2002) Convergent evolution sheds light on the anti-β-elimination mechanism common to family 1 and 10 polysaccharide lyases. Proc Natl Acad Sci U S A 99:12067–12072. doi:10.1073/pnas.182431199
7. Abbott DW, Boraston AB (2007) A family 2 pectate lyase displays a rare fold and transition metal-assisted B-elimination. J Biol Chem 282:35328–35336. doi:10.1074/jbc.M705511200
8. Yip VLY, Withers SG (2006) Breakdown of oligosaccharides by the process of elimination. Curr Opin Chem Biol 10:147–155. doi:10.1016/j.cbpa. 2006.02.005
9. Gacesa P (1987) Alginate-modifying enzymes. FEBS Lett 212:199–202. doi:10.1016/0014-5793(87)81344-3
10. Abbott DW, Gilbert HJ, Boraston AB (2010) The active site of oligogalacturonate lyase provides unique insights into cytoplasmic oligogalacturonate β-elimination. J Biol Chem 285:39029–39038. doi:10.1074/jbc.M110.153981
11. Hepler PK, Winship LJ (2010) Calcium at the cell wall-cytoplast interface. J Integr Plant Biol 52:147–160. doi:10.1111/j.1744-7909.2010.00923.x
12. Shevchik VE, Condemine G, Robert-Baudouy J, Hugouvieux-Cotte-Pattat N (1999) The exopolygalacturonate lyase PelW and the oligogalacturonate lyase Ogl, two cytoplasmic enzymes of pectin catabolism in Erwinia chrysanthemi 3937. J Bacteriol 181:3912–3919
13. Abbott DW, Thomas D, Pluvinage B, Boraston AB (2013) An ancestral member of the polysaccharide lyase family 2 displays endolytic activity and magnesium dependence. Appl Biochem Biotechnol 171:1911–1923. doi:10.1007/s12010-013-0483-9
14. McLean R, Hobbs JK, Suits MD et al (2015) Functional analyses of resurrected and contemporary enzymes illuminate an evolutionary path for the emergence of exolysis in polysaccharide lyase family 2. J Biol Chem 290:21231–21243. doi:10.1074/jbc.M115.664847
15. Harding MM (2001) Geometry of metal - ligand interactions in proteins. Acta Crystallogr Sect D Biol Crystallogr 57:401–411. doi:10.1107/S0907444999007374
16. Gasteiger E, Hoogland C, Gattiker A, et al (2005) Protein identification and analysis tools on the ExPASy Server. Proteomics protocols handbook. pp 571–607. doi:10.1385/1592598900

Chapter 5

Colorimetric Detection of Acetyl Xylan Esterase Activities

Galina Mai-Gisondi and Emma R. Master

Abstract

Colorimetric detection of reaction products is typically preferred for initial surveys of acetyl xylan esterase (AcXE) activity. This chapter will describe common colorimetric methods, and variations thereof, for measuring AcXE activities on commercial, synthesized, and natural substrates. Whereas assays using *p*NP-acetate, α-naphthyl acetate, and 4-methylumbelliferyl acetate (4MUA) are emphasized, common methods used to measure AcXE activity towards carbohydrate analogs (e.g., acetylated *p*-nitrophenyl β-D-xylopyranosides) and various acetylated xylans are also described. Strengths and limitations of the colorimetric assays are highlighted.

Key words Acetyl xylan esterase, Colorimetric assays, *p*NP-acetate, *α*-Naphthyl acetate, 4-Methylumbelliferyl acetate, Acetylated xylooligosaccharides, β-Xylosidase-coupled assay

1 Introduction

Within the carbohydrate-active enzyme (CAZy) database (www.cazy.org), acetyl xylan esterases are currently classified into carbohydrate esterase (CE) families 1–7, and 16, as well as the recently proposed family CE17 [1–3]. Most of AcXEs employ a Ser-His-Asp catalytic triad to catalyze the deacylation of substituted sugars [1], although family CE4 includes enzymes that employ a metal-ion dependent hydrolysis mechanism [4]. Given the lack of low cost and well characterized acetylated xylo-oligosaccharides and xylans, broad screens for AcXE activity often begin by using *p*-nitrophenyl acetate (*p*NP-acetate), α-naphthyl acetate, or 4MUA [5–7]. Slight variations to the standard assay are summarized in Table 1. Whereas continuous product detection is possible through assays that use *p*NP-acetate, α-naphthyl acetate, and 4MUA, main limitations to using *p*NP-acetate is instability at alkaline pH [8]. Because 4MUA is more stable under these pH conditions, it is often the preferred substrate for determination of pH optima [7, 8].

The original version of this chapter was revised. The erratum to this chapter is available at:
DOI 10.1007/978-1-4939-6899-2_23

D. Wade Abbott and Alicia Lammerts van Bueren (eds.), *Protein-Carbohydrate Interactions: Methods and Protocols*, Methods in Molecular Biology, vol. 1588, DOI 10.1007/978-1-4939-6899-2_5, © Springer Science+Business Media LLC 2017

Table 1
Summary of reported methods for reaction termination and product detection in assays containing *p*NP-acetate, α-naphthyl acetate, and 4MUA

Substrate	Developing reagent	Product detection	Reference (s)
p-Nitrophenyl acetate (*p*NP-acetate) 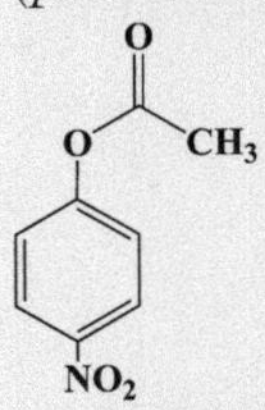	None	*p*-nitrophenol at 410 nm	[5, 9–12]
	None	*p*-nitrophenol at 420 nm	[6, 13–16]
	None	*p*-nitrophenol at 405 nm	[17–22]
	Sodium carbonate (Na_2CO_3)	*p*-nitrophenol at 405 nm	[23]
	none	Acetic acid detection kit from Boehringer Mannheim	[8]
α-naphthyl acetate 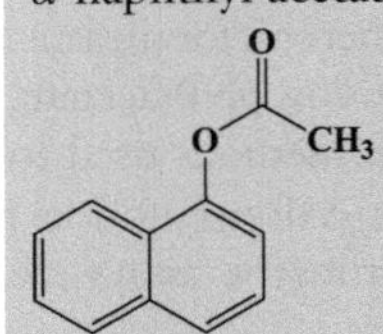	Fast Garnet GBC in sodium dodecyl sulfate; incubate at room temperature for 15 min	α-naphthol in complex with developing reagent at 560 nm	[16–18, 24]
	Fast Corinth V salt in sodium acetate buffer (pH 4.3) containing Tween 20; incubate at room temperature for 10 min	α-naphthol in complex with developing reagent at 535 nm	[25–27]
	none	α-naphthol directly at 321 nm	[28]
4-methylumbelliferyl acetate (4MUA) 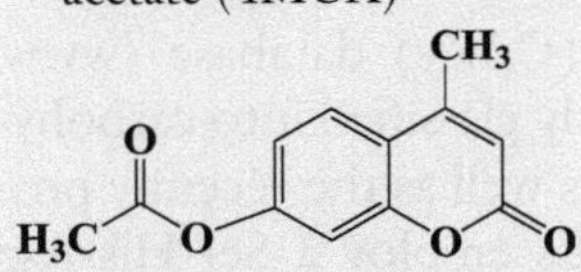	Citric acid to decrease pH to 2–3	4-methylumbelliferone (4-MU) at 354 nm	[7]

Certain AcXEs, such as those belonging to CE4 family, exhibit no activity towards the chromogenic substrates described above [1], which has motivated the application of alternative commercial compounds (Table 2) and synthesis of additional substrate analogs (Table 3). These substrates have also been used to uncover the regioselectivity of AcXEs [9–14].

Following initial activity screens and determination of pH optimum, the specific activity and kinetic parameters of AcXEs can be determined using natural acetylated oligosaccharides and polysaccharides. For example, AcXE activity has been measured using native, acetylated xylan from birchwood [5, 14, 17], beechwood [17], and oat spelt [11], whereas oligosaccharides can be prepared through β-xylanase treatment of xylans [38–40]. Alkali extracted xylans can also be chemically acetylated prior to their use in activity assays [23, 41]. The release of acetic acids from these substrates is

Table 2
Summary of noncommercial, synthesized carbohydrate analogs used to detect AcXE activity

Substrate	Reported approach to product detection	Reference(s)
Acetylated methyl β-D-xylopyranosides (e.g., 2,3,4-tri-O-acetylated methyl-β-xylopyranoside) 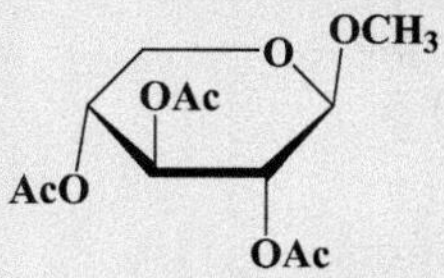 Acetylated xylobiose 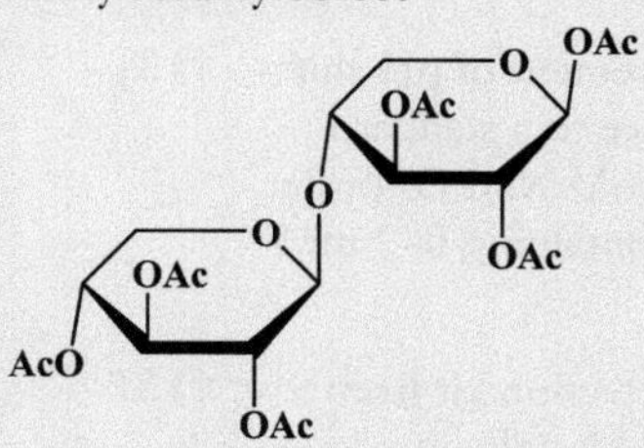	• Gas-liquid chromatography (GLC-MS) • Acetic acid detection with K-ACETRM acetic acid kit from Megazyme • TLC detection of released sugars • ^{1}H NMR determination of regioselectivity	[2, 9, 29–31]
Deoxy and fluoro derivatives of methyl β-D-xylopyranoside diacetates (e.g., 2-deoxy-2-fluoro-3,4-diacetylated methyl β-D-xylopyranoside) 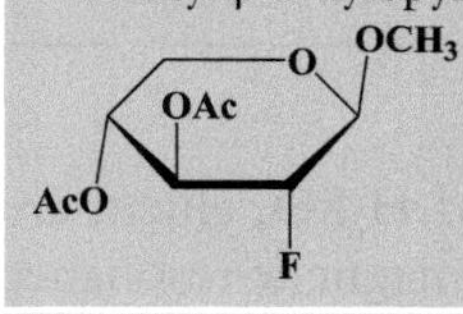	• TLC detection of released sugars	[32, 33]
Monoacetylated *p*-nitrophenyl β-D-xylopyranosides (e.g., 2-O-acetyl nitrophenyl β-D-xylopyranoside) 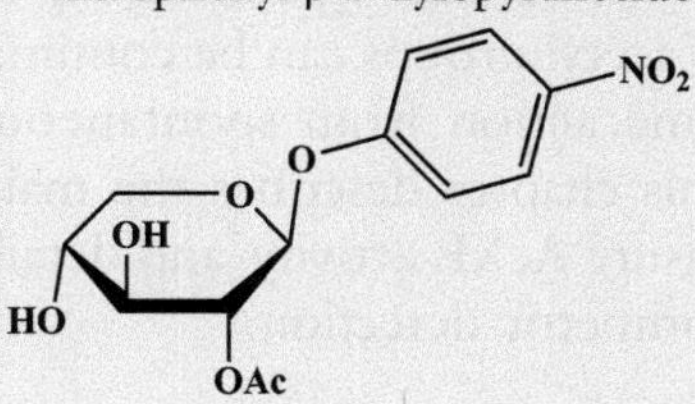	Addition of $Na_2B_4O_7$ (or N_2CO_3) followed by *p*-nitrophenol detection at 405 nm	[34–36]

typically measured by monitoring the drop in pH of reaction mixtures either directly [13, 42], or indirectly using a colorimetric pH indicator such as bromothymol blue (BTB) [43]. Such measurements have been facilitated by the establishment of commercial kits that also permit indirect, spectrophotometric detection of acetic acid release (e.g., acetic acid assay kit from Megazyme (K-ACET) or the acetate colorimetric assay kit from Sigma (MAK086). Alternatively, high performance liquid chromatography (HPLC) is widely used for direct, quantitative measurement of acetic acid

Table 3
Summary of reported methods for reaction termination and product detection using natural substrates

Substrate	Developing reagent	Production detection	Reference
Acetylated birchwood xylan	Addition of ice-cold ethanol and incubation on ice to precipitate the xylan; removal of precipitate by centrifugation; freeze and lyophilize supernatant	The residue is suspended in water, acidified by addition of phosphoric acid, and then acetate is measured by gas–liquid chromatography	[18]
Acetylated birchwood xylan	None	Acetic acid detection kit from Boehringer Mannheim	[10]
Acetylated beechwood xylan	None	Direct measurement of pH shift and reference to a standard curve of 2 mM Na_2PO_4 buffer (pH 7.0) containing 0–2 mM acetic acid	[13]
Acetylated oat spelt xylan	Not mentioned	Acetic acid detection kit from Boehringer Mannheim	[13]
Acetyl glucuronoxylan (oligosaccharides)	Denaturation of enzyme at 100 °C for 5 min	Carbohydrate detection by MALDI ToF and TLC following GH10 xylanase and esterase treatment	[37]

release, where an Aminex HPX-87 column and H_2SO_4 eluent are commonly used [1, 5, 17, 44]. Finally, nuclear magnetic resonance spectroscopy (NMR), including ^{1}H NMR and 2D-NMR, can be used to detect deacetylation products [45]. In particular, ^{1}H NMR has been the method of choice to confirm the regioselectivity of AcXEs [46, 47], whereby release of acetyl groups can be continuously tracked to distinguish enzyme action from spontaneous acetyl group migration [1, 34]. This chapter describes the main substrates and methods used to measure AcXE activity, and details five common methods for their colorimetric detection.

2 Materials

Common buffers can be used unless otherwise specified. The stock buffers described here are prepared as 500 mM solutions.

2.1 Assay Using pNP-Acetate

1. Substrate stock solution: 50 mM *p*-nitrophenyl acetate (*p*NP-acetate) in dimethyl sulfoxide DMSO (*see* **Note 1**). Weigh 0.09 g of *p*NP-acetate (181.15 g/mol) using a digital balance (0.1 mg readability), transfer to a 10 mL volumetric flask, and add DMSO up to the mark.

2. Product stock solution: 50 mM *p*-nitrophenol (*p*NP) in DMSO (*see* **Note 2**). Weigh 0.07 g of *p*NP (139.11 g/mol) using a digital balance (0.1 mg readability), transfer to a 10 mL volumetric flask, and add DMSO up to the mark.
3. Stock solution of the test enzyme in water or reaction buffer (*see* **Note 3**).

2.2 Assay Using α-Naphthyl Acetate

1. Substrate stock solution: 100 mM α-naphthyl acetate in DMSO (*see* **Note 4**). Weigh 0.19 g of α-naphthyl acetate (186.21 g/mol) using a digital balance (0.1 mg readability), transfer to a 10 mL volumetric flask, and add DMSO up to the mark.
2. Product stock solution: 100 mM α-naphthol in DMSO. Weigh 0.14 g of α-naphthol (144.17 g/mol) using a digital balance (0.1 mg readability), transfer to a 10 mL volumetric flask, and add DMSO up to the mark.
3. Fast Corinth Salt stopping and dye solution (*see* **Note 5**): 0.01% Fast Corinth V Salt (zinc chloride double salt) in 1 M sodium acetate buffer (pH 4.3) containing 10% Tween 20. To prepare 1 M sodium acetate buffer, prepare separately 1 M sodium acetate and 1 M acetic acid solution and adjust pH of sodium acetate with acetic acid.
4. Stock solution of the test enzyme in water or reaction buffer (*see* **Note 3**).

2.3 Assay Using 4MUA

1. Substrate stock solution: 100 mM **4**MUA in dimethyl sulfoxide (DMSO) (*see* **Note 6**). Weigh 0.22 g of **4**MUA (218.2 g/mol) using a digital balance (0.1 mg readability), transfer to a 10 mL volumetric flask, and add DMSO up to the mark.
2. Product stock solution: 100 mM 4MU in DMSO (*see* **Note 7**). Weigh 0.18 g of **4**MU (176.17 g/mol) using a digital balance (0.1 mg readability), transfer to a 10 mL volumetric flask, and add DMSO up to the mark.
3. Stopping solution: 50 mM citric acid (pH 2.2). Weigh 4.8 g of citric acid salt (192.12 g/mol), transfer to a graduated cylinder, and make up to 500 mL with milliQ water. Filter through 0.45 μM filter and transfer into a glass bottle.
4. Stock solution of the test enzyme in water or reaction buffer (*see* **Note 3**).

2.4 Assay Using Acetylated Natural Substrates

1. Substrate stock solution: 3% of acetylated oligosaccharides or polysaccharides prepared in milliQ water. Weigh 0.3 g of substrate, transfer to a 10 mL volumetric flask, and add water up to the mark.
2. Stopping solution: 0.33 M H_2SO_4 prepared in milliQ water. Transfer 50 mL of milliQ water to a graduated cylinder. Weight

3.23 g of concentrated H_2SO_4 (98.08 g/mol), transfer into the cylinder, and fill up to 100 mL with milliQ water.

3. Detection method: commercially available acetic acid kits (e.g., Acetic Acid Assay Kit (K-ACET; Megazyme) or Acetate Colorimetric Assay Kit **(MAK086; Sigma) or HPLC** [5, 48].
4. Stock solution of the test enzyme in water or reaction buffer (*see* **Note 3**).
5. Water bath with shaker.

2.5 The β-Xylosidase-Coupled Assay with 4-Nitrophenyl β-D-Xylopyranosides

1. Substrate stock solutions (200 mM) in DMSO: 2-*O*-, 3-*O*-, or 4-*O*-acetyl *p*-nitrophenyl β-D-xylopyranoside, and *p*-nitrophenyl β-D-xylopyranoside. Weigh 0.60 g of **mono-***O*-acetyl *p*-nitrophenyl β-D-xylopyranoside ($C_{12}H_{15}NO_8$; calculated molecular weight: 301.25 g/mol) or 0.54 g of *p*-nitrophenyl β-D-xylopyranoside (271.22 g/mol), using a digital balance (0.1 mg readability), transfer to a 10 mL volumetric flask, and add DMSO up to the mark.
2. Product stock solution: 50 mM *p*-nitrophenol (*p*NP) in DMSO. Weigh 0.07 g of *p*NP (139.11 g/mol) using a digital balance (0.1 mg readability), transfer to a 10 mL volumetric flask, and add DMSO up to the mark.
3. Enzyme mixture: comprising the test enzyme (acetyl xylan esterase) and 1.5 nkat (equates to ~0.09 U or 1.5 nmol/s) of a β-xylosidase determined using *p*-nitrophenyl β-D-xylopyranoside (e.g., recombinant XlnD from *Aspergillus niger* (AnGH3) produced in *Saccharomyces cerevisiae* Y293 [49–51] or β-xylosidase XloA (locus tag: TM0076) from *T. maritima* introduced into pET24d vector and expressed in *E. coli* DL41 [36]. Commercial β-D-xylosidases are also available (e.g., E-BXSR from Megazyme; X3504 from Sigma) (*see* **Note 8**).
4. Stopping solution: saturated $Na_2B_4O_7$ (~0.12 M). Dissolve 15 g of $Na_2B_4O_7 \bullet 10\ H_2O$ in 350 mL milliQ water, stir for several hours, and then decant the solution from the residual solid.

3 Methods

3.1 Assay Using pNP-Acetate

1. To prepare a final reaction volume of 1 mL containing 50 mM buffer pH 6.5 (or lower pH) and 1 mM *p*NP-acetate (*see* **Note 9**), transfer 100 μL of a 500 mM stock-buffer (e.g., sodium citrate buffer) solution to 780 μL of milliQ water and add 20 μL of the 50 mM *p*NP-acetate stock solution.
2. Control samples are prepared as above without adding enzyme.
3. Mix and incubate the solution in a preheated water bath for 5 min.

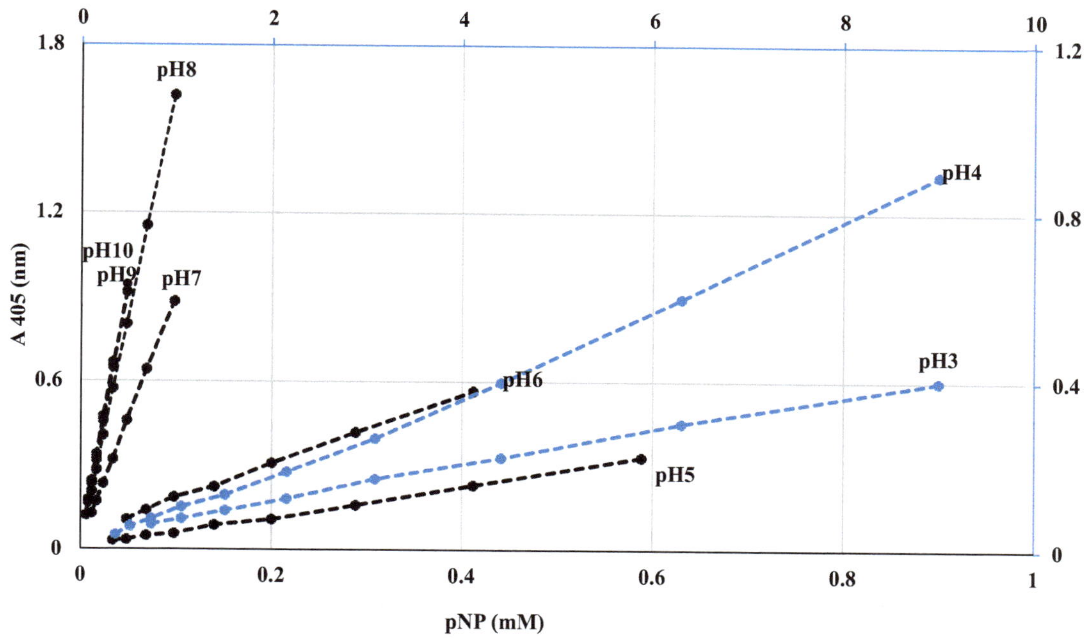

Fig. 1 Impact of pH on standard curves generated using *p*-nitrophenol (*p*NP)

4. To initiate the reaction, add 100 μL of enzyme solution at an appropriate dilution for product detection. Negative control samples will substitute the enzyme solution for 100 μL of water or storage buffer for the enzyme. If measuring reaction rates, the enzyme dose must also ensure a linear relationship between reaction time and product release (*see* **Note 10**).
5. Directly measure *p*-nitrophenol formation at *A*405 nm using a 1.5 mL cuvette; use milliQ water as the blank (*see* **Note 11**).
6. The absorbance of *p*-nitrophenol is pH dependent (Fig. 1). Therefore, a standard curve is required for each tested pH condition. The following *p*-nitrophenol concentrations are recommended for 1 mL cuvette (i.e., 1 cm path length) at the specified pH value: pH 3: 0.75–9 mM; pH 4: 0.3–9 mM; pH 5: 0.03–0.6 mM; pH 6: 0.05–0.5 mM; pH 7: 0.01–0.1 mM; pH 8: 0.01–0.1 mM; pH 9: 0.01–0.05 mM; pH 10: 0.005–0.5 mM.

3.2 Assay Using α-Naphthyl Acetate

1. To prepare a final reaction volume of 2 mL containing 1 mM α-naphtyl acetate in final 50 mM buffer, pre-equilibrate 20 μL of the 100 mM substrate stock solution with 200 μL of 500 mM buffer and 1.58 mL milliQ water at the reaction temperature.
2. Control samples are prepared as above without adding enzyme.
3. Initiate the reaction by adding 200 μL of enzyme solution at an appropriate dilution for product detection (*see* **Note 10**). Negative control samples will substitute the enzyme solution for 200 μL of water or storage buffer for the enzyme.

4. Incubate the reaction mix for 10 min at a specified temperature.
5. Add 1 mL of 0.01 Fast Corinth Salt stopping and dye solution.
6. Read absorbance at A535 nm exactly 10 min after addition of the dye; milliQ water containing dye is used as the blank.
7. To prepare a standard curve, dilute 100 mM α-naphthol 1:10 with milliQ water and use 10 mM solution for dilution series containing 0.01–0.1 mM α-naphtol in cuvette with 1 cm path length after addition of stopping reagent.

3.3 Assay Using 4MUA

1. To prepare a final reaction volume of 400 μL containing 100 mM buffer and 1 mM 4MUA, transfer 80 μL of a 500 mM stock buffer solution to 306 μL milliQ water and add 4 μL of the 100 mM 4MUA stock solution (*see* **Note 12**).
2. Control samples are prepared as above without adding enzyme.
3. Mix and incubate the solution in a preheated water bath for 5 min.
4. To initiate the reaction, add 10 μL of enzyme solution at an appropriate dilution for product detection. If measuring reaction rates, the enzyme dose must also ensure a linear relation between reaction time and product release. Negative control samples will substitute the enzyme solution for 10 μL of water or storage buffer for the enzyme.
5. Incubate the reaction mix for 10 min at a specified temperature (*see* **Note 10**).
6. After 10 min, add 600 μL of 50 mM citric acid (pH 2.2) and vortex rigorously to stop the reaction.
7. Filter the samples through a 0.2 μm GHP Acrodiscs 13 filter (PALL) to remove insoluble particles prior to measurement (*see* **Note 13**).
8. Transfer filtered samples into disposable 1.5 mL polystyrene cuvettes and read the absorbance at 354 nm; milliQ water is used as the blank.
9. To prepare a standard curve, dilute 100 mM 4 MU stock solution to 10 mM in DMSO. Use 0.01–1 mM of 4 MU for 1 cm path length after addition of stopping solution (for 96-well plate use 0.05–2.5 mM of 4 MU without adding of stopping solution).

3.4 Assay Using Acetylated Natural Substrates

1. To prepare a final reaction volume of 200 μL containing 50 mM buffer and 1.5% (w/v) oligosaccharide, mix 20 μL of 500 mM buffer, 30 μL water, and 100 μL of a 3% (w/v) oligosaccharide to stock solution (*see* **Note 14**).
2. Control samples are prepared as above without adding enzyme.

3. Preheat the tubes containing substrate–buffer–water mix to the reaction temperature for 5 min.
4. Initiate reaction by adding 50 μL enzyme dilution. Negative control samples will substitute the enzyme solution for 50 μL of water or storage buffer for the enzyme.
5. Incubate with shaking (80–200 rpm) at a temperature and duration suitable for measuring reaction rates (*see* **Note 15**).
6. Detect acetic acids released using a commercial acetic acid kit or by HPLC after stopping the reaction with 40 μL of 0.33 M H_2SO_4 or by heating at 100 °C for 10–20 min (*see* **Note 16**).

3.5 The β-Xylosidase-Coupled Assay with 4-Nitrophenyl β-D-Xylopyranosides

1. To prepare a final reaction volume of 250 μL containing 100 mM buffer (pH 5–6) (*see* **Note 17**) and 4 mM substrate, pre-warm 50 μL of 500 mM buffer together with 145 μL milliQ water and add 5 μL of 200 mM mono-acetylated substrate stock solution.
2. Control samples are prepared as above without adding enzyme.
3. Initiate the reaction by adding 50 μL of enzyme mixture directly after the substrate addition. Negative control samples will substitute the enzyme solution for 50 μL of water or storage buffer for the enzyme. Approximately 0.16 nkat of AcXE is reported to correspond to an A_{405} of 1.75 in 1 cm light path cuvette; absorbance values over 0.1 are recommended [35].
4. Incubate the reaction mix for 10 min at 40 °C or less (*see* **Note 18**).
5. Add 800 μL of saturated $Na_2B_4O_7$ solution (*see* **Note 19**) and measure absorbance at 405 nm.
6. To prepare a standard curve use 0.003–0.3 mM 4-nitrophenol for both 1 cm light path cuvette or 96 well plate after addition of stopping solution.

4 Notes

1. Acetonitrile was alternatively used to prepare 80 mM *p*NP-acetate stock solution [28]. In the product sheet of Sigma-Aldrich, methanol is recommended. Methanol (52%) was confirmed experimentally in our group to prepare a 10 mM *p*NP-acetate solution. Higher *p*NP-acetate concentrations or less than 52% of methanol led to precipitation. In methanol, stock solutions can be stored for about one week at 2–8 °C with only a small increase in free nitrophenol. By contrast, aqueous solutions should be freshly prepared each day (product sheet of Sigma-Aldrich).
2. 10 mM *p*NP can also be dissolved in 40% methanol (determined experimentally by our group).

3. The concentration of an enzyme stock solution is dependent on the enzyme preparation and dilution required in the assay.
4. A 100 mM stock solution of α-naphthyl acetate in DMSO is insoluble if diluted to 10 mM with water. Alternative methods describe using 1 mM substrate in buffer (buffer is dependent on enzyme used) [25] or 5 mM stock solution in methanol [24]. The maximal concentration of α-naphthyl acetate that is soluble in aqueous solution is reported to be 2 mM [14].
5. Fast Corinth Salt stopping solution contains insoluble particles. However, filtration can reduce product detection. This developing procedure is also pH sensitive.
6. Alternatively, 4MUA is reported to solubilize as a 5 mM acetone solution [28].
7. Use of DMSO to dissolve 4MU is recommended in the product information sheet provided by Sigma-Aldrich. 4MU is soluble at 1 mM in solutions containing 1% DMSO.
8. β-Xylosidase should not be capable of hydrolyzing the monoacetates of 4-nitrophenyl β-D-xylopyranosides (NPh-Xyl), should be free of deacetylating esterases, and activity on 4-nitrophenyl β-D-xylopyranosides should not be inhibited by acetic acid released by AcXEs.
9. Concentrations of 1–2 mM of *p*NP-acetate are described for this assay.
10. When determining rates of reaction, the reaction time is typically between 5–20 min, to minimize the chance of enzyme denaturation and conversion of more than 10% substrate.
11. If absorbance of substrate is high, solutions containing substrate but no enzyme should also be prepared as a negative reference.
12. The 4MUA assay reported by Shao et al. [7] begins by mixing the enzyme sample with the reaction buffer prior to substrate addition. This may be because the substrate can precipitate once suspended in aqueous solutions, depending on substrate concentration and reaction temperature.
13. Depending on the reaction temperature, 1 mM of substrate in aqueous solvent may lead to precipitation, which can interfere with product detection. In this case, residual substrate can be removed by filtration prior to absorbance readings.
14. Acetate buffer should not be used in this assay. The final substrate concentration will depend on the sensitivity of the detection method used.
15. 200 rpm is indicated in Juturu et al. [23]. The time needed for the reaction depends on the substrate concentration, enzyme performance, and sensitivity of the acetic acid detection method.

16. Heating to stop the reaction is preferred in the cases where pH changes caused by addition of H_2SO_4 are not desirable. If the reaction products will also be analyzed by NMR to determine the regioselectivity of the enzyme, then the precautions against acetyl group migration that are described in **Notes 17** and **18** should also be applied.
17. Acetyl group migration is avoided at pH 4, and is very low below pH 6 [34].
18. Acetyl group migration is also mitigated by running reactions below 40 °C and for short incubation times (i.e., less than 30 min) [34].
19. Saturated $Na_2B_4O_7$ solution intensifies the yellow color of product but can also increase background detection of remaining substrate.

Acknowledgments

This work was supported by a grant to G.M. from Ella and Georg Ehrnrooth foundation, Finland and an ERC Consolidator Grant to E.M. (BHIVE—648925). We thank Professor M. Tenkanen for her critical review of the manuscript.

References

1. Biely P (2012) Microbial carbohydrate esterases deacetylating plant polysaccharides. Biotechnol Adv 30:1575–1588. doi:10.1016/j.biotechadv.2012.04.010
2. Alalouf O, Balazs Y, Volkinshtein M et al (2011) A new family of carbohydrate esterases is represented by a GDSL hydrolase/acetylxylan esterase from *Geobacillus stearothermophilus*. J Biol Chem 286:41993–42001. doi:10.1074/jbc.M111.301051
3. Lombard V, Golaconda Ramulu H, Drula E et al (2014) The carbohydrate-active enzymes database (CAZy) in 2013. Nucleic Acids Res 42:D490–D495
4. Taylor EJ, Gloster TM, Turkenburg JP et al (2006) Structure and activity of two metal ion-dependent acetylxylan esterases involved in plant cell wall degradation reveals a close similarity to peptidoglycan deacetylases. J Biol Chem 281:10968–10975. doi:10.1074/jbc.M513066200
5. Biely P, Puls J, Schneider H (1985) Acetyl xylan esterases in fungal cellulolytic systems. FEBS Lett 186:80–84. doi:10.1016/0014-5793(85)81343-0
6. Johnson KG, Fontana JD, MacKenzie CR (1988) Measurement of acetylxylan esterase in *Streptomyces*. Methods Enzymol 160:551–560. doi:10.1016/0076-6879(88)60168-6
7. Shao W, Wiegel J (1995) Purification and characterization of two thermostable acetyl xylan esterases from *Thermoanaerobacterium sp.* strain JW/SL-YS485. Appl Environ Microbiol 61:729–733
8. Christakopoulos P, Mamma D, Kekos D et al (1999) Enhanced acetyl esterase production by *Fusarium oxysporum*. World J Microbiol Biotechnol 15:443–446. doi:10.1023/A:1008936204368
9. Biely P, Côté G, Kremnický L et al (1996) Substrate specificity of acetylxylan esterase from *Schizophyllum commune*: mode of action on acetylated carbohydrates. Biochim Biophys Acta 1298:209–222. doi:10.1016/S0167-4838(96)00132-X
10. Lee H, To RJ, Latta RK et al (1987) Some properties of extracellular acetylxylan esterase produced by the yeast *Rhodotorula mucilaginosa*. Appl Environ Microbiol 53:2831–2834

11. Degrassi G, Okeke BC, Bruschi CV et al (1998) Purification and characterization of an acetyl xylan esterase from *Bacillus pumilus*. Appl Environ Microbiol 64:789–792
12. Chung HJ, Park SM, Kim HR et al (2002) Cloning the gene encoding acetyl xylan esterase from *Aspergillus ficuum* and its expression in *Pichia pastoris*. Enzyme Microb Technol 31:384–391. doi:10.1016/S0141-0229(02)00122-9
13. Halgasová N, Kutejová E, Timko J (1994) Purification and some characteristics of the acetylxylan esterase from *Schizophyllum commune*. Biochem J 298(Pt 3):751–755
14. McDermid KP, Forsberg CW, Mackenzie CR (1990) Purification and properties of an acetylxylan esterase from *Fibrobacter succinogenes* S85. Appl Environ Microbiol 56:3805–3810. doi:10.1016/j.enzmictec.2007.09.007
15. Merino-Trigo A, Sampedro L, Rodrguez-Berrocal FJ et al (1999) Activity and partial characterisation of xylanolytic enzymes in the earthworm *Eisenia andrei* fed on organic wastes. Soil Biol Biochem 31:1735–1740. doi:10.1016/S0038-0717(99)00092-9
16. Bauer S, Vasu P, Persson S et al (2006) Development and application of a suite of polysaccharide-degrading enzymes for analyzing plant cell walls. Proc Natl Acad Sci USA 103:11417–11422. doi:10.1073/pnas.0604632103
17. Chungool W, Thongkam W, Raweesri P et al (2008) Production, purification, and characterization of acetyl esterase from *Streptomyces sp.* PC22 and its action in cooperation with xylanolytic enzymes on xylan degradation. World J Microbiol Biotechnol 24:549–556. doi:10.1007/s11274-007-9509-1
18. Hespell RB, O'Bryan-Shah PJ (1988) Esterase activities in *Butyrivibrio fibrisolvens* strains. Appl Environ Microbiol 54:1917–1922
19. Blum DL, Li XL, Chen H, Ljungdahl LG (1999) Characterization of an acetyl xylan esterase from the anaerobic fungus *Orpinomyces sp.* strain PC-2. Appl Environ Microbiol 65:3990–3995
20. Westlake K, Mackie RI, Dutton MF (1987) T-2 toxin metabolism by ruminal bacteria and its effect on their growth. Appl Environ Microbiol 53:587–592
21. Navarro-Fernández J, Martínez-Martínez I, Montoro-García S et al (2008) Characterization of a new rhamnogalacturonan acetyl esterase from *Bacillus halodurans* C-125 with a new putative carbohydrate binding domain. J Bacteriol 190:1375–1382. doi:10.1128/JB.01104-07
22. Neumueller KG, Streekstra H, Gruppen H et al (2014) *Trichoderma longibrachiatum* acetyl xylan esterase 1 enhances hemicellulolytic preparations to degrade corn silage polysaccharides. Bioresour Technol 163:64–73. doi:10.1016/j.biortech.2014.04.001
23. Juturu V, Aust C, Wu J (2013) Heterologous expression and biochemical characterization of acetyl xylan esterase from *Coprinopsis cinerea*. World J Microbiol Biotechnol 29:597–605. doi:10.1007/s11274-012-1215-y
24. Koseki T, Furuse S, Iwano K et al (1997) An *Aspergillus awamori* acetylesterase: purification of the enzyme, and cloning and sequencing of the gene. Biochem J 326:485–490
25. Poutanen K, Sundberg M (1988) An acetyl esterase of *Trichoderma reesei* and its role in the hydrolysis of acetyl xylans. Appl Microbiol Biotechnol 28:419–424
26. Poutanen K, Sundberg M, Korte H et al (1990) Deacetylation of xylans by acetyl esterases of *Trichoderma reesei*. Appl Microbiol Biotechnol 33:506–510
27. He X (2003) A continuous spectrophotometric assay for the determination of diamondback moth esterase activity. Arch Insect Biochem Physiol 54:68–76. doi:10.1002/arch.10103
28. Garcia-Sastre A, Villar E, Manuguerra JC et al (1991) Activity of influenza C virus O-acetylesterase with O-acetyl-containing compounds. Biochem J 273(Pt2):435–441
29. Biely P, Côté G, Kremnický L et al (1996) Substrate specificity and mode of action of acetylxylan esterase from *Streptomyces lividans*. FEBS Lett 396:257–260. doi:10.1016/0014-5793(96)01080-0
30. Biely P, Mastihubová M, Côté GL et al (2003) Mode of action of acetylxylan esterase from *Streptomyces lividans*: a study with deoxy and deoxy-fluoro analogues of acetylated methyl β-d-xylopyranoside. Biochim Biophys Acta 1622:82–88. doi:10.1016/S0304-4165(03)00130-2
31. Horton D, Lauterback JH (1969) Relative reactivities of hydroxyl groups in carbohydrate derivatives. Specific NMR spectral assignments of acetyl groups in methyl tetra-O-acetyl-alpha-D-glucopyranoside and related derivatives. J Org Chem 34:86–92. doi:10.1021/jo00838a021
32. Mastihubová M, Biely P (2001) A common access to 2- and 3-substituted methyl β-D-xylopyranosides. Tetrahedron Lett 42:9065–9067. doi:10.1016/S0040-4039(01)01957-8
33. Mastihubová M, Biely P (2004) Deoxy and deoxyfluoro analogues of acetylated methyl beta-D-xylopyranoside-substrates for acetylxylan esterases. Carbohydr Res 339:2101–2110. doi:10.1016/j.carres.2004.06.001
34. Mastihubová M, Biely P (2004) Lipase-catalysed preparation of acetates of

4-nitrophenyl β-D-xylopyranoside and their use in kinetic studies of acetyl migration. Carbohydr Res 339:1353–1360. doi:10.1016/j.carres.2004.02.016

35. Biely P, Mastihubová M, la Grange DC et al (2004) Enzyme-coupled assay of acetylxylan esterases on monoacetylated 4-nitrophenyl beta-D-xylopyranosides. Anal Biochem 332:109–115. doi:10.1016/j.ab.2004.04.022
36. Levisson M, Han GW, Deller MC et al (2012) Functional and structural characterization of a thermostable acetyl esterase from *Thermotoga maritima*. Proteins 80:1545–1559. doi:10.1002/prot.24041
37. Biely P, Cziszárová M, Agger JW et al (2014) *Trichoderma reesei* CE16 acetyl esterase and its role in enzymatic degradation of acetylated hemicellulose. Biochim Biophys Acta 1840:516–525. doi:10.1016/j.bbagen.2013.10.008
38. Biely P, Cziszárová M, Uhliariková I et al (2013) Mode of action of acetylxylan esterases on acetyl glucuronoxylan and acetylated oligosaccharides generated by a GH10 endoxylanase. Biochim Biophys Acta 1830:5075–5086. doi:10.1016/j.bbagen.2013.07.018
39. Rantanen H, Virkki L, Tuomainen P et al (2007) Preparation of arabinoxylobiose from rye xylan using family 10 *Aspergillus aculeatus* endo-1,4-β-d-xylanase. Carbohydr Polym 68:350–359. doi:10.1016/j.carbpol.2006.11.022
40. Pastell H, Tuomainen P, Virkki L et al (2008) Step-wise enzymatic preparation and structural characterization of singly and doubly substituted arabinoxylo-oligosaccharides with non-reducing end terminal branches. Carbohydr Res 343:3049–3057. doi:10.1016/j.carres.2008.09.013
41. Dalrymple BP, Cybinski DH, Layton I et al (1997) Three *Neocallimastix patriciarum* esterases associated with the degradation of complex polysaccharides are members of a new family of hydrolases. Microbiology 143:2605–2614. doi:10.1099/00221287-143-8-2605
42. Pouvreau L, Jonathan MC, Kabel MA et al (2011) Characterization and mode of action of two acetyl xylan esterases from *Chrysosporium lucknowense* C1 active towards acetylated xylans. Enzyme Microb Technol 49:312–320. doi:10.1016/j.enzmictec.2011.05.010
43. Martínez-Martínez I, Montoro-García S, Lozada-Ramírez JD et al (2007) A colorimetric assay for the determination of acetyl xylan esterase or cephalosporin C acetyl esterase activities using 7-amino cephalosporanic acid, cephalosporin C, or acetylated xylan as substrate. Anal Biochem 369:210–217. doi:10.1016/j.ab.2007.06.030
44. Cybinski DH, Layton I, Lowry JB et al (1999) An acetylxylan esterase and a xylanase expressed from genes cloned from the ruminal fungus *Neocallimastix patriciarum* act synergistically to degrade acetylated xylans. Appl Microbiol Biotechnol 52:221–225
45. Chong SL, Virkki L, Maaheimo H et al (2014) O-Acetylation of glucuronoxylan in Arabidopsis thaliana wild type and its change in xylan biosynthesis mutants. Glycobiology 24:494–506. doi:10.1093/glycob/cwu017
46. Uhliariková I, Vršanská M, McCleary BV et al (2013) Positional specifity of acetylxylan esterases on natural polysaccharide: an NMR study. Biochim Biophys Acta 1830:3365–3372. doi:10.1016/j.bbagen.2013.01.011
47. Neumüller KG, de Souza AC, van Rijn JH et al (2015) Positional preferences of acetyl esterases from different CE families towards acetylated 4-O-methyl glucuronic acid-substituted xylo-oligosaccharides. Biotechnol Biofuels 8:1–11. doi:10.1186/s13068-014-0187-6
48. Neumüller KG, de Souza AC, Van Rijn J et al (2013) Fast and robust method to determine phenoyl and acetyl esters of polysaccharides by quantitative 1H NMR. J Agric Food Chem 61:6282–6287. doi:10.1021/jf401393c
49. Biely P, Mastihubová M, Tenkanen M et al (2011) Action of xylan deacetylating enzymes on monoacetyl derivatives of 4-nitrophenyl glycosides of β-D-xylopyranose and α-L-arabinofuranose. J Biotechnol 151:137–142. doi:10.1016/j.jbiotec.2010.10.074
50. Biely P, Hirsch J, la Grange DC et al (2000) A chromogenic substrate for a beta-xylosidase-coupled assay of alpha-glucuronidase. Anal Biochem 286:289–294. doi:10.1006/abio.2000.4810
51. Topakas E, Kyriakopoulos S, Biely P et al (2010) Carbohydrate esterases of family 2 are 6-O-deacetylases. FEBS Lett 584:543–548. doi:10.1016/j.febslet.2009.11.095

Chapter 6

Methods for Determining Glycosyltransferase Kinetics

Maria Ngo and Michael D.L. Suits

Abstract

Glycosyltransferases are a class of biosynthetic enzymes that transfer individual activated monosaccharide units to specific acceptors. Colorimetric assays using the detection of released products such as para-nitrophenol and coupled assays for inorganic phosphate detection allow for convenient and quantifiable kinetic characterization. These techniques may be applied to determine the enzymatic activity of glycosyltransferases by indirectly measuring the transfer of nucleotide-activated donor carbohydrate units to various cognate acceptor molecules. In addition to an overview of these methods, the protocol for quantifying the glycosyltransferase activity used for the characterization of penicillin-binding proteins (PBPs) involving the transfer of lipid II to form elongated murein chains during bacterial cell wall synthesis is described herein.

Key words Glycosyltransferases, Enzyme kinetics, Malachite green dye, *para*-Nitrophenol, Colorimetric assay, Lipid II transfer

1 Introduction

Glycosyltransferases (GTs) are biosynthetic enzymes that catalyze the transfer of phosphate-activated carbohydrate donors to specific acceptor molecules, resulting in the formation of glycosidic bonds (*see* Carbohydrate Active Enzymes; http://www.cazy.org/) [1]. For instance, in the case of glycosyltransferases, donor glucosyl units in the form of uridine diphosphate glucose (UDP-Glu), a nucleotide-activated sugar, are conjugated to specific acceptor molecules to form products [2, 3]. While we overview three approaches for performing kinetic analyses with specific GTs, the methods outlined are broadly applicable for the kinetic characterization of GT systems in general. Specifically, we overview kinetic characterization of GTs through para-nitrophenol depletion, malachite green dye inorganic phosphate detection methods, and fluorescamine-labeled quantification via high-performance liquid chromatography (HPLC).

The determination of optimal condition for kinetic characterization, such as temperature and pH in which a GT reaction occurs may initially be determined using the relative absorbance of

D. Wade Abbott and Alicia Lammerts van Bueren (eds.), *Protein-Carbohydrate Interactions: Methods and Protocols*, Methods in Molecular Biology, vol. 1588, DOI 10.1007/978-1-4939-6899-2_6,

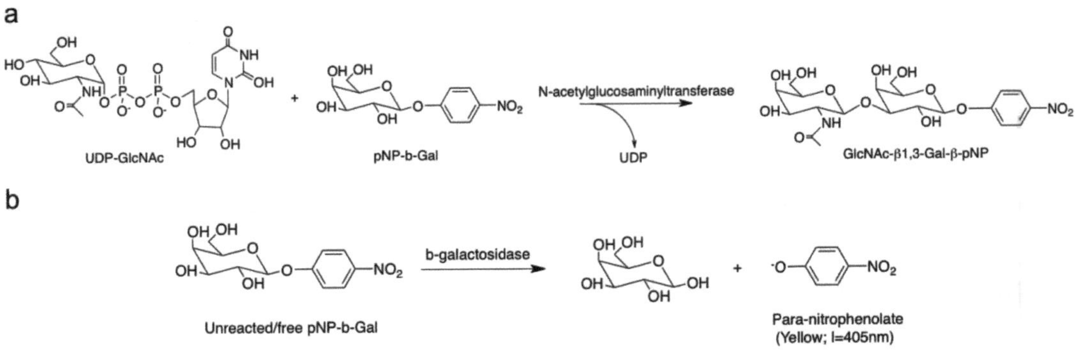

Fig. 1 (**a**) The overall reaction employed for the pNP- and malachite green assay in the following protocol. The incubation of UDP-GlcNAc (glycosyl donor) with pNP-β-Gal (glycosyl acceptor) in the presence of the enzyme will result in the transfer of the single glucosyl unit forming a glycosidic bond. (**b**) The analysis of enzyme activity from the pNP-assay requires an additional step of adding a β-galactosidase to cleave the galactosyl unit from any unreacted/free pNP-β-Gal. The release of para-nitrophenolate from the glycosyl acceptor, emits a *yellow* color and will allow an inverse quantification (λ = 405 nm) of how much product was formed. A high signal of para-nitrophenolate indicates that there is low enzyme activity due to the amount of free glycosyl acceptors still available for reaction

para-nitrophenol (pNP) at a wavelength of 405 nm under basic conditions. This methodology was utilized for the kinetic characterization of LgtA, an N-acetylglucosaminyltransferase from *Neisseria meningitidis*, which transfers the glucosyl unit from the glycosyl donor, uridine diphosphate N-acetylglucosamine (UDP-GlcNAc), to the glycosyl acceptor, 4-nitrophenyl-β-D-galactopyranoside (pNP-β-Gal) (Fig. 1a) [4–6]. After a suitable incubation period for catalysis to progress, an exolytic β-galactosidase is added to the reaction to hydrolyze any remaining pNP-β-Gal to D-galactoside and p-nitrophenol. Under basic conditions, the resonance structure of p-nitrophenol to the p-nitrophenolate ion results in the development of a yellow color that can be quantified at 405 nm, which correlates to the amount of unreacted pNP-β-Gal remaining (Fig. 1b). As not all GTs are active under basic conditions, this type of analysis may involve quenching the reaction through the addition of NaOH as an endpoint assay. A lower absorbance value will inversely correlate with a higher level of activity from the enzyme because there is a lower amount of available pNP-β-Gal remaining in the reaction mixture [7]. In this way, the pNP-colorimetric assay is useful for efficiently comparing the relative absorbance values, and thus, relative enzymatic activity. However, this approach is not typically recommended for the characterization of enzyme kinetics for extended incubation periods due to its susceptibility to auto-hydrolysis that can give greater phenolate signals than expected [8].

Although auto-hydrolysis of pNP substrates may be largely corrected against a suitable blank (buffer, control protein, acceptor, and UDP-donor only) [9], another method that provides a more sensitive and reliable measurement of GT activity is the malachite green

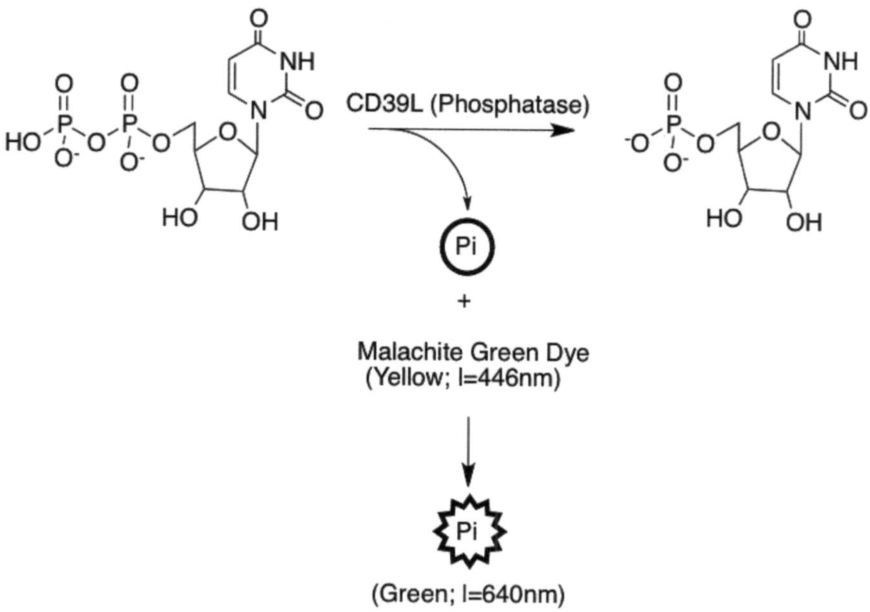

Fig. 2 Following the same reaction as seen in Fig. 1a, the enzyme activity is measured differently in the malachite green assay than in the pNP-assay (*see* Fig. 1b). Rather than detecting the quantity of free glycosyl acceptor, the malachite green assay employs the use of CD39L, a phosphatase, to cleave the end of UDP to release an inorganic phosphate. With the addition of the malachite green dye reagent (originally yellow in color and quantifiable at 446 nm), the reaction will yield a malachite green–phosphate complex molecule indicated by a color change to green, which can also be quantified at 640 nm. The detection of released inorganic phosphate from the glycosyl donor directly correlates to the amount of product formed

assay [10]. This assay typically employs the optimal conditions that were determined from initial pNP-colorimetric assays. In this approach the malachite green dye reagent acts indirectly in that it provides a measure of the release of inorganic phosphate from reacted glycosyl donors [11, 12]. When the donor sugar unit is transferred to the glycosyl acceptor, the remaining moiety of the glycosyl donor is UDP that has a diphosphate attached to the nucleotide. A phosphatase can be added which will cleave the inorganic phosphate from the free end of UDP (Fig. 2). The addition of malachite green dye reagent, which is initially yellow in color and can be read at 446 nm, will interact with one unit of phosphate to produce a large complex molecule concomitantly with a change to a green colour that can be quantified at 640 nm. A standard curve generated with inorganic phosphate from KH_2PO_4 will allow for the quantification of free phosphate that remained following the GT reaction.

Lastly, the glycosyltransfer reaction at the core of the biosynthesis of murein or peptidoglycan, is widely studied and involves a three-stage process of: (1) N-acetylglucosamine (NAG) conversion to N-acetylmuramic acid (NAM) in the cytosol, (2) the addition of undecaprenyl phosphate to NAM to produce Lipid I, and (3) the elongation of murein through the glycosyltransfer of the

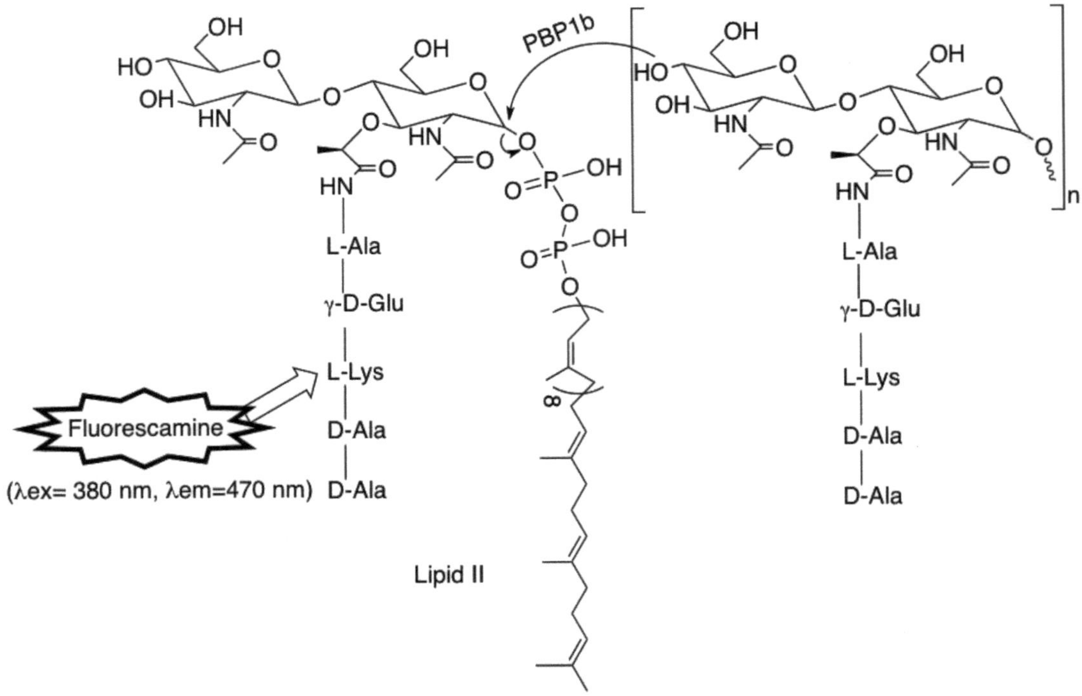

Fig. 3 Generally known as stage 3 of the biosynthetic process of the bacterial cell wall, penicillin-binding protein (PBP1b) elongates the formation of murein by transferring Lipid II to an extension of the carbohydrate chain. Here, PBP1b catalyzes the O-glycosidic bond between NAG and NAM units, which also releases undecaprenyl phosphate from the NAM unit. Fluorescamine is used to label the primary amine group belonging to lysine for quantification at an excitation wavelength of 380 nm, and emission wavelength of 470 nm

carbohydrate units (Lipid II), fluorescamine labeling and transpeptidation reaction by penicillin-binding proteins (PBPs) [13, 14] (Fig. 3). The protocol described [15] involves more sophisticated equipment for detection than the two aforementioned methods. However, the detection of fluorescamine-labeled Lipid II provides a sensitive method for the characterization of glycosyltransferase activity at membrane surfaces where substrate decoration is biologically significant, but the abundance of catalytic products is low. The reaction consists of reaction buffer, Lipid II (substrate), and PBP1b (GT-enzyme). In this reaction, the GT activity of PBP1b links units of Lipid II to produce an elongated chain. Lipid II is then labeled with fluorescamine for detection [16] and purified using an anion-exchange column through HPLC [17, 18]. The phosphate groups give Lipid II a net negative charge [19], allowing purification using anion-exchange chromatography. The NAM moiety also provides the short peptide bond in which fluorescamine can react with side group of lysine by adjusting the pH of the reaction above 9 (fluorescamine reacts with primary amine groups; lysine $pK_{a2} = 8.95$). Finally, the measurement of free Lipid

II is read at the excitation and emission wavelengths of 380 nm and 470 nm, respectively. Theoretically, the aliquots retrieved from the longer periods of incubation should result in a lower intensity of fluorescence for Lipid II as there would be less available substrate for the GT reaction. Also, because only Lipid II is labeled, the detection of the peak should appear at the same retention time from any collected aliquots. Integration of the peak areas and comparison of an N-acetyl lysine standard curve may also be used to determine concentration of remaining Lipid II [20].

2 Materials

Prepare all reagents with MilliQ water and store at room temperature unless otherwise stated.

2.1 General Glycosyltransferase Assays

1. Reaction buffer: 50 mM Tris–HCl, pH 7.5, 150 mM NaCl. Add about 50 mL of water to a glass beaker (*see* **Note 1**). Weigh 0.6057 g of Tris–HCl and use stir bar to mix solution while adjusting the pH to 7.5 (*see* **Note 2**). Dilute the buffer up to a final volume of 100 mL.
2. Glycosyl donor substrate: 500 μM of uridine 5′-diphospho-N-acetylglucosamine sodium salt (UDP-GlcNAc) in reaction buffer. Store at 4 °C.
3. Glycosyl acceptor substrate: 500 μM of 4-nitrophenyl β-D-galactopyranoside (pNP-β-Gal) in reaction buffer. Store at 4 °C (*see* **Note 3**).
4. Reaction enzyme: 50 nM glycosyltransferase in reaction buffer. Store at 4 °C (*see* **Note 4**).
5. 25 mM β-galactosidase in water. Store at 4 °C.
6. 1 M NaOH in water.
7. Clear 96-well microplate (*see* **Note 5**).
8. Metals: 10 mM $MgCl_2$; 10 mM $MnCl_2$; and 10 mM $CaCl_2$ (*see* **Note 6**) in water.
9. 10 mM ethylenediaminetetraacetic acid (EDTA) in water.
10. Buffers: 50 mM MES, pH 5.5, 50 mM MES, 6.0, and 50 mM MES, 6.5. Add 5 mL of water to a 10-mL conical tube. Weigh 0.0976 g of MES and use stir bar to mix solution while adjusting the pH to 5.5. Dilute to a final volume of 10 mL. Repeat this for pH 5 and 6.5.

    50 mM PIPES, pH 6.5 and 50 mM PIPES, pH 7.0. Add 5 mL of water to a 10-mL conical tube. Weigh 0.1512 g of PIPES and use stir bar to mix solution while adjusting the pH to 6.5. Repeat this for pH 7.0.

50 mM HEPES pH 7.0, 50 mM HEPES pH 7.0, 50 mM HEPES pH 7.0, 7.5, and 50 mM HEPES pH 7.0, 8.0. Add 5 mL of water to a 10-mL conical tube. Weigh 0.1192 g of HEPES and use stir bar to mix solution while adjusting the pH to 7.0. Repeat this for pH 7.5 and 8.0.

50 mM Tris–HCl, pH 8.0 and 50 mM Tris–HCl, 8.5. Add 5 mL of water to a 10-mL conical tube. Weigh 0.0606 g of Tris–HCl and use stir bar to mix solution while adjusting the pH to 8.0. Repeat synthesis for pH 8.5.

11. Phosphate standard curve: Use 1 M stock phosphate solution to prepare standard concentrations 0, 20, 40, 60, 80, 100, 120, 140, 160, 180, 200, 300, 400, 500, 600, 800, and 1000 μM in water.
12. Phosphatase: 0.1 μg/μL CD39L in water. Store at 4 °C.
13. Malachite Green Dye Reagent (*see* **Note 7**): 0.045% (w/v) malachite green, 4.2% (w/v) ammonium molybdate, 1% Triton X-100.

2.2 Anion-Exchange HPLC

1. Substrate: 2 μM Lipid II. Store at 4 °C.
2. Enzyme: 60 nM PBP1b. Store at 4 °C.
3. Reaction Buffer: 50 mM HEPES buffer, pH 7.5, 0.085% (w/v) decyl-polyethylene glycol (PEG), 10% (v/v) dimethyl Sulfoxide (DMSO), 10 mM $MgCl_2$. Add ~50 mL water to a glass beaker. Weigh 1.192 g of HEPES and use a stir bar to mix solution while adjusting the pH to 7.5. Add 0.0850 g of decyl-PEG (*see* **Note 8**), 10 mL of DMSO, and 0.0952 g of $MgCl_2$ to the buffer solution. Dilute to a final volume of 100 mL with water.
4. Quench buffer: 1 M bicine, pH 10.2. Add 5 mL of water to a 10-mL conical tube. Weigh 1.6320 g of bicine and use stir bar to mix solution while adjusting the pH to 10.2 (*see* **Note 9**). Dilute to a final volume of 10 mL with water.
5. Label: 10 mM fluorescamine in acetone. Protect from light with aluminum foil until time of use.
6. Elution buffer: 0.5 M ammonium acetate in 80% (v/v) methanol. Measure 1 L of 80% (v/v) methanol and slowly add 38.540 g of ammonium acetate while mixing. Filter-sterilize the solution.
7. Standard curve: Use 0.5 M stock N-acetyl lysine solution to prepare standard concentrations 0, 0.1, 0.5, 1, 5, 10, 20, 40, 60, 80, and 100 μM in water.
8. Anion-exchange HPLC with fluorescence detector.

3 Methods

3.1 Optimizing Conditions Using pNP Spectroscopic Properties

3.1.1 Effect of Temperature

1. Prepare 100 μL reactions in a 96-well microplate: Produce triplicates of each (*see* **Note 10**):
   (a) Blank—Pipette 10 μL of 500 μM of UDP-GlcNAc, 10 μL of 500 μM pNP-β-Gal, and dilute with 80 μL of reaction buffer.
   (b) Sample—Pipette 10 μL of 500 μM of UDP-GlcNAc, 10 μL of 500 μM pNP-β-Gal, 1 μL of 50 nM of reaction enzyme (*see* **Note 11**), and dilute with 79 μL of reaction buffer.
2. Mix reactions by pipetting up and down slowly (*see* **Note 12**).
3. Incubate at room temperature for 30 min.
4. Add 1uL of 25 mM of β-galactosidase to all the reactions. Incubate at 37 °C for 5 min.
5. Quench the reaction with 1uL of 1 M NaOH.
6. Read the microplate at 405 nm.
7. Repeat **steps 1–5** but for **step 3**, incubate the reactions at 37 °C for 30 min.
8. Proceed to the next section with the temperature that yielded the lower absorbance value as the optimal condition.

3.1.2 Effect of Metal Ions

1. Prepare 100 μL reactions in a 96-well microplate: Produce triplicates of each (*see* **Note 13**):
   (a) Blanks—Pipette 1μL of 10 mM of metal ion (Mg^{2+}, Mn^{2+}, Ca^{2+}) into 79 μL of 50 mM reaction buffer. Add 10 μL of 500 μM of UDP-GlcNAc, and 10 μL of 500 μM of pNP-β-Gal.
   (b) Samples with metal ions—Pipette 1 μL of 10 mM of metal ion (Mg^{2+}, Mn^{2+}, Ca^{2+}) into 78 μL of 50 mM reaction buffer. Add 10 μL of 500 μM of UDP-GlcNAc, and 10 μL of 500 μM of pNP-β-Gal. Add 1 μL of 50 nM of reaction enzyme last.
   (c) Blank with no metal—Pipette 1 μL of 10 mM of EDTA into 78 μL of 50 mM of reaction buffer. Add 10 μL of 500 μL of UDP-GlcNAc, and 10 μL of 500 μM of pNP-β-Gal.
   (d) Samples with no metals—Pipette 1 μL of 10 mM of EDTA into 78 μL of 50 mM of reaction buffer. Add 10 μL of 500 μM of UDP-GlcNAc, and 10 μL of 500 μM of pNP-β-Gal. Add 1 μL of 50 nM of reaction enzyme last.
2. Mix reactions by pipetting up and down slowly.
3. Incubate at optimized temperature for 30 min.

4. Add 1 μL of 25 nM of β-galactosidase to all reactions. Incubate at 37 °C for 5 min.
5. Quench the reaction with 1uL of 1 M NaOH.
6. Read the microplate at 405 nm.
7. Proceed to the next section with the optimal temperature and metal ion (or none) conditions that yielded the lower absorbance values.

3.1.3 Effect of pH

1. Prepare 100 μL reactions in a 96-well microplate:
   Produce triplicates of each (*see* **Note 14**):
   (a) Blanks—Pipette 1 μL of 10 mM of optimal metal ion (or none), 10 μL of 500 μM of UDP-GlcNAc, and 10 μL of 500 μM of pNP-β-Gal. Dilute to final volume of 100 μL with 50 mM MES pH 5.5. Reproduce blank setup in the next well, changing only the buffer at different pH (MES pH 5.5, 6.0, 6.5, PIPES pH 6.5, 7.0, HEPES pH 7.0, 7.5, 8.0, bicine pH 8.0, 8.5).
   (b) Samples—Pipette 1 μL of 10 mM of optimal metal ion (or none), 10 μL of 500 μM of UDP-GlcNAc, and 10 μL of 500 μM of pNP-β-Gal. Dilute to a final volume of 99 μL with 50 mM MES pH 5.5. Add 1 μL of 50 nM of reaction enzyme last. Reproduce sample set-up in the next well, changing only the buffer at different pH (MES pH 5.5, 6.0, 6.5, PIPES pH 6.5, 7.0, HEPES pH 7.0, 7.5, 8.0, bicine pH 8.0, 8.5).
2. Mix reactions by pipetting up and down slowly.
3. Incubate at optimized temperature for 30 min.
4. Add 1 μL of 50 mM of β-galactosidase to all reactions. Incubate at 37 °C for 5 min.
5. Quench the reaction with 1 μL of 1 M NaOH.
6. Read the microplate at 405 nm.
7. Proceed to kinetic characterization with the optimal temperature, metal ion (or none), and pH conditions (*see* **Note 15**).

3.2 Kinetic Activity with Malachite Green Assay

3.2.1 Phosphate Standard Curve

1. Dilute from a 1 M stock phosphate solution to prepare 100 μL standards in triplicate of concentrations 0, 20, 40, 60, 80, 100, 120, 140, 160, 180, 200, 300, 400, 500, 600, 800, 1000 μM in a microplate.
2. Add 200 μL of malachite green reagent dye to each standard. Mix the reaction by pipetting up and down slowly. Increasingly darker shade of green should appear as standard phosphate concentration increases.
3. Record absorbance values at 620 nm.
4. Repeat **steps 1–3** until linear plot (concentration vs. absorbance) is produced.

3.2.2 Kinetic Characterization at Optimal Conditions

1. Prepare 1 mL reactions in clean, sterile 1.5-mL centrifuge tubes:
   Produce triplicates of each:
   (a) 1 μM substrate blanks—Add 10 μL of optimal 10 mM of metal ion (or none), 2 μL of 500 μM of UDP-GlcNAc, and 100 μL of 500 μM of pNP-β-Gal. Dilute to final volume of 1 mL with 50 mM of reaction buffer at optimal pH.
   (b) 1 μM substrate samples—Add 10 μL of 10 mM of metal ion (or none), 2 μL of 500 μM of UDP-GlcNAc, and 100 μL of 500 μM of pNP-β-Gal. Dilute to 1 mL with 50 mM of reaction buffer at optimal pH. Add 10 μL of 50 nM of reaction enzyme last.
2. Mix reactions by pipetting up and down slowly.
3. Incubate the reactions at the optimal temperature for 60 min in total. Take a 100 μL aliquot from the reaction mix and transfer into the 96-well microplate. This is time = 0 measurement. Add 1 μL of 0.1 μg/μL of CD39L3 to the aliquot.
4. Immediately end the reaction (*see* **Note 16**) from the aliquot by adding 200 μL of malachite green reagent dye. Mix the reaction by pipetting up and down slowly.
5. Retrieve aliquots at 5, 10, 15, 20, 30, 40, 50, and 60 min marks. Repeat **steps 3** and **4** for each aliquot.
6. Read the microplate at 620 nm. Absorbance values should fall between the ranges of the phosphate standard curve. If not, dilute samples by tenfold or more until an appropriate absorbance is obtained.
7. Repeat **steps 1–8** for each increasing final UDP-GlcNAc concentrations (within the range of 1–300 μM) to produce a rectangular hyperbolic, Michaelis–Menten plot.

3.3 Anion-Exchange HPLC of Lipid II Reaction

3.3.1 N-Acetyl Lysine Standard Curve

1. Dilute from a 0.5 M stock N-acetyl lysine solution to prepare 100 μL standards of concentrations 0, 0.1, 0.5, 1, 5, 10, 20, 40, 60, 80, and 100 μM in a microplate.
2. Add 20 μL of 1 M bicine pH 10.2 to each standard concentration. Then label each standard with 12 μL of 10 mM fluorescamine. Mix the reaction by pipetting up and down slowly.
3. Collect fluorescence values of each standard at an excitation wavelength of 380 nm and emission wavelength of 470 nm.
4. Produce a standard curve (concentration vs. fluorescence units) for sample analysis.

3.3.2 For PBD Specific Glycosyltransferase-Type Reactions Using Lipid II Substrate

1. Prepare 1 mL reactions in clean 1.5-mL centrifuge tubes:
   Perform in triplicate:
   (a) Sample—Pipette 10 μL of 2uM Lipid II, 1 μL of 60 nM PBP1b, and dilute to a final volume of 1 mL with the reaction buffer.

2. Mix reactions by pipetting up and down slowly.
3. Incubate reaction at 30 °C. Collect a 30 μL aliquot from the sample reaction and transfer into a new tube labeled as time = 0.
4. Adjust the pH above 9 with 6 μL of 1 M bicine pH 10.2. Then label the samples with 3.6 μL of 10 mM fluorescamine.
5. Retrieve aliquots at t = 5, 10, 15, 30, 30, 45, 60 min marks. Repeat **steps 3–4** for each collected aliquot.
6. Inject (*see* **Note 17**) 20 μL of the labeled t = 0 sample onto an anion-exchange HPLC and subject it to a linear gradient of the elution buffer.
7. Collect data with an excitation wavelength of 380 nm and emission wavelength of 470 nm.
8. Repeat **steps 6–7** with the remaining labeled samples (t = 5, 10, 15, 30, 45, 60 min). The peaks from each sample should overlay around the same time of elution with the only change being a decrease in intensity of fluorescence as time increases.

4 Notes

1. Adding a small amount of water to the beaker prior to addition of dry masses helps to dissolve chemicals easier and prevents clumping at the bottom.
2. Use HCl or NaOH to adjust pH of buffer stocks appropriately.
3. Prepare fresh stock of glycosyl acceptor if solution appears yellow due to auto-hydrolysis.
4. Concentration of reaction enzyme may be adjusted higher or lower depending on the rate of activity.
5. No preference for the type of microplate; however, round-bottoms allow better mixing for assay-type reactions, while flat-bottoms allow for uniform surface and subsequent readings.
6. Listed are only a few commonly used divalent ions that may affect activity of enzyme.
7. Malachite green dye reagent should appear yellow in color in the absence of inorganic phosphate.
8. Heating solution slightly may help improve suspension of polyethylene glycol.
9. The pH does not have to be exact, as long as it is above pH 9.0 because the pK_a of lysine's side chain is ~8.95 for the ability of fluorescamine to bind.
10. In total, six wells will be used: three blanks and three samples incubated at room temperature, and three blanks and three samples at 37 °C.

11. Always add the enzyme last.
12. Slightly tapping the microplate on a flat surface without splashing may also help to mix reaction. Or a "shake" method can be applied from the microplate reader before readings if available.
13. In total, 24 wells will be used: three blanks for each metal ion (Mg^{2+}, Mn^{2+}, Ca^{2+}), three samples for each metal ion, three blanks for testing the absence of metal with EDTA, and three samples with EDTA.
14. In total, 60 wells will be used: three blanks and three samples for each pH (50 mM MES pH 5.5, 6.0, 6.5, PIPES pH 6.5, 7.0, HEPES pH 7.0, 7.5, 8.0, bicine pH 8.0, 8.5).
15. Reaction buffer may be changed if a better yield of activity is produced than the default buffer (50 mM Tris–HCl, pH 7.5).
16. Aliquots may be frozen with liquid nitrogen to prevent further activity while waiting for other time-aliquots. Or readings should be taken immediately after adding malachite green dye reagent to aliquot.
17. Ensure no air bubbles are injected along with sample to prevent flow rate or detection signal.

Acknowledgment

This work was supported by NSERC grant RGPIN-2014-05018 and the Faculty of Science, Wilfrid Laurier University.

References

1. Lombard V, Bernard T, Rancurel C et al (2010) A hierarchical classfication of polysaccharide lyases for glycogenomics. Biochem J 432:437–444
2. Campbell JA, Davies GJ, Bulone V, Henrissat B (1997) A classification of nucleotide-diphospho-sugar glycosyltransferases based on amino acid sequence similarities. Biochem J 326:929–942
3. Coutinho PM, Deleury E, Davies GJ, Henrissat B (2003) An evolving hierarchical family classification for glycosyltransferases. J Mol Biol 328:307–317
4. Liu Z, Lu Y, Zhang J et al (2003) Glycosylation reactions Using recombinant *Escherichia coli*. Appl Environ Microbiol 69:2110–2115
5. Liu XW, Xia C, Li L et al (2009) = O55:H7. Bioorg Med Chem 17:4910–4915
6. Zhang J, Kowal P, Fang J et al (2002) Efficient chemoenzymatic synthesis of globotriose and its derivatives with a recombinant alpha(1-4)-galactosyltransferase. Carbohydr Res 337: 969–976
7. Shen R, Wang S, Ma X et al (2010) An easy colorimetric assay for glycosyltransferases. Biochem 75:944–950
8. Bertram M, Manschot-Lawrence C, Floter E, Bornscheuer UT (2007) A microtiter plate-based assay method to determine fat quality. Eur J Lipid Sci Technol 109:180–185
9. Solhtalab M, Karbalaei-Heidari HR, Absalan G (2015) Tuning of hydrophilic ionic liquids concentration: a way to prevent enzyme instability. J Mol Catal B: Enzym 122: 125–130
10. Wu ZL, Ethen CM, Prather B et al (2011) Universal phosphatase-coupled glycosyltransferase assay. Glycobiology 21:727–733
11. Baykov AA, Evtushenko OA, Avaeva SM (1988) A malachite green procedure for

orthophosphate determination and its use in alkaline phosphatase-based enzyme immunoassay. Anal Biochem 171:266–270

12. Tedaldi L, Evitt A, Göös N et al (2014) A practical glycosyltransferase assay for the identification of new inhibitor chemotypes. MedChemComm 5:1193–1201
13. Schwartz B, Markwalder JA, Wang Y (2001) Lipid II: total synthesis of the bacterial cell wall precursor and utilization as a substrate for glycosyltransfer and transpeptidation by penicillin binding protein (PBP) 1b of *Escherichia coli*. J Am Chem Soc 123: 11638–11643
14. Terrak M, Nguyen-distèche M (2006) Kinetic characterization of the monofunctional glycosyltransferase from *Staphylococcus aureus*. J Bateriol 188:2528–2532
15. Schwartz B, Markwalder JA, Seitz SP et al (2002) A kinetic characterization of the glycosyltransferase activity of *Escherichia coli* PBP1b and development of a continuous fluorescence assay. Biochemistry 41:12552–12561
16. Blanchaert B, Adams E, Van Schepdael A (2014) An overview of analytical methods for monitoring bacterial transglycosylation. Anal Methods 6:7590–7596
17. Borisova SA, Zhao L, Melancon CE et al (2004) Characterization of the glycosyltransferase activity of DesVII: analysis of and implications for the biosynthesis of macrolide antibiotics. J Am Chem Soc 126:6534–6535
18. Cartwright AM, Lim EK, Kleanthous C, Bowles DJ (2008) A kinetic analysis of regiospecific glucosylation by two glycosyltransferases of *Arabidopsis thaliana*: domain swapping to introduce new activities. J Biol Chem 283:15724–15731
19. Barrett D, Barrett D, Leimkuhler C et al (2005) Kinetic characterization of the glycosyltransferase module of *Staphylococcus aureus* PBP2. Microbiology 187:2215–2217
20. Weis M, Lim EK, Bruce NC, Bowles DJ (2008) Engineering and kinetic characterisation of two glucosyltransferases from *Arabidopsis thaliana*. Biochimie 90:830–834

Chapter 7

Analyzing Activities of Lytic Polysaccharide Monooxygenases by Liquid Chromatography and Mass Spectrometry

Bjørge Westereng, Magnus Ø. Arntzen, Jane Wittrup Agger, Gustav Vaaje-Kolstad, and Vincent G.H. Eijsink

Abstract

Lytic polysaccharide monooxygenases perform oxidative cleavage of glycosidic bonds in various polysaccharides. The majority of LMPOs studied so far possess activity on either cellulose or chitin and analysis of these activities is therefore the main focus of this review. Notably, however, the number of LPMOs that are active on other polysaccharides is increasing. The products generated by LPMOs from cellulose are either oxidized in the downstream end (at C1) or upstream end (at C4), or at both ends. These modifications only result in small structural changes, which makes both chromatographic separation and product identification by mass spectrometry challenging. The changes in physicochemical properties that are associated with oxidation need to be considered when choosing analytical approaches. C1 oxidation leads to a sugar that is no longer reducing but instead has an acidic functionality, whereas C4 oxidation leads to products that are inherently labile at high and low pH and that exist in a keto-gemdiol equilibrium that is strongly shifted toward the gemdiol in aqueous solutions. Partial degradation of C4-oxidized products leads to the formation of native products, which could explain why some authors claim to have observed glycoside hydrolase activity for LPMOs. Notably, apparent glycoside hydrolase activity may also be due to small amounts of contaminating glycoside hydrolases since these normally have much higher catalytic rates than LPMOs. The low catalytic turnover rates of LPMOs necessitate the use of sensitive product detection methods, which limits the analytical possibilities considerably. Modern liquid chromatography and mass spectrometry have become essential tools for evaluating LPMO activity, and this chapter provides an overview of available methods together with a few novel tools. The methods described constitute a suite of techniques for analyzing oxidized carbohydrate products, which can be applied to LPMOs as well as other carbohydrate-active redox enzymes.

Key words Lytic polysaccharide monooxygenase, High-performance anion-exchange chromatography, Porous graphitized carbon, Aldonic acid, Gemdiol, Hydrophilic interaction liquid chromatography

1 Introduction

1.1 Chitin

The fact that LPMOs (previously known as CBM33 and GH61) are enzymes was discovered in 2010 by Vaaje-Kolstad et al. [1]. The first described activity for a LPMO was CBP21 (or *Sm*LPMO10A),

D. Wade Abbott and Alicia Lammerts van Bueren (eds.), *Protein-Carbohydrate Interactions: Methods and Protocols*, Methods in Molecular Biology, vol. 1588, DOI 10.1007/978-1-4939-6899-2_7, © Springer Science+Business Media LLC 2017

a chitin-active C1-oxidizing bacterial LPMO. The formation of oxidized chito-oligosaccharides was analyzed at high resolution using hydrophilic interaction liquid chromatography (HILIC) (Fig.1a) and MALDI-ToF (using both ion doping and isotope labeling; Fig. 1b). The HILIC method was developed for separating native chito-oligosaccharides and chitoaldonic acids in the same analysis. Whereas native chito-oligosaccharides retain well in acetonitrile–water, proper retention of aldonic acids (i.e., charged carboxylic acids) requires increased ionic strength and pH of the elution buffer.

In this early work, detection of oxidized oligomeric products by mass spectrometry was shown to be very useful, but also challenging, due to the equilibrium between the aldonic acids and the corresponding lactones; and the overlapping masses of the sodium adducts of oxidation products and the commonly observed K-adducts of the corresponding native oligosaccharides. In this respect, the combination of MALDI-ToF MS with isotope labeling and/or metal doping is very useful [1] (Fig. 1b, c). Metal doping is simple and should probably be used routinely. The use of labeling techniques is more complicated, but also a powerful tool for the identification of oxidations, as shown in Fig. 1b.

1.2 Cellulose

As predicted upon the discovery of LPMO activity on chitin in 2010, LPMOs acting on cellulose were described soon after, in 2011, by several groups [3–6]. Oxidized cello-oligosaccharides were analyzed by high-performance anion-exchange chromatography (HPAEC; Fig. 2) and MALDI-ToF MS. Analysis of cello-oligosaccharides requires different detection methods (e.g., pulsed amperometric detection, charged aerosol detection or ESI-MS) compared to chito-oligosaccharides because the former do not

Fig. 1 Analysis of C1-oxidized chito-oligosaccharides. (**a**) UHPLC-HILIC analysis of oxidized chito-oligosaccharides. Some ionic strength (15 mM Tris–HCl, pH 8.0) was essential to obtain retention of aldonic acids. Note that α and β-anomers would be separated under these chromatographic conditions if the oligosaccharide would have a normal reducing end (as in **c**, below). The lack of such separation thus indicates that the reducing end is modified. (**b**) (*i*) Equilibrium between the lactone form and the aldonic acid form of oxidized chitobiose. (*ii*) MS analysis at lower pH (promoting the lactone form) without metal doping, showing both the lactone (1257 for the sodium adduct) and the aldonic acid form (1275 for the sodium adduct) of the hexameric C1-oxidized product, as well as the distribution of sodium and potassium adducts. Note that the mass difference between sodium (23) and potassium (39) is 16. Furthermore, note that aldonic acids form diagnostic sodium and potassium salts, meaning that one proton is replaced by Na^+ or K^+. (*iii*) Analysis of the sample of (*ii*) at higher pH (almost no lactone) and (*iv*) after lithium (7) doping gives a simpler spectrum representing the chitoaldonic acids: *m/z* 1259 for the lithium adduct and *m/z* 1265, for the lithium salt of the lithium adduct. (*v*) Fragmentation mass spectra of chitohexaaldonic acid (indicated above the spectrum) formed in reactions with $H_2{}^{16}O$ (*black*) or $H_2{}^{18}O$ (*red*). Only the Y ions show *m/z* +2 for reactions run in $H_2{}^{18}O$, showing that the oxidation is in the down-stream end. (**c**) Analysis of $GlcNAc_2$ in its native and oxidized (aldonic acid) form. The oxidized disaccharide was generated by reaction of $GlcNAc_2$ with a chito-oligosaccharide oxidase called ChitO (*blue* chromatogram; *see* [2] for details). Note that native $GlcNAc_2$ (*green* chromatogram) elutes earlier, and with anomer separation. (**a**) and (**b**) were reproduced from [1] with permission from AAAS; (**c**) was reproduced from [2]

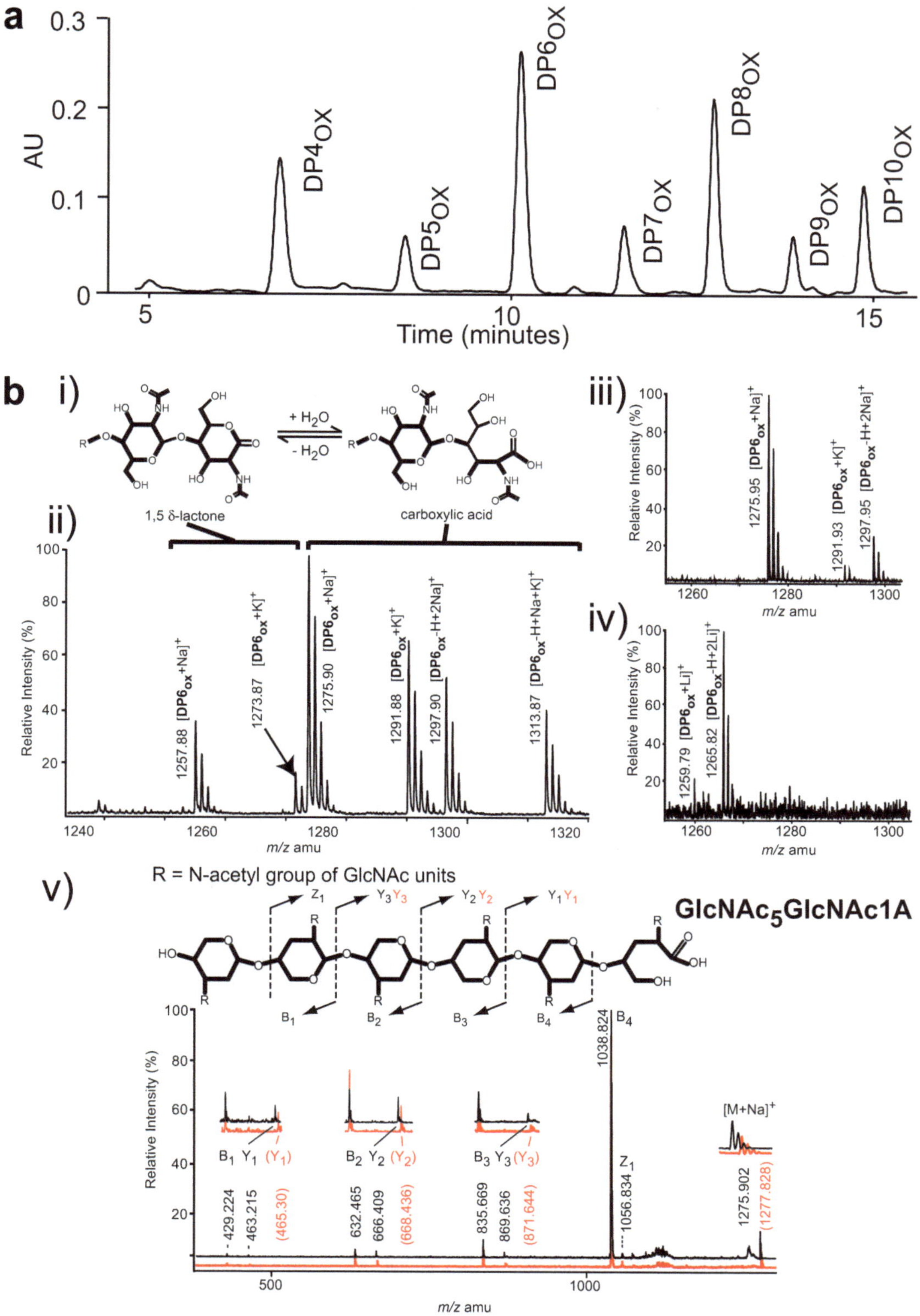
a
AU
Time (minutes)
DP4OX
DP5OX
DP6OX
DP7OX
DP8OX
DP9OX
DP10OX
b
i)
+ H2O
- H2O
1,5 δ-lactone
carboxylic acid
ii)
Relative Intensity (%)
1257.88 [DP6ox+Na]+
1273.87 [DP6ox+K]+
1275.90 [DP6ox+Na]+
1291.88 [DP6ox+K]+
1297.90 [DP6ox-H+2Na]+
1313.87 [DP6ox-H+Na+K]+
m/z amu
iii)
1275.95 [DP6ox+Na]+
1291.93 [DP6ox+K]+
1297.95 [DP6ox-H+2Na]+
iv)
1259.79 [DP6ox+Li]+
1265.82 [DP6ox-H+2Li]+
v)
R = N-acetyl group of GlcNAc units
GlcNAc5GlcNAc1A
Z1
Y3 Y3
Y2 Y2
Y1 Y1
B1
B2
B3
B4
[M+Na]+
429.224
463.215
(465.30)
632.465
666.409
(668.436)
835.669
869.636
(871.644)
1038.824
1056.834
1275.902
(1277.828)

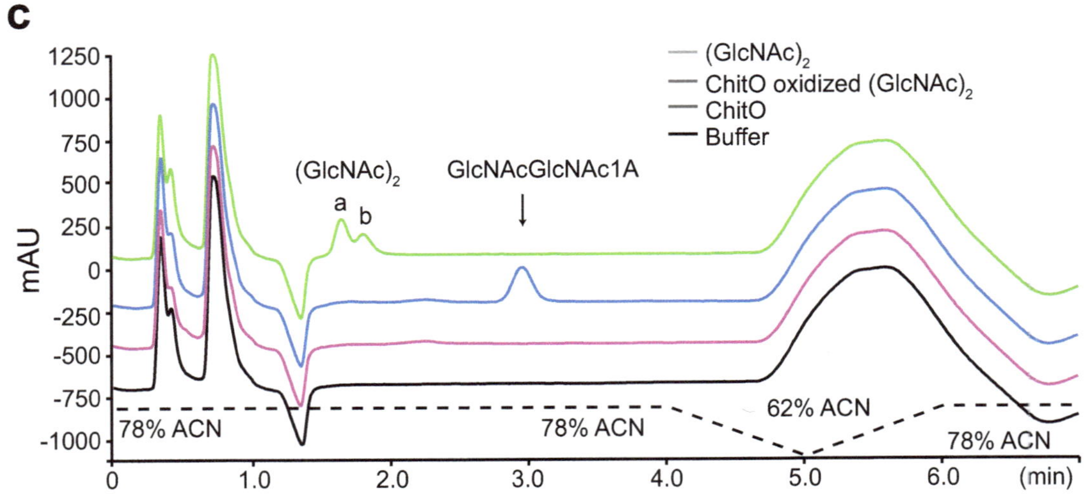

Fig. 1 (continued)

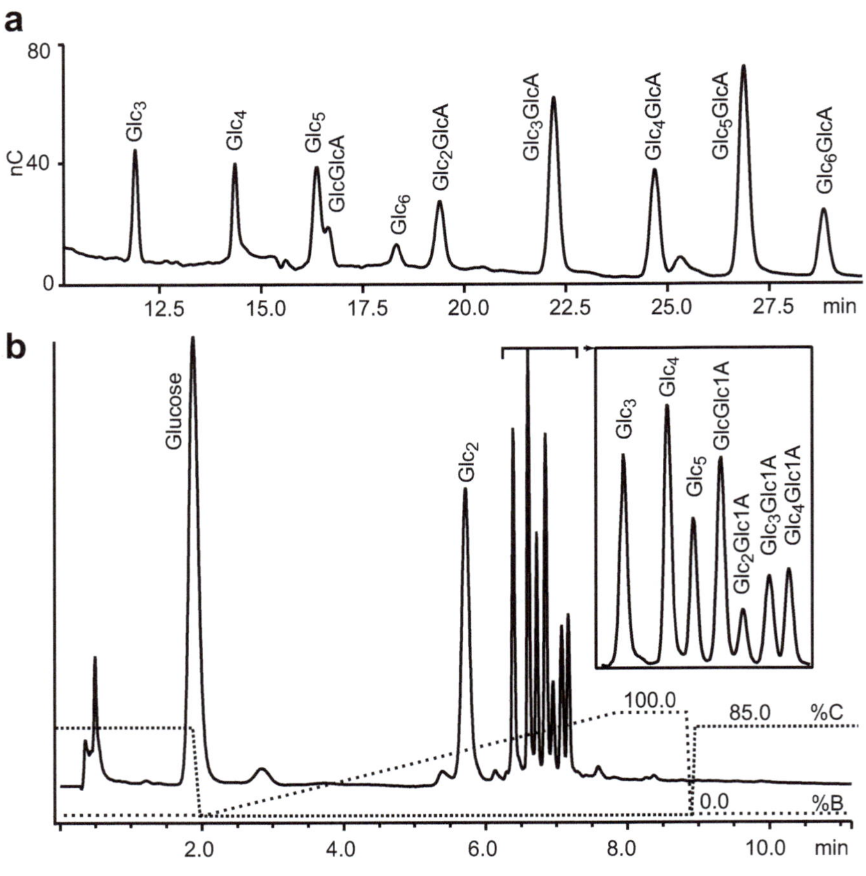

Fig. 2 HPAEC analysis of C1 oxidized cello-oligosaccharides. (**a**) Standard procedure (*see* Subheading 3.2); native cello-oligosaccharides elute first followed by the aldonic acids. There is a slight overlap between the two product clusters, which implies that the C1-oxidized monomer and dimer elute among the late eluting native oligosaccharides. (**b**) the faster procedure (*see* **Note 2**); by shortening the column length to a guard column, run times of approximately 10 min are sufficient, allowing high throughput HPAEC, if reduced resolution is acceptable. (**a**) is reproduced from [3] with permission from John Wiley and Sons; (**b**) is reproduced from ref. [7] with permission from Elsevier

absorb UV light. Aldonic acids are stable at high pH and HPAEC is therefore a suitable method for analyzing C1-oxidized cellodextrins. In later work, it was shown that the normally time-consuming HPAEC procedure could be made much faster by only using a guard column that provides sufficient separation of aldonic acids and native species with a total run time of only 10 min (Fig. 2b; [7]). At the alkaline pH during the HPAEC analysis, the equilibrium between the lactone and acid is strongly shifted toward the aldonic acid, and this makes HPAEC ideal for analysis of C1-oxidized products (the pK_a of cellobionic acid is 3.5 [8]).

In contrast to the chemically stable aldonic acids, oxidation in the non-reducing end (C4-oxidation) results in products that are much more prone to decomposition at extreme pH. It was recently shown that the gemdiols undergo on-column decomposition during HPAEC [8] (Fig. 3), leading to products with additional oxidations and, most importantly, native cello-oligosaccharides that have one less glucose that the original C4-oxidized product.

Due to this undesirable effect during HPAEC an alternative method based on porous graphitized carbon (PGC) chromatography has been developed to enable simultaneous screening of C1 and C4 oxidized cellodextrins (Fig. 4; [8]). PGC chromatography may be combined with charged aerosol detection (CAD), where sufficiently high sensitivity can be obtained by employing low ion-strength eluents (low nanomole range; [8]). While CAD

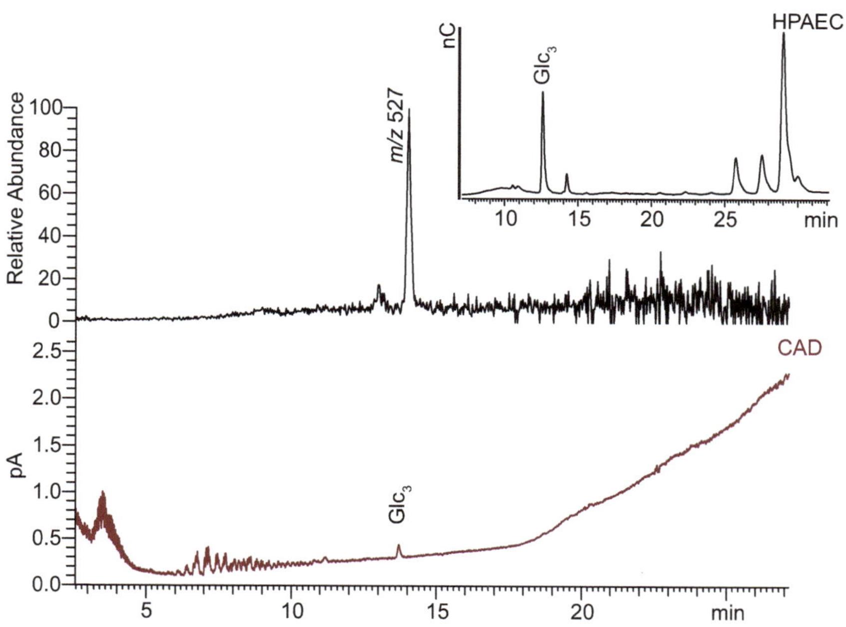

Fig. 3 Decomposition of C4-oxidized cellodextrins during HPAEC. A purified C4 oxidized tetramer was subjected to standard HPAEC (*upper right chromatogram*) and the peak eluting at 12.8 min (labeled Glc_3) was collected and reinjected on a PGC column where it coelutes with cellotriose (*lower choromatogram*; 13.8 min). Mass spectrometry analysis of the compound confirms that it is a native trimer (*m/z* = 527; sodium adduct). This figure was reproduced from [8] with permission from Elsevier

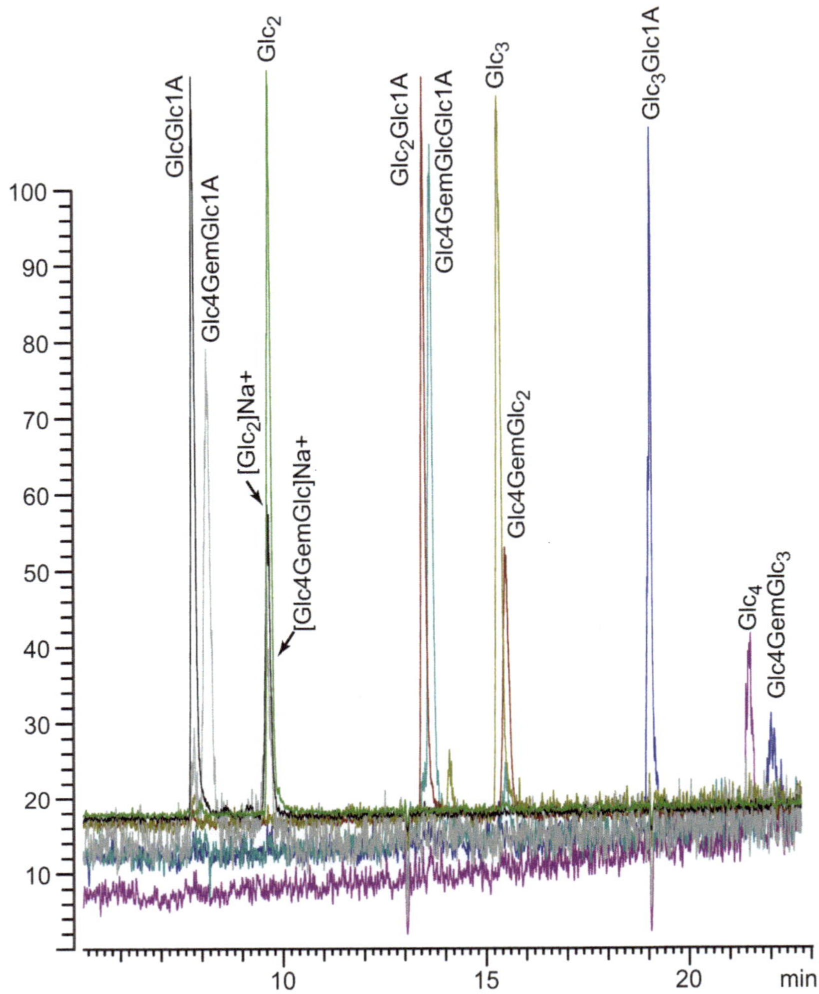

Fig. 4 Porous graphitized carbon chromatography of a mixture of C1-oxidized, C4-oxidized, double oxidized, and native cello-oligosaccharides. The chromatogram comprises 12 overlayed Extracted Ion Chromatogram (EIC) traces (lithium adducts if not otherwise indicated). The traces show that the PGC column offers superb separation between C1 and C4 oxidized cello-oligosaccharides, whereas native and C4-oxidized as well as C1-oxidized and double-oxidized species partially co-elute. This method may easily be combined with a universal detector like charged aerosol detection (CAD) for quantification purposes. The figure was reproduced from [8] with permission from Elsevier

detection provides sensitivity sufficient to enable kinetic analysis, limitations in product separation are such that the method is only suitable for oligosaccharides up to DP5, which needs to be taken into consideration when working with LPMOs that release higher DP products. Since the oxidative modifications in the non-reducing end render products with a high degree of similarity to native cello-oligosaccharides several products tend to co-elute. The big advantage here is that the PGC-CAD method, in contrast to HPAEC, can easily be combined with MS detection which allows discrimination between co-eluting species. Furthermore, native products may be removed by treatment with a beta-glucosidase (which acts from the non-reducing end and will only work on the native compounds). Obviously, because of the partial co-elution of C4-oxidized products and their native counterparts, it is of utmost importance to ensure that there is no background formation of native oligosaccharides by contaminating cellulases.

1.3 Do LPMOs Have Glycoside Hydrolase Side Activities?

The question whether or not LPMOs possess additional glycosyl hydrolase activity has been discussed repeatedly, due to the inherent appearance of native oligosaccharides during product analysis, for both C1 and C4 oxidizing LPMOs. Native oligosaccharides are formed from the substrate when an oxidation event occurs close to the reducing end (in the case of C1 oxidation) or non-reducing end (in the case of C4 oxidation; Fig. 5). In order to investigate

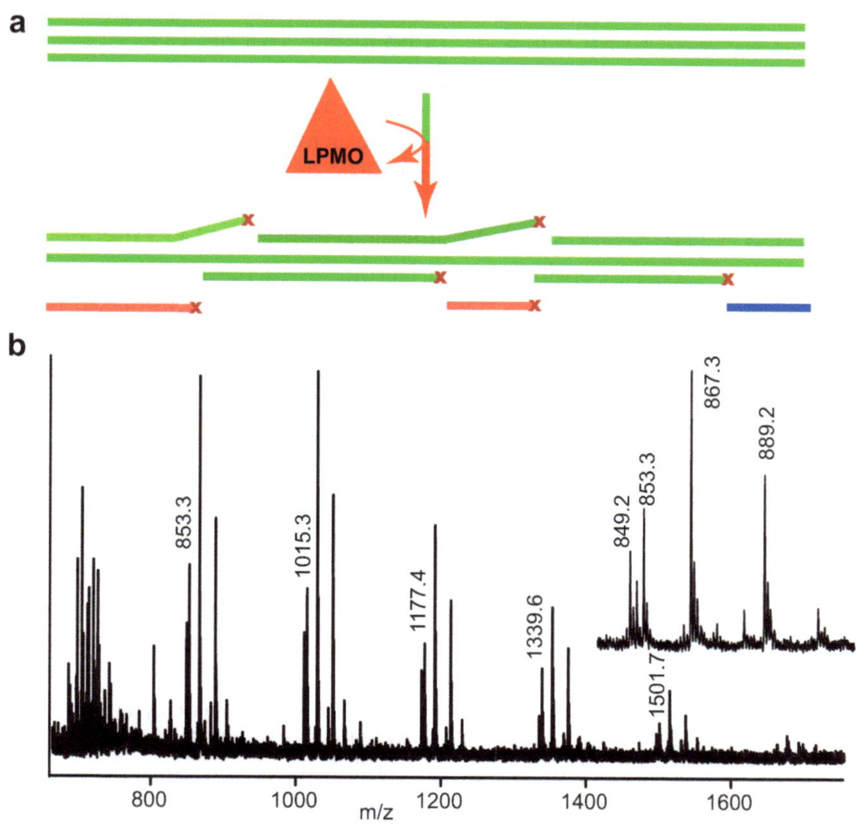

Fig. 5 LPMO activity on reduced cellulose. (**a**) Schematic presentation of LPMO activity (C1 oxidation) on a normal cellulose fiber (*green*). Oxidation events are marked by *red crosses*. Oxidation events may result in soluble shorter products (*red* and *blue*) and longer insoluble products (*green*). Soluble oxidized oligosaccharides are marked as *red lines* with *red crosses*. The *blue line* indicates the release of a native product, which may happen when the oxidation event occurs close to the reducing end of the substrate. (**b**) MALDI-ToF spectra of reduced PASC treated with a C1-oxidizing LPMO (*Pc*GH61D). The spectra show that the vast majority of the released native oligosaccharides are reduced (*m/z* values corresponding to reduced celloligosaccharides are 853.3, 1015.3, 1177.4, 1339.6, and 1501.7). The *inset* shows details for the pentamer (sodium adducts labeled): 849, DP5-lactone; 853, DP5, reduced; 867, DP5ox, alodonic acid; 889, DP5ox, sodium salt of the aldonic acid. (**a**) is reproduced from [7] with permission from Elsevier; (**b**) is reproduced from [4])

the true occurrence of glycosyl hydrolase activity, the substrate may be reduced to its corresponding glucitol in the downstream end. Upon reaction with LPMO, any glycoside hydrolase activity would reveal itself by high release of native oligosaccharides compared to the amount of reduced oligosaccharides. This approach was applied to C1-oxidizing *Pc*GH61D (or *Pc*LPMO9D) which resulted in increasing amounts of oxidized oligosaccharides compared to glucitol oligosaccharides over time (3:1 after 4 h; 13:1 after 20 h). Only minor amounts of native oligosaccharides were released, indicating that there is no significant glycoside hydrolase activity (Fig. 5; [4]).

C4-oxidizing LPMOs seem to produce larger amounts of native products, but this is due to chemcial modfications during the analytical process, as discussed to above. A nice overview over the (apparent) production of native cellodextrins by varying types of LPMOs may be found in Fig. 1 of ref. [7].

1.4 Analyzing MS Data and Differentiating Between C1 and C4 Oxidized Products

When analyzing C1 and C4 oxidized products using MALDI, a general feature for aldonic acids is that they form salts of their adducts, and this formation of double adducts is typical for carboxylic acids ([9, 10]). Since the aldonic acid to lactone equilibrium is favored toward the aldonic acid under MALDI conditions, signals corresponding to the lactone form tend to be weak. For C4 oxidized products, the 4-keto to gemdiol equilibrium is less skewed, and, due to efficient dehydration during spotting of MALDI sample plates, the keto signal (with m/z −2 compared to the native), is much more pronounced than the lactone signal (also m/z −2 compared to the native) for C1 oxidized products. Thus, despite similar masses of the products (note that the aldonic acid and the gemdiol have identical masses too), mass spectra will show characteristic differences that relate to C1 vs C4 oxidation. MS-MS approaches will yield different fragmentation patterns for C1 and C4 oxidized species [11]. In short, C4 oxidized species tend to show double water loss and dominant ring fragmentation, while C1 oxidized species do not exhibit these features during fragmentation but instead show diagnostic decarboxylation. For details on fragmentation, readers are directed to Isaksen et al. [11]. Extensive fragmentation data on xyloglucan is presented in Agger et al. [12].

In mass spectrometry, analyzing products from LPMO reactions is a major challenge because of the overlapping masses of common species. The mass difference of oxidized and native sugars is m/z 16 while the mass difference between sodium and potassium adducts is also m/z 16. In most experimental conditions both sodium and potassium adducts may be present, meaning that the native-potassium $[M+K]^+$ and the oxidized-sodium $[M+Na]^+$ species will have overlapping m/z values. This poses considerable problems in interpreting MS data and ion doping (*see* Subheading 3) is regularly used to reveal the true nature of the products.

Table 1
Theoretical and observed masses of native and oxidized cello-oligomers harboring potassium and sodium adducts, respectively

	Native $[M+K]^+$			Oxidized $[M+Na]^+$			
	Theoretical (m/z)	Observed (m/z)	Error (ppm)	Theoretical (m/z)	Observed (m/z)	Error (ppm)	Required resolution
Glc_2	381.0794	381.0787	1.84	381.1003	381.1000	0.79	18,234
Glc_3	543.1322	543.1315	1.29	543.1532	543.1533	−0.18	25,864
Glc_4	705.1850	705.1835	2.13	705.2060	705.2058	0.28	33,581
Glc_5	867.2378	867.2362	1.84	867.2588	867.2585	0.35	41,298
Glc_6	1029.2907	1029.2892	1.46	1029.3116	1029.3118	−0.19	49,249

The observed data were achieved using a high-resolution Q-Exactive mass spectrometer with the resolution set to 140,000. The required resolution was calculated as $R = M/\Delta M$, where ΔM is the difference between the two masses that one wants to separate

Importantly, modern mass spectrometers can achieve resolutions up to several hundred thousand using the orbitrap principle and Fourier transformation, while time-of-flight mass spectrometers typically achieve resolutions of up to around 40,000. By employing the newest high resolution methods it is actually possible to discriminate between species with "overlapping" m/z values. Table 1 shows relevant adduct pairs with overlapping m/z values and the resolution required for being able to unambiguously discriminate between those. Figure 6 illustrates the separation of these ion pairs using a Q-Exactive mass spectrometer employing the orbitrap principle and demonstrates that differentiation between native potassium adducts and oxidized sodium adducts can be achieved.

1.5 Verification of Product Identity and Product Stability

Product identities have been addressed in various manners, as discussed above. Notably, the identities of both C1-oxidized cello-oligosaccharides [7] and C4-oxidized products have also been verified by NMR [11]. Initially, it was proposed that some LPMOs could oxidize C6 [6], but there is little proof to support this oxidation mode, which, notably, would likely not lead to cleavage. This being said, it is not unlikely that other oxidations may occur, either directly by the LPMO, possibly as a side reaction, or indirectly, via tautomerization.

1.6 Recent Developments

Continuous developments in research on LPMOs and other carbohydrate-active redox enzymes will require an expanded repertoire of screening methods capturing a wider range of products. Today, LPMOs have been shown to be active on hemicelluloses (glucomannan, mixed linked beta-glucan and xyloglucan) [12, 13],

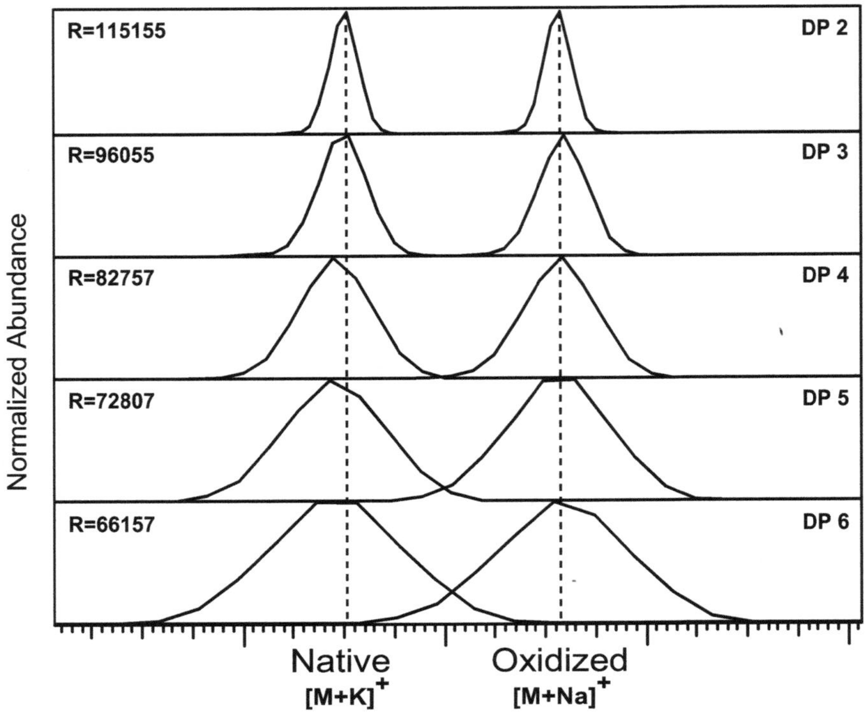

Fig. 6 The power of high-resolution mass spectrometry. The figure shows native (K adduct) and oxidized cello-oligosaccharides (Na adduct) in the range of DP 2-6. MS-acquisitions were done using a Q-Exactive mass spectrometer in profile mode with 140,000 set resolution on two separate samples: (I) 0.005 g/L Glc_{2-6} standard in 1.5 mM KCl and (II) products of an LPMO reaction. Mass spectra in a fixed relevant range (average *m/z* of native and oxidized ± 0.04 Da) of samples I and II were overlaid and the intensities normalized. The *X*-axis represents this *m/z* window of 0.08 Da in total. *R* indicates the peak width resolution as reported by the Thermo Xcalibur software

starch [14, 15], and xylan [16]. Some examples of LPMO action on more complex, natural samples have been shown in studies on xyloglucan and mannan by Agger et al. [12] and studies on xylan by Frommhagen et al. [16]. Another factor potentially contributing to sample complexity is the wide range of reductants that can activate LPMOs ([17, 18]). Certain reductants can cause challenges in analytics since sample compounds may give interfering signals. Since the effect of sample background varies depending on the choice of method (e.g., ascorbic acid has interfering peaks with C4 oxidized products in HPAEC, but not with C1 oxidized products) the effect of sample background must be evaluated for each individual analytical case. Furthermore, some reductants and the use of extreme reaction conditions may cause side reactions [19] which unavoidably results in even more complex product mixtures. Below, we will outline several of the techniques currently used to detect the activity of LMPOs.

2 Materials

2.1 HILIC Chromatography

1. Equipment: UHPLC system (Agilent 1290) with a diode array UV detector.
2. Columns: BEH amide column (2.1 × 150mm) and a BEH Amide VanGuard pre column (2.1 × 5 mm) both having a column material particle size of 1.7 μm.
3. Acetonitrile (ACN) HPLC grade.
4. Tris–HCl (15 mM, pH 8). Dissolve 2.36 g of Tris–HCl in approx 950 mL of Milli-Q water. Adjust to pH 8 with HCl, fill up to a final volume of 1 L with Milli-Q water.

2.2 HPAEC

1. Equipment: Ion exchange chromatography system with pulsed amperometric detection (PAD) (ICS3000, Dionex).
2. Columns: CarboPac PA1 (2 × 250 mm) and a CarboPac PA1 guard (2 × 50 mm) columns (Dionex, Thermo).
3. MilliQ water. Measure the desired volume of Milli-Q water (Type I, 18.2 MΩ·cm) directly in a dedicated HPAEC mobile phase bottle. Sonicate for 20 min to remove dissolved carbon dioxide and transfer immediately hereafter to the HPAEC system and store under N_2-saturated headspace.
4. Sodium Hydroxide (0.1 M). Measure exactly 2 L of Milli-Q water (Type I, 18.2 MΩ·cm) directly in a dedicated HPAEC mobile phase bottle. Sonicate for 20 min to remove dissolved carbon dioxide and transfer immediately hereafter to the HPAEC system and store under N_2-saturated headspace. Add 10.4 mL of NaOH from a 50% liquid solution. Do not use NaOH pellets. Close the mobile phase bottle and swirl gently to ensure proper mixing. Maintain N_2-saturated headspace until the mobile phase is discarded.
5. Sodium acetate (1 M in 0.1 M NaOH). Dissolve 82.03 g of anhydrous sodium acetate (≥99% purity) in 1 L of Milli-Q water (Type I, 18.2 MΩ·cm). Filter the solution through no less than a 0.45 μm filter directly into a dedicated HPAEC mobile phase bottle. Sonicate for 20 min to remove dissolved carbon dioxide and transfer immediately hereafter to the HPAEC system and store under N_2-saturated headspace. Add 5.2 mL of NaOH from a 50% liquid solution. Do not use NaOH pellets. Close the mobile phase bottle and swirl gently to ensure proper mixing. Maintain N_2-saturated headspace until the mobile phase is discarded.

2.3 PGC Chromatography

1. Equipment: UHPLC system (Ultimate3000RS, Dionex) set up with charged aerosol detection (Corona ultra) and an ESI-MS detector (Velos pro).

2. Columns: Porous graphitized carbon columns; Hypercarb (2.1 × 150mm; 3 μm) and a Hypercarb guard (2.1 × 10 mm; 3 μm) from Thermo Electron Corporation, San José, USA.
3. Ammonium acetate (10 mM, pH 8). Dissolve 0.771 g of ammonium acetate in approx 950 mL of Milli-Q water. Adjust pH with ammonia until pH 8 and fill up to a final volume of 1 L with Milli-Q water.
4. Acetonitrile (ACN) HPLC grade.
5. Sodium chloride (1 μM NaCl, no buffer). Dissolve 0.058 g of NaCl in 1 L Milli-Q water (1 mM). From this 1 mM solution, mix 1 mL with Milli-Q water to a final volume of 1 L.

2.4 MALDI-ToF Analysis and Lithium Doping

1. Equipment: Bruker Ultraflex MALDI-TOF/TOF instrumentwith a Nitrogen 337 nm laser beam (Bruker Daltonics GmbH, Bremen, Germany).
2. Lithium chloride solution (the LiCl concentration should be approximately twice the concentration of the buffer used in the LPMO reaction). Dissolve the desired amount of LiCl in Milli-Q water.
3. 2,5-dihydroxybenzoic acid (DHB) solution: dissolve 4.5 mg DHB (Bruker Daltonics) in 150 μL acetonitrile and 350 μL water.
4. MTP 384 target plate ground steel TF from Bruker Daltonics (or equivalent).

2.5 High Resolution MS to Discriminate Between Potassium and Sodium Adducts by Direct Infusion Q-Exactive MS

1. Q-Exactive hybrid quadrupole orbitrap mass spectrometer (Thermo Scientific, Bremen, Germany).
2. MilliQ water.
3. Potassium chloride (KCl, 1.5 mM). Dissolve 111.8 mg KCl in MilliQ water and adjust to a final volume of 1 L.

2.6 Isotope Labeling

1. Tris–HCl (*see* Subheading 2.1).
2. Ascorbic acid (AA, 1 mM), prepare a fresh 100 mM stock solution of reduced ascorbic acid by dissolving 17.6 mg AA in 1 mL MilliQ water.
3. N_2(g).
4. $H_2{}^{18}O$ and $^{18}O_2$ from Cambridge Isotope laboratories (catalogue numbers OLM-240-97-1 and OLM-212-PK, respectively).

2.7 Reduction of Cellulose

1. Phosphorous acid swollen cellulose prepared from Avicel PH-101 cellulose from Fluka analytical (Sigma-Aldrich, St. Louis, USA).
2. Sodium hydroxide (NaOH, 12.5 mM).
3. Sodium borohydride ($NaBH_4$) anhydrous.
4. Glacial acetic acid.

3 Methods

The protocols provided below cover most available methods for characterizing LPMOs and analyzing oxidized products by HPLC and mass spectrometry. For more details readers are directed to the original publications related to the method in question. Where applicable, notes are appended in Subheading 4.

3.1 HILIC Chromatography for the Analysis of C1 Oxidized Chitin Oligosaccharides

1. Use an instrumental setup as mentioned in Subheading 2.1 or similar.
2. Dissolve sample in 72% (v/v) acetonitrile (ACN). Inject 5 μL sample (*see* **Note 1**).
3. Operate the system at 30 °C (column temperature) and a flowrate of 0.4 mL/min.
4. Keep starting conditions 72% ACN (A):28% 15 mM Tris–HCl pH 8.0 (B) for 4 min, then use an 11 min linear gradient to 62% A: 38% B, which is held for 3 min.
5. Recondition column by applying a 2 min gradient to initial conditions and subsequent operate at initial conditions for 5 min.
6. Monitor eluted oligosaccharides by recording UV absorption at 205 nm (*see* **Note 1**).

3.2 HPAEC-PAD for the Analysis of Various Cello-Oligosaccharides

1. Use an instrumental setup as mentioned in Subheading 2.2 or similar.
2. Centrifuge samples for 3 min in an Eppendorf centrifuge at maximum speed and transfer supernatants to HPLC vials without any further adjustments.
3. Set column temperature 30 °C and use 0.25 mL/min flow rate.
4. Use mobile phases containing 0.1 M NaOH (A) and 0.1 M NaOH, 1 M sodium acetate (B) (*see* **Note 2A**).
5. Use the following gradient: a 10 min linear gradient from 100% A (starting condition) to 10% B, a 15 min linear gradient to 30% B, a 5 min exponential gradient (Dionex curve 6) to 100% B.
6. Recondition column by running initial conditions (100%A) for 9 min [3] (*see* **Note 2A**).
7. For other applications and mass spectrometry adaptations (*see* **Note 2B** and **C**).

3.3 Simultaneous Analysis of Aldonic Acids and C4-Oxidized Cello-Oligosaccharides by Porous graphitized Carbon (PGC) Chromatography

1. Use an instrumental setup as written in Subheading 2.3 or similar.
2. Centrifuge samples for 3 min in an Eppendorf centrifuge at maximum speed and transfer supernatant to HPLC vials without any further adjustments.
3. Operate the column at 0.4 mL/min and 70 °C.

4. Use the following gradient: 0–1 min, 100% eluent A (10 mM NH_4-acetate, pH 8.0); 1–15 min, linear gradient to 27.5% eluent B (acetonitrile); 15–28 min, linear gradient to 60% B; 28–35 min, isocratic at 60% B [7] (*see* **Note 3**).
5. Recondition the column by applying 100% eluent A for 9 min.
6. The chromatography system was composed as follows: Dionex UltiMate3000 RSLC set up with detection by electrospray ionization-MS (VelosPro LTQ linear iontrap, Thermo Scientific) or optionally with parallel use of MS with a Charged Aerosol Detector (CAD) (ESA inc., Dionex, Sunnyvale, USA). ESI-MS detection is used for qualitative detection, whereas the CAD is used for quantitation. The CAD detector is a universal detector, where the response is independent of the analyte, making it possible to prepare calibration curves with easily accessible standards that are structurally similar, but not identical, to the analyte. Such standards need to have the same elution time as the analyte to prevent gradient effects (*see* **Note 3**).

3.4 MALDI-ToF Analysis and Lithium Doping

1. To prepare samples for MALDI-ToF analysis reactions should be run at low buffer concentrations (as a rule of thumb, less than 50 mM, but lower is better), and no MS-incompatible ions like phosphate/nitrate should be used.
2. Centrifuge samples in an Eppendorf centrifuge at maximum speed for 2 min at room temperature.
3. Apply 2 μL saturated DHB solution to a MALDI plate.
4. Apply 1 μL sample, and mix with 3).
5. Dry the spot under a stream of warm air.
6. Analyze the sample on a MALDI-ToF instrument (*see* **Note 4A**).
7. Mix 1 μL sample with 9 μL LiCl solution and vortex for 5 s.
8. Apply 2 μL saturated DHB solution to a MALDI plate. DHB is the standard matrix used for all MALDI experiments, but other matrices may work equally well.
9. Add 1 μL of the lithium-doped sample from (1) to (2) and mix.
10. Dry the spot under a stream of warm air.
11. Analyze the sample on a MALDI-ToF instrument (*see* **Note 4B**).

3.5 High Resolution MS to Discriminate Between Potassium and Sodium Adducts by Direct Infusion Q-Exactive MS

The following procedure describes the manual analysis of oligosaccharides using direct injections on a Q-Exactive hybrid quadrupole orbitrap mass spectrometer. Other high-resolution mass spectrometers equipped with a nano-flow ion source can also be used with minor adaptations of this protocol.

1. Mount the Nanospray Flex ion source to the Q-Exactive, and change the nano head to the Offline nano ES head to allow manual use of tapered capillary emitters.

2. Prepare the sample by diluting with water or potassium chloride (1.5 mM) when conducting potassium doping.
3. Cut the emitter to a suitable length and load 3 μL sample directly into the emitter using a gel-loader tip. Avoid bubbles. Use a new emitter for every sample.
4. Assemble the emitter in the ion source and position the tip in front of the skimmer and apply positive pressure using a syringe to start the liquid flow. If no flow is observed, a gentle "crash" into the skimmer to scratch the emitter tip may be necessary. The optimal flow lays between 50 and 300 nL/min and, although this is difficult to control, the flow rate can be estimated by the time it takes for the complete sample to be injected (i.e., 3 μL fully injected in 10 min gives a flow of 300 nL/min).
5. Apply the settings from Table 2 and switch on the Q-Exactive. Adjust the emitter position and the spray voltage, if needed, to achieve a stable spray.
6. Acquire MS full scan data for 20 s. Optionally, the Q-Exactive can be set to cycle between MS full scans and MS/MS fragment scans by providing an inclusion list of selected precursor ions. Fragmentation can be achieved using stepped normalized collision energy from 25 to 48. In order to achieve high quality MS/MS spectra, the number of microscans should be at least three and the maximum injection time set to 800 ms. (*see* **Note 5**).

Table 2
Q-Exactive parameters for direct injections

Sheet gas flow rate	0
Aux gas flow rate	0
Sweep gas flow rate	0
Spray voltage (kV)	0.9–1.5[a]
Capillary temperature	250 °C
S-lens RF level	50
Scan range	150–2000 *m/z*
Resolution	140,000
Polarity	Positive
AGC target	5e6
Maximum inject time	100 ms
Spectrum scan mode	Profile
Micro scans	1

[a]The spray voltage normally needs to be adjusted to the needle distance and the sample concentration

3.6 Isotope Labeling

Stable isotope reagents such as $H_2{}^{18}O$ and $^{18}O_2$ can be used in LPMO reactions to demonstrate the incorporation of molecular oxygen and water in the products formed by these enzymes. Identification of products containing the ^{18}O isotope is achieved by mass spectrometry where products have m/z +2 compared to products formed in ^{16}O conditions. Such experiments and product analysis have been described in detail in Vaaje-Kolstad et al. [1] and the protocols used by these authors are outlined below. The reaction volume, substrate concentrations, etc. described are optimal for demonstrating activity of a chitin-active LPMO toward chitin, but may need optimization if the methods are used for other substrates and enzymes.

3.6.1 Reactions in Buffered $H_2{}^{18}O$

1. Suspend 2.0 mg of dry substrate in 1.0 mL pure $H_2{}^{18}O$ in a 2.0 mL glass vial. This leaves a headspace of approximately 1 mL when the vial is sealed. Seal the vial airtight and mix thoroughly. Let the substrate suspension hydrate overnight at room temperature.
2. Dissolve a sufficient amount of reducing agent (e.g., ascorbic acid) in an appropriate volume of pure $H_2{}^{18}O$ to yield a final concentration of 1.0 M. Keep the solution in an aluminum foil wrapped test tube (to shield from light) on ice.
3. In order to achieve the correct pH in the $H_2{}^{18}O$ reaction solution, transfer 10 μL of a 1.0 M nonvolatile buffer (e.g., Tris–HCl pH 8.0) to a 2.0 mL glass vial and evaporate off the liquid by heating with dry air (approximately 60 °C).
4. Transfer 498 μL of the substrate suspension to the glass vial containing the dried buffer and mix thoroughly to dissolve the buffer components. Subsequently, add 0.5 μL of the reducing agent solution (dissolved in $H_2{}^{18}O$) and 0.75 μL of a 660 μM solution of the LPMO (dissolved in $H_2{}^{16}O$) to the buffered substrate suspension to yield final concentrations of 1 mM reducing agent and 1 μM enzyme. A high concentration of the enzyme stock solution is desirable in order to keep $H_2{}^{16}O$ contamination at a minimum (replacing the $H_2{}^{16}O$ in the enzyme stock solution to $H_2{}^{18}O$ is possible, but is not considered as an option due to the high cost of pure $H_2{}^{18}O$).
5. Seal the glass vial airtight and incubate the reaction for an appropriate time (usually 1 h or more) at an appropriate temperature (enzyme dependent) and with vigorous mixing (e.g., 1000 rpm in an Eppendorf Thermomixer).analyze products by MS.

3.6.2 Reactions in a Solution Saturated with $^{18}O_2$

1. Prepare a buffered LPMO reaction solution (e.g., 20 mM Tris–HCL pH 8.0) containing 2.0 mg/mL substrate and 1.0 mM reducing agent (e.g., ascorbic acid) in a glass vial that can be closed airtight with a screw cap containing a Teflon coated rubber septum. Make sure that the reaction volume only represents approximately 50% of the vial volume.

2. Close the vial tightly with the screw cap and connect to a Schlenk line (for details, *see* procedure described under the heading "molecular oxygen free reaction" in Vaaje-Kolstad et al. [1] to remove oxygen from the headspace and dissolved oxygen from the solution. This is achieved by performing five cycles of degassing and N_2 filling. It is important to have a slight N_2 over pressure after the final N_2 filling in order to avoid contamination of the head space with air when removing the vial from the Schlenk line.
3. Remove the vial from the Schlenk line and perforate the septum with the needle of a Hamilton syringe preloaded with a concentrated LPMO solution. Add the LPMO to the reaction mixture by injecting an appropriate volume (as low as possible in order to minimize addition of dissolved $^{16}O_2$) to a final concentration of 1.0 μM. Withdraw the Hamilton syringe from the vial.
4. Connect a gas cylinder containing compressed $^{18}O_2$ gas to the vial by pushing a needle fitted to the outlet of the gas cylinder through the vial septum.
5. Using the Schlenk line, place the vial under vacuum in order to remove atmospheric gas residing in the tubing connected to the $^{18}O_2$ gas container and the headspace of the vial.
6. Disconnect the vial from the Schlenk line needle and fill the head space of the vial with $^{18}O_2$ gas by slowly opening the gas cylinder regulator.
7. After 30 s, close the gas cylinder regulator and carefully remove the needle from the vial.
8. Incubate the vial containing the LPMO reaction mixture for an appropriate number of hours (usually 1–24) at the desired temperature with vigorous mixing (e.g., 1000 rpm in an Eppendorf Thermomixer) and analyze products by MS.
9. The lactone—aldonic acid equilibrium will lead to exchange of oxygen atoms (*see* **Note 6**).

3.7 Reduction of Cellulose

Reduced phosphoric acid swollen cellulose (PASC) can be prepared with the following procedure:

1. Use a 2 mL 2% (w/v) PASC suspension in water and centrifuge for 3 minutes at 21,000 × *g*. Remove the supernatant and resuspend the pellet in 1 mL MilliQ H_2O. Centrifuge again for 3 minutes at 21,000 × *g* and remove the supernatant.
2. Resuspend the pellet in 4 mL 12.5 mM NaOH.
3. Add 25 mg $NaBH_4$ and leave the tubes at ambient temperature overnight with occasional stirring.
4. Quench the reaction by neutralizing with 100 μL glacial acetic acid, followed by centrifugation as described above.

5. Wash the pellet four times with MilliQ H_2O and finally resuspend in MilliQ H_2O to obtain a 2% (w/v) solution of reduced PASC.

3.8 Future Perspectives

In this chapter we outline several crucial aspects of carbohydrate analysis that can be applied to analyze soluble products generated by LPMOs. More work is needed for developing effective methods for monitoring the insoluble products, i.e., oxidations on the insoluble material. Insight into oxidations on insoluble products may in some case be obtained by completely solubilizing LPMO-treated material with hydrolases and then analyze soluble oxidized products. Less quantitative methods based on labeling oxidized chain ends and microscopy are also available [20]. In addition to this, the field of size exclusion chromatography (SEC) in both the aqueous and nonaqueous mode has seen large improvements in the past years. In particular, several column producers today design SEC columns for UHPLC conditions and this enables higher throughput, higher resolution, and smaller injection volumes than what we have seen so far. Using SEC in ionic liquid mode for analyzing molecular distributions in cellulose [21] as well as for studying the molecular distribution of product mixtures after enzymatic treatments has a large potential to broaden our understanding of the effects of LPMO treatments.

Research on LPMOs has only just begun and, despite major achievements [22–24], there is much exciting research ahead. The analytical tools described above will be invaluable for further unravelling of LPMO function in nature and in the biorefinery.

4 Notes

1. Samples must have the same proportion of organic solvent as in the chromatographic starting conditions, if not, this is likely to compromise resolution. Some ionic strength (provided by the added Tris–HCl) is needed in order to obtain retention of the aldonic acids. Furthermore, an adapted version of this method that is more suitable for the shortest products appears in [2]. This study also describes a method for enzymatically generating chitoaldonic acid standards using a chito-oligosaccharide oxidase [2].
2. (A) When eluents are prepared note that when approx half of the 50% NaOH solution has been used, discard the remaining for the purpose of mobile phase preparation due to risk of carbonate contamination. It is critical to follow this procedure for mobile phase preparation or to follow equivalent recommendations by the instrument vendor, in order to achieve satisfactory quality of analysis. The most important things to pay attention to are (a) water and chemical quality, (b) sufficient

degassing for removal of dissolved carbon dioxide, (c) storage in atmospheres with reduced content of carbon dioxide (N_2 or He-saturated headspace), (d) regular change of mobile phases (2–3 days shelf life), and (e) to avoid all kinds of detergents in mobile phases, hence no detergent washing of mobile phase bottles between eluent preparations. Restrict cleaning to rinsing with Milli-Q water (Type I, 18.2 MΩ cm). Extensive exchange of mobile phases on the column and regeneration after each eluent changes is also important in order to remove accumulation of carbonate contaminations on the column which compromise resolution. (B) This method is used for native and oxidized cello-oligosaccharides, and may be adapted to be used for xyloglucan fragments as described in [12]. If there is a need for higher throughput, a 10 min method for separation and detection of aldonic acids may be used [7]. (C) For validation purposes HPAEC may be coupled to ESI-MS as explained in [8], but this requires anion suppressor, additional pumps and flow splitting. If you do not have access to online MS detection with your HPAEC a simpler approach that does not require a complex instrument setup and is based on offline MS is described in [8]. In short the latter procedure implies manual fractionation, desalting and MS analysis by MALDI-ToF, direct infusion ESI-MS, or injection onto another, simpler LC-MS system.

3. The elution of products may vary slightly depending on which UHPLC system you are using. This is due to for example varying dead volumes/gradient mixing in the hardware that is used. Gradient needed may PGC allows simultaneous detection of C1 and C4 oxidized products only in the range from DP2-5. *See* details in Westereng et al. [8]. Native cello-oligosaccharides co-elute with C4-oxidized cello-oligosaccharides and double oxidized compounds co-elute with C1-oxidized oligosaccharides.

   Due to incompatibility between an alkaline mobile phase and the detection principle of CAD, it is beneficial to lower the pH of eluent A to 6.5 in cases where CAD is applied [8]. Sensitivity in the low nanomole range is usually needed and improved sensitivity with CAD may be achieved by lowering the ionic strength. Analysis of uncharged compounds may be achieved at ionic strengths down to 1 μM NaCl and this enables product quantification with a sensitivity that is suitable for characterizing LPMO activities [8].

4. (A) Under standard conditions, more than one type of adduct is commonly observed during MALDI analysis. The most dominant adduct is sodium, but also potassium, hydrogen, and NH_4 adducts can occasionally be observed. One simple way of overcoming this multiplicity of signals, which hampers

product identification, is ion doping to force the adduct composition to a defined adduct type. An example of lithium doping is given in Fig. 1, which shows complete lithium adduct formation. (B) Normally the mixing of the sample and the LiCl solution (or other doping reagent) as indicated is sufficient to achieve complete doping. Testing two to three different concentrations of doping reagent may be necessary to find conditions that provide sufficient doping. Note that adding too much of doping reagent can give problems due to ion suppression. It is important to run LPMO reactions with low buffer concentrations in order to avoid ion suppression and extensive background signals. Doping with other ion salts, such as NaCl and KCl, may be performed in a similar manner as outlined above, but note that adduct formation efficiency varies between ions. Details on adduct formation may be found in [25].

5. The actual flow rate in the emitter is difficult to control and is dependent on several factors such as sample viscosity, emitter opening size, applied positive pressure, distance from needle to skimmer, and spray voltage. Some trial and error must be expected to optimize these parameters. Typically, without spray voltage applied, the positive pressure alone should be able to form small droplets at the emitter tip. If no liquid appears, a small "crash" into the skimmer may scratch the emitter open. If too much liquid comes out, the emitter opening is too large, and a new emitter needs to be installed. This method is not applicable for high-throughput analysis or quantitative measurements as a new emitter, including manual optimization, is needed for each sample.

6. Aldonic acids dissolved in aqueous buffers are in a pH-dependent equilibrium with the corresponding δ-lactone. The δ-lactone is formed by dehydration of the aldonic acid. The aldonic acids is re-formed by hydrolysis of the δ-lactone. Thus, when performing ^{18}O–isotope labeling experiments, care must be taken to avoid (a) exchange of the incorporated ^{18}O atom with ^{16}O $^{18}O_2$ experiments and (b) incorporation of a second ^{18}O atom in $H_2{}^{18}O$ experiments. Since the aldonic acid-δ-lactone equilibrium is strongly dominated by the aldonic acid at alkaline pH, it is preferable to conduct isotope labeling experiments at pH>7. The isotope labeling experiments performed by Vaaje-Kolstad et al. [1] were all conducted at pH 8.0. At this pH the δ-lactone forms of the C1 oxidized products are not observed using MALDI-ToF MS and the equilibrium is such that there is enough time to carry out product analyses before the exchange of oxygen atoms becomes noticeable.

Acknowledgments

This work was supported by the Norwegian Research Council through grants and 193817, 214138, 214613, 216162, 243663, and 244259.

References

1. Vaaje-Kolstad G, Westereng B, Horn SJ, Liu ZL, Zhai H, Sorlie M, Eijsink VGH (2010) An oxidative enzyme boosting the enzymatic conversion of recalcitrant polysaccharides. Science 330(6001):219–222. doi:10.1126/science.1192231
2. Loose JS, Forsberg Z, Fraaije MW, Eijsink VG, Vaaje-Kolstad G (2014) A rapid quantitative activity assay shows that the *Vibrio cholerae* colonization factor GbpA is an active lytic polysaccharide monooxygenase. FEBS Lett 588(18):3435–3440. doi:10.1016/j.febslet.2014.07.036
3. Forsberg Z, Vaaje-Kolstad G, Westereng B, Bunaes AC, Stenstrom Y, MacKenzie A, Sorlie M, Horn SJ, Eijsink VGH (2011) Cleavage of cellulose by a CBM33 protein. Protein Sci 20(9):1479–1483. doi:10.1002/Pro.689
4. Westereng B, Ishida T, Vaaje-Kolstad G, Wu M, Eijsink VG, Igarashi K, Samejima M, Stahlberg J, Horn SJ, Sandgren M (2011) The putative endoglucanase PcGH61D from phanerochaete chrysosporium is a metal-dependent oxidative enzyme that cleaves cellulose. PLoS One 6(11):e27807. doi:10.1371/journal.pone.0027807
5. Phillips CM, Beeson WT, Cate JH, Marletta MA (2011) Cellobiose dehydrogenase and a copper-dependent polysaccharide monooxygenase potentiate cellulose degradation by neurospora crassa. ACS Chem Biol 6(12):1399–1406. doi:10.1021/cb200351y
6. Quinlan RJ, Sweeney MD, Lo Leggio L, Otten H, Poulsen JCN, Johansen KS, Krogh KBRM, Jorgensen CI, Tovborg M, Anthonsen A, Tryfona T, Walter CP, Dupree P, Xu F, Davies GJ, Walton PH (2011) Insights into the oxidative degradation of cellulose by a copper metalloenzyme that exploits biomass components. Proc Natl Acad Sci U S A 108(37):15079–15084. doi:10.1073/pnas.1105776108
7. Westereng B, Agger JW, Horn SJ, Vaaje-Kolstad G, Aachmann FL, Stenstrom YH, Eijsink VG (2013) Efficient separation of oxidized cello-oligosaccharides generated by cellulose degrading lytic polysaccharide monooxygenases. J Chromatogr A 1271(1):144–152. doi:10.1016/j.chroma.2012.11.048
8. Westereng B, Arntzen MØ, Aachmann FL, Várnai A, Eijsink VGH, Agger JW (2016) Simultaneous analysis of C1 and C4 oxidized oligosaccharides, the products of lytic polysaccharide monooxygenases acting on cellulose. J Chromatogr A 1445:46–54 http://dx.doi.org/10.1016/j.chroma.2016.03.064
9. Coenen GJ, Bakx EJ, Verhoef RP, Schols HA, Voragen AGJ (2007) Identification of the connecting linkage between homo- or xylogalacturonan and rhamnogalacturonan type I. Carbohydr Polym 70(2):224–235
10. Westereng B, Coenen GJ, Michaelsen TE, Voragen AGJ, Samuelsen AB, Schols HA, Knutsen SH (2009) Release and characterization of single side chains of white cabbage pectin and their complement-fixing activity. Mol Nutr Food Res 53(6):780–789. doi:10.1002/mnfr.200800199
11. Isaksen T, Westereng B, Aachmann FL, Agger JW, Kracher D, Kittl R, Ludwig R, Haltrich D, Eijsink VG, Horn SJ (2014) A C4-oxidizing lytic polysaccharide monooxygenase cleaving both cellulose and cello-oligosaccharides. J Biol Chem 289(5):2632–2642. doi:10.1074/jbc.M113.530196
12. Agger JW, Isaksen T, Varnai A, Vidal-Melgosa S, Willats WG, Ludwig R, Horn SJ, Eijsink VG, Westereng B (2014) Discovery of LPMO activity on hemicelluloses shows the importance of oxidative processes in plant cell wall degradation. Proc Natl Acad Sci U S A 111(17):6287–6292. doi:10.1073/pnas.1323629111
13. Bennati-Granier C, Garajova S, Champion C, Grisel S, Haon M, Zhou S, Fanuel M, Ropartz D, Rogniaux H, Gimbert I, Record E, Berrin JG (2015) Substrate specificity and regioselectivity of fungal AA9 lytic polysaccharide monooxygenases secreted by podospora anserina. Biotechnol Biofuels 8:90. doi:10.1186/s13068-015-0274-3
14. Vu VV, Beeson WT, Span EA, Farquhar ER, Marletta MA (2014) A family of starch-active polysaccharide monooxygenases. Proc Natl Acad Sci U S A 111(38):13822–13827. doi:10.1073/pnas.1408090111
15. Lo Leggio L, Simmons TJ, Poulsen JC, Frandsen KE, Hemsworth GR, Stringer MA,

von Freiesleben P, Tovborg M, Johansen KS, De Maria L, Harris PV, Soong CL, Dupree P, Tryfona T, Lenfant N, Henrissat B, Davies GJ, Walton PH (2015) Structure and boosting activity of a starch-degrading lytic polysaccharide monooxygenase. Nat Commun 6:5961. doi:10.1038/ncomms6961

16. Frommhagen M, Sforza S, Westphal AH, Visser J, Hinz SWA, Koetsier MJ, van Berkel WJH, Gruppen H, Kabel MA (2015) Discovery of the combined oxidative cleavage of plant xylan and cellulose by a new fungal polysaccharide monooxygenase. Biotechnol Biofuels 8:12. doi:10.1186/s13068-015-0284-1
17. Kracher D, Scheiblbrandner S, Felice AKG, Breslmayr E, Preims M, Ludwicka K, Haltrich D, Eijsink VGH, Ludwig R (2016) Extracellular electron transfer systems fuel cellulose oxidative degradation. Science. doi:10.1126/science.aaf3165
18. Westereng B, Cannella D, Agger JW, Jorgensen H, Andersen ML, Eijsink VGH, Felby C (2015) Enzymatic cellulose oxidation is linked to lignin by long-range electron transfer. Sci Rep-Uk 5. doi: 10.1038/srep18561 ARTN 18561
19. Cannella D, Mollers KB, Frigaard NU, Jensen PE, Bjerrum MJ, Johansen KS, Felby C (2016) Light-driven oxidation of polysaccharides by photosynthetic pigments and a metalloenzyme. Nat Commun 7. doi:10.1038/ncomms11134
20. Eibinger M, Ganner T, Bubner P, Rosker S, Kracher D, Haltrich D, Ludwig R, Plank H, Nidetzky B (2014) Cellulose surface degradation by a lytic polysaccharide monooxygenase and its effect on cellulase hydrolytic efficiency. J Biol Chem 289(52):35929–35938. doi:10.1074/jbc.M114.602227
21. Potthast A, Radosta S, Saake B, Lebioda S, Heinze T, Henniges U, Isogai A, Koschella A, Kosma P, Rosenau T, Schiehser S, Sixta H, Strlic M, Strobin G, Vorwerg W, Wetzel H (2015) Comparison testing of methods for gel permeation chromatography of cellulose: coming closer to a standard protocol. Cellul 22(3):1591–1613. doi:10.1007/s10570-015-0586-2
22. Beeson WT, Vu VV, Span EA, Phillips CM, Marletta MA (2015) Cellulose degradation by polysaccharide monooxygenases. Annu Rev Biochem 84:923–946. doi:10.1146/annurev-biochem-060614-034439
23. Hemsworth GR, Johnston EM, Davies GJ, Walton PH (2015) Lytic polysaccharide monooxygenases in biomass conversion. Trends Biotechnol 33(12):747–761. doi:10.1016/j.tibtech.2015.09.006
24. Hemsworth GR, Davies GJ, Walton PH (2013) Recent insights into copper-containing lytic polysaccharide mono-oxygenases. Curr Opin Struct Biol 23(5):660–668. doi:10.1016/j.sbi.2013.05.006
25. Cancilla MT, Wang AW, Voss LR, Lebrilla CB (1999) Fragmentation reactions in the mass spectrometry analysis of neutral oligosaccharides. Anal Chem 71(15):3206–3218. doi:10.1021/Ac9813484

Chapter 8

Carbohydrate Depolymerization by Intricate Cellulosomal Systems

Johanna Stern, Lior Artzi, Sarah Moraïs, Carlos M.G.A. Fontes, and Edward A. Bayer

Abstract

Cellulosomes are multi-enzymatic nanomachines that have been fine-tuned through evolution to efficiently deconstruct plant biomass. Integration of cellulosomal components occurs via highly ordered protein–protein interactions between the various enzyme-borne dockerin modules and the multiple copies of the cohesin modules located on the scaffoldin subunit. Recently, designer cellulosome technology has been established to provide insights into the architectural role of catalytic (enzymatic) and structural (scaffoldin) cellulosomal constituents for the efficient degradation of plant cell wall polysaccharides. Owing to advances in genomics and proteomics, highly structured cellulosome complexes have recently been unraveled, and the information gained has inspired the development of designer cellulosome technology to new levels of complex organization. These higher-order designer cellulosomes have in turn fostered our capacity to enhance the catalytic potential of artificial cellulolytic complexes. In this chapter, methods to produce and employ such intricate cellulosomal complexes are reported.

Key words Cellulosome, Cellulose, Cellulase, Xylanase, Multi-enzymatic complex

1 Introduction

Plant cell wall polysaccharides, primarily cellulose and hemicellulose, comprise a major reservoir of carbon and energy. The complex construction of the plant cell wall restricts accessibility of its carbohydrate components to enzymatic attack, and, consequently, the recycling of photosynthetically fixed carbon is a relatively inefficient biological process of increasing biotechnological importance. Limited numbers of microorganisms are capable of deconstructing and utilizing recalcitrant carbohydrates as carbon sources. Enzyme systems produced by these microorganisms comprise diverse and extensive consortia of multimodular enzymes that either act freely or are organized in high-molecular-mass multi-enzyme complexes termed cellulosomes [1]. The most efficient natural cellulolytic system known is produced by the anaerobic thermophilic bacterium,

D. Wade Abbott and Alicia Lammerts van Bueren (eds.), *Protein-Carbohydrate Interactions: Methods and Protocols*, Methods in Molecular Biology, vol. 1588, DOI 10.1007/978-1-4939-6899-2_8, © Springer Science+Business Media LLC 2017

Clostridium thermocellum, which possesses a well characterized cellulosome system [2].

The major player in cellulosome organization is a non-catalytic primary scaffoldin subunit, which binds the entire complex to its substrate via a carbohydrate-binding module (CBM) and incorporates different enzymes via specific high-affinity multi-modular cohesin–dockerin interactions [3, 4]. The primary scaffoldin anchors the cellulosome onto the bacterial cell surface through the interaction of its own C-terminal dockerin to cohesins located in cell-bound anchoring scaffoldins [3].

Cellulosomal systems of *Acetivibrio cellulolyticus* [5, 6], *Ruminococcus flavefaciens* [7], and more recently *Clostridium clariflavum* [8–10] (Fig. 1) contain adaptor scaffoldins which mediate the binding between the enzyme-bearing primary scaffoldin and cell surface-anchoring scaffoldins. The intricacy and reticulated nature of these cellulosomal architectures is believed to improve cellulosome flexibility through the amplification of possible

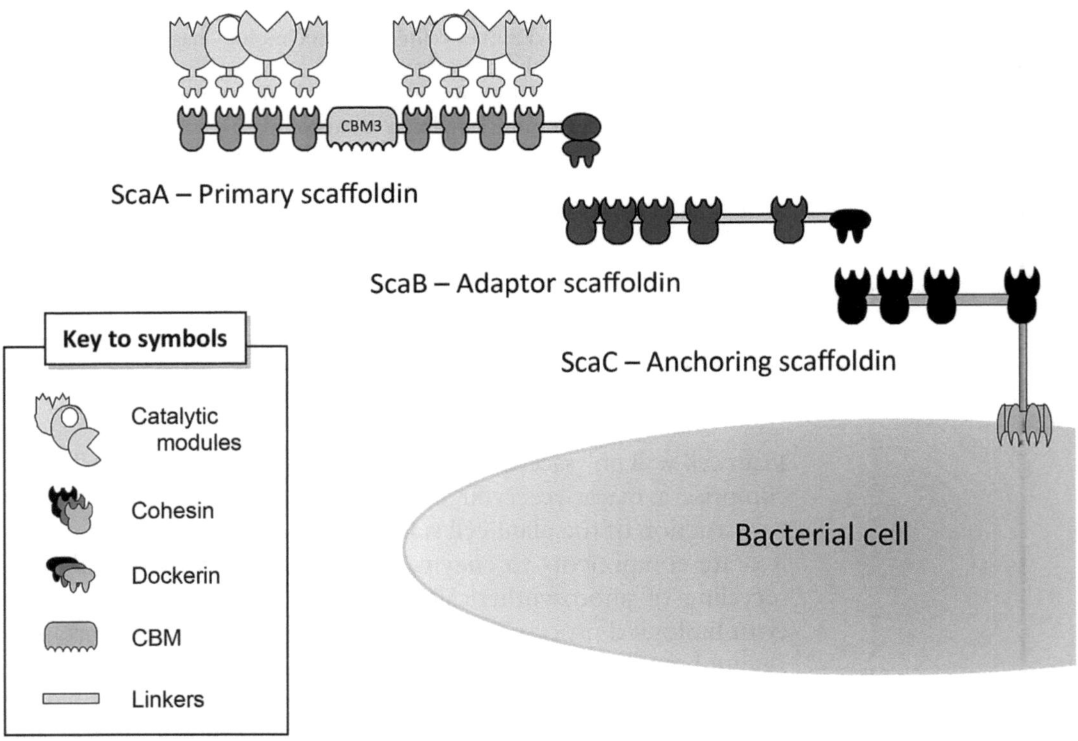

Fig. 1 Organization of the *Clostridium clariflavum* cellulosome. The cellulosome of *C. clariflavum* is the largest cellulosomal complex known up to date, constructed of an octavalent (eight cohesins) primary scaffoldin (ScaA) that interacts with the pentavalent adaptor scaffoldin (ScaB), which, in turn, interacts with the tetravalent anchoring scaffoldin (ScaC) that attaches the complex to the bacterial cell surface. Altogether, this complex can contain up to 160 enzymatic subunits. In addition to the trifold arrangement of scaffoldins and enzymes presented in the figure, the bacterium produces ten additional scaffoldins. Thus, a variety of other cellulosome assemblies are produced by *C. clariflavum*, involving the different scaffoldins and dockerin-bearing enzymes

enzymatic compositions, which allows the bacterium to easily adapt to environmental changes related to substrate composition. Recently, the key elements cellulosome have been uncovered. Proteomic approaches have been used to characterize cellulosome complexes of *C. clariflavum* grown on different cellulosic substrates [10], revealing how cellulosome composition and enzymatic activity is adapted as a function of the carbon source.

The Lego-like nature of the cellulosomal subunits may be transformed by mixing and matching different cohesin–dockerin pairs to assemble artificial cellulosome-like systems, referred to as designer cellulosomes [11]. These artificial nanodevices comprise a chimaeric scaffoldin with cohesin modules of divergent specificity, usually derived from different bacteria, and allow the controlled incorporation of recombinant matching dockerin-bearing enzymes that might benefit from enzyme proximity (Fig. 2). The designer cellulosome concept is based on the high affinity and specific interaction between a cohesin and a dockerin [12].

While the overall architecture of designer cellulosomes is generally quite simple, a more elaborate type of artificial cellulosome exists, characterized by the presence of an adaptor chimaeric scaffoldin [13]. The adaptor scaffoldin enables both enhanced

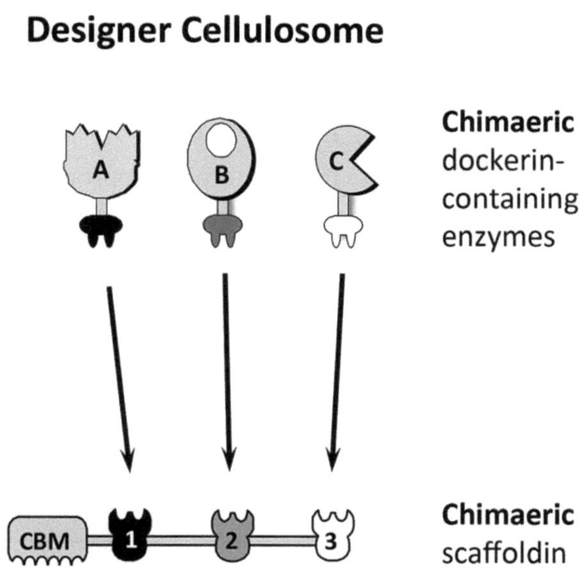

Fig. 2 Schematic representation of the designer cellulosome concept. The chimaeric scaffoldin is composed of a carbohydrate-binding module (CBM) with three cohesins (1, 2, and 3) from various bacterial systems with different specificities. The chimaeric containing-enzymes are composed of a catalytic module (*A*, *B*, and *C*), each of which bears an appended dockerin module that binds specifically to the chimaeric scaffoldin. The catalytic modules may be from the same bacterium or from totally different species

diversity and amplified numbers of catalytic subunits that can be incorporated into the system.

In this chapter, we describe methodologies to produce such intricate cellulosomal structures to improve the efficiency of carbohydrate depolymerization. Additionally, we outline the methods used to isolate *C. clariflavum* cellulosomes as an example of the technology that may be used for the purification of high-molecular-weight natural cellulosomes. These artificial cellulosomes may be generated from a natural complex cellulosomal system, such as *C. clariflavum*, or by constructing elaborate artificial cellulosomal assemblies. Regarding the production of natural cellulosomal complexes, this chapter includes the conditions for proper fermentation of *C. clariflavum*, followed by the methodology for isolation and purification of the different cellulosomal fractions. For the production of elaborate designer cellulosomes, we provide methods for genetic engineering of the different chimaeric components and analysis of the resulting chimaeric proteins. Finally, strategies to probe the enzymatic activity of cellulosomes specialized in the degradation of different cellulosic substrates are detailed.

2 Materials

2.1 Purification of Natural Cellulosomes

2.1.1 GS-2 Medium for Growth of C. clariflavum (for 1 L)

1. 0.5 g K_2HPO_4.
2. 0.5 g $MgCl_2 \cdot 6H_2O$.
3. 0.5 g KH_2PO_4.
4. 1.3 g NH_4SO_4.
5. 10.5 g morpholinopropane sulfonic acid (MOPS).
6. 5 g yeast extract.
7. 2 mg resazurin.
8. Mineral solution [16].

2.2 Preparation of Designer Cellulosomes

2.2.1 Expression of Chimaeric Scaffoldins

1. LB medium

Yeast extract	0.5% (w/v)
Tryptone	1.0%
NaCl	0.5%

2. Isopropyl β-D-1-thiogalactopyranoside (IPTG) 1 M (10 mL) Weigh 2.383 g, and dissolve in 10 mL water. Filter the solution (filters: 0.2–0.45 mM). Dispense into 1 mL aliquots. Store at −20 °C.

2.2.2 Purification of Chimaeric Scaffoldins

1. Protease inhibitors 200× (10 mL). Weigh 350 mg phenylmethylsulfonyl fluoride (PMSF), 120 mg benzamidine, and 15 mg benzamide. Resuspend with 10 mL ethanol. Store at −20 °C.

2. Wash Buffer #1: TBS, 1 M NaCl (250 mL). Weigh 12.6 g NaCl. Transfer to a glass beaker. Introduce 25 mL of TBS 10×. Complete with water to 250 mL.
3. Elution Buffer: 1% triethylamine (50 mL). Prepare fresh for each experiment. In hood, introduce 0.5 mL triethylamine into a 50 mL falcon tube. Complete with water to 50 mL.
4. Neutralization Buffer: 1 M MES (2-(N-morpholino)ethanesulfonic acid), pH 5.5 (50 mL). Weigh 9.76 g of MES and place in 50 mL falcon tube. Add about 40 mL water. Adjust to pH to 5.5 with NaOH. Complete with water to 50 mL.

2.2.3 SDS-PAGE

1. SDS-Resolving Buffer: 1.5 M Tris pH 8.8 with sodium dodecyl sulfate (SDS): Prepare 500 mL. Weigh 90.75 g of Tris (hydroxymethyl) amino methane and 1.78 g SDS. Transfer to a glass beaker. Mix and adjust pH with HCl solution (*see* **Note 1**). Make up to 500 mL with water. Store at 4 °C.
2. SDS-Stacking Buffer: 1 M Tris pH 6.8 with SDS (500 mL) Weigh 60.5 g of Tris (hydroxymethyl) amino methane and 3.7 g SDS. Transfer to a glass beaker. Mix and adjust pH with HCl solution. Bring to 500 mL with water. Store at 4 °C.
3. SDS-Running Buffer 10× (5 L). Weigh 151.43 g Tris (hydroxymethyl) amino methane, 20.7 g glycine and 50 g SDS. Complete to 5 L with water. Store at room temperature.
4. SDS-Sample Buffer 3× (100 mL). Mix 5 mL β-mercaptoethanol, 10 mg Bromophenol blue, 3× SDS, 10 mL glycerin, and 6.25 mL of Stacking Buffer. Complete to 100 mL with water. Store at room temperature.

2.3 Analysis of Cohesin–Dockerin Interactions

2.3.1 Modified ELISA-Based Assay

1. Coating Solution: 0.1 M sodium carbonate, pH 9. Prepare 500 mL of 5× stock solution. Weigh 264.972 g of sodium carbonate, transfer to a glass beaker, and add 400 mL water. Mix and adjust pH with HCl solution (*see* **Note 1**). Bring to 500 mL with water. Store at 4 °C.
2. Tris-Buffered Saline (TBS): Prepare 1 L of 10× stock solution. Weigh 80 g NaCl, 2 g KCl and 30 g Tris (hydroxymethyl) amino methane. Transfer to a glass beaker and add 900 mL water. Mix and adjust pH to 7.4 with HCl (*see* **Note 1**). Bring to 1 L with water. Store at 4 °C
3. Wash Buffer #2: TBS supplemented with 10 mM $CaCl_2$ and 0.05% Tween 20. For each experiment prepare 500 mL of fresh Wash Buffer. Prepare a volume of 1 L of stock solution of 2 M $CaCl_2$ (*see* **Note 2**). Weigh 221.96 g of $CaCl_2$, transfer to a glass beaker and add 1 L water. Mix and store at 4 °C. Introduce 50 mL of 10× TBS, 2.5 mL of 2 M $CaCl_2$, and 250 μL of Tween 20. Make up to 500 mL with water. Store at 4 °C.

4. Blocking Buffer: TBS 10 mM $CaCl_2$ 2% BSA 0.05% Tween 20: For each experiment prepare 50 mL of fresh Blocking Buffer. Introduce 5 mL of TBS 10×, 250 μL of 2 M $CaCl_2$ solution, 25 μL of Tween 20, and 1 g of BSA into a 50 mL falcon tube. Make up to 50 mL. Mix vigorously, and store at 4 °C.
5. Primary antibody preparation: rabbit anti-xylanase T-6 antibody (*see* Ref. [14] for details), diluted 1:10,000 in Blocking Buffer.
6. Secondary antibody preparation: HRP-labeled anti-rabbit antibody diluted 1:10,000 in Blocking Buffer.
7. TMB (3,3′-5,5′-tetramethylbenzidine) substrate-chromogen (Dako A/S, Glostrup, DK): 1 M H_2SO_4.
8. CBM-fused cohesins (CBM-Cohs) and xylanase-fused dockerins (Xyn-Docs) were prepared as described in Barak et al. 2005 [15]. CBM-fused cohesins are composed of the cellulose-binding CBM3a of CipA from *C. thermocellum* fused to a cohesin module with the selected specificity. Xyn-Docs are composed of xylanase T-6 from *Geobacillus stearothermophilus* fused to a dockerin module with the selected specificity.

2.3.2 Non-denaturing Gel Electrophoresis

1. Dilution Buffer: same as Wash Buffer #2 (*see* Subheading 2.3.1).
2. ND-Resolving Buffer (100 mL): Weigh 18.2 g Tris (hydroxymethyl) amino methane and place in a glass beaker with 90 mL water. Adjust to pH 8.9 with HCl. Complete with water to 100 mL. Store at 4 °C.
3. ND-Stacking Buffer (100 mL): Weigh 5.7 g Tris (hydroxymethyl) amino methane and put in glass beaker with 90 mL water. Adjust to pH to 6.7 with H_3PO_4. Complete with water to 100 mL. Store at 4 °C.
4. ND-Running Buffer 10× (250 mL): Weigh 7.5 g Tris (hydroxymethyl) amino methane, 36 g glycine and place in glass beaker. Complete with water to 250 mL.
5. ND-Sample Buffer 3× (10 mL): Introduce 3 mL glycerol, 3 mL 10× Running Buffer and 4 mL of water. Finally introduce 0.2 mg of bromophenol blue.

2.3.3 Affinity Pull-Down Assay

1. Cellobiose suspension (50 mL): 10% cellobiose (w/v) (Sigma-Aldrich) suspended in 50 mM acetate buffer, pH 5. Store at 4 °C.
2. Wash Buffer #3 (10 mL): 50 mM acetate buffer, pH 5, supplemented with 0.05% Tween 20 (v/v). Store at 4 °C.

2.3.4 Assembly of Extended Designer Cellulosome

1. Wash buffer #2 (as for Modified ELISA-based assay Subheading 2.3.1).

2.4 Enzymatic Activity Assays

2.4.1 Substrate Degradation

1. Wash Buffer #2: *see* above (*see* Subheading 2.3.1).
2. Reaction Buffer: 50 mM acetate buffer, pH 5.0, 24 mM $CaCl_2$, 4 mM EDTA.
   (a) Phosphoric Acid Swollen Cellulose (PASC) suspension (1 L): In hood, place 6 g of Avicel with 300 mL phosphoric acid into a 3 L glass beaker, and mix for 3 h. Add 1.5 L of water and mix for 3 min. Centrifuge at 14,000 × *g* for 30 min, and resuspend with water. Repeat three times. Resuspend with 800 mL water. Adjust pH to 7 with NaOH 10 M. Use at a final concentration of 0.75% (w/v).
   (b) Avicel Suspension (20 mL): Avicel 10–20% (w/v) (Sigma-Aldrich) suspended in 100 mM acetate buffer, pH 5.0. Store at 4 °C.
   (c) Xylan Solution (5 mL): 2% xylan (birchwood, beechwood, or oat spelt from Sigma-Aldrich), suspended in 50 mM citrate buffer, pH 6.0. Prepare it fresh for each experiment.
   (d) Switchgrass, provided by Nott Farms (Ont.) Ltd. and Switch Energy Corp (Ontario, Canada) is treated as follows: 100 g of switchgrass is suspended in 700 mL of 5% (w/v) H_2SO_4. The suspension containing the switchgrass is boiled with stirring for 1 h. The treated switchgrass is washed with water and filtered with a vacuum glass filter until neutral pH is reached. The switchgrass is concentrated to a final solid content of 27% (w/w).
   (e) Hatched Wheat Straw (0.2–0.8 mm), provided by Valagro Carbone Renouvelable Poitou-Charentes (Poitiers, France), is treated as follows: The crude substrate is incubated in distilled water under mild stirring for 3 h at room temperature, vacuum filtered on a 2.7-mm glass filter, resuspended in water, and incubated for 16 h under mild stirring at 4 °C. The suspension is filtered and washed three times with distilled water. A sample is dried overnight at 100 °C for estimation of dry weight.

2.4.2 DNS Assay for Determination of Reducing Sugars

1. Weigh 40 g DNS (3,5-dinitro-salicylic acid), 8 g phenol, 2 g sodium sulfite (Na_2SO_3), 800 g Na-K tartarate (Rochelle salts).
2. Add all the above to 2 L 2% NaOH.
3. Mix with a magnetic stirrer overnight at room temperature, cover with aluminum foil.
4. Complete with water to a final volume of 4 L.
5. Store in a dark bottle at 4 °C.

3 Methods

3.1 Purification of Natural Cellulosomes

In nature, cellulosomes are produced by a restricted subset of anaerobic bacterial species that live in a variety of habitats under different environmental conditions [3, 17]. The structure and composition of cellulosomes varies within each bacterial species, and the number of enzymes in each complex is controlled by the molecular architecture of the scaffoldins that assemble the complex. Cellulosomes are generally attached to the bacterial cell surface [4, 18–21], although they may be released from the cell and function as cell-free complexes [10, 22].

The cellulosome of *C. thermocellum* is recruited to the cell surface, although the complex is released into the extracellular medium in the late stationary phase of growth. This facilitates the collection of these complexes and their further purification [4, 23]. Purification of cellulosomes can be performed using various approaches, depending on the properties required for their application. These include gel filtration for high-molecular-weight complexes [10]; affinity-digestion for highly cellulolytic cellulosomes that adhere to cellulose and hydrolyze it efficiently [23, 24]; ion exchange chromatography [25] and ammonium sulfate precipitation [26].

In this chapter, we provide the methods used to isolate *C. clariflavum* cellulosomes as an example of the technology that may be used for the purification of high-molecular-weight natural cellulosomes.

C. clariflavum is a thermophilic bacterium that grows under strict anaerobic conditions. Bioinformatic and biochemical analyses of its cellulosomal system revealed that it possesses one of the largest cellulosome complexes known up to date, comprising up to 160 dockerin-bearing enzymes [9]. *C. clariflavum* was cultivated for the subsequent isolation of different cellulosome complexes. Components of purified cellulosomes were characterized by mass spectrometry and the ability of native cellulosomes to degrade different substrates was evaluated.

3.1.1 Fermentation of C. Clariflavum

1. In order to produce large amounts of natural cellulosomes from *C. clariflavum*, grow the bacteria in large scale (e.g., 10 L fermenter) under anaerobic conditions at 55 °C on GS-2 medium. Make sure to sparge the medium with nitrogen during the entire fermentation process.
2. Adjust the solution to pH 7.2 with 10 M NaOH.
3. Add the desired carbon source. For example: 8 g/L cellobiose (Sigma-Aldrich), 2 g/L Avicel or 2 g/L acid-pretreated switchgrass to induce expression of genes associated with cellulosome production.

4. Grow the bacterium until late stationary phase.
5. Assess growth phase by measuring optical density (A_{600nm}) at different time points when the substrate is soluble (cellobiose). Centrifuge the cells of each sample, and resuspend in TBS. Extract 1 mL sample, and measure the optical density. When cells are grown on insoluble carbon sources, growth is measured by enzymatic activity of samples collected at different time points on carboxymethyl cellulose (CMC; VWR International Ltd., Poole, England).
6. After cultivation, collect the spent growth medium, and remove cells by centrifugation at 13,690 × *g* for 15 min.

3.1.2 Isolation of High-Molecular-Weight Cellulosomes

C. clariflavum cellulosomes are predicted to have an exceptionally high molecular weight. Therefore, a separation method based on size exclusion is desirable.

1. Concentrate the media supernatant fluids by 100-fold using a peristaltic pump (MasterFlex l/S pump system, Easy-Load II pump head [Cole-Parmer, Vernon Hills, IL]) with a 500-kDa-cutoff Pellicon 2 membrane (Millipore, Darmstadt, Germany).
2. Separate the protein contents of the concentrated solution by preparative gel filtration chromatography, using a chromatography system for laboratory-scale protein purification (Äkta start; GE Healthcare, Uppsala, Sweden). Use a gel filtration column HiPrep 26/60 Sephacryl S-500 HR (GE Healthcare, Little Chalfont, UK), with a separation range of 40 kDa to 20 MDa. Use TBS as the Running Buffer. Chromatographic separation of the proteins should result in the elution of two major high-molecular-weight peaks, with a calculated molecular mass of ~ 400 and ~ 1200 kDa, for each growth condition.
3. Subject the different fractions comprising the two peaks to SDS-PAGE (*see* Fig. 3 for an example of SDS-PAGE of the cellulosome fractions).

3.2 Preparation of Designer Cellulosomes

The development of the designer cellulosome concept proceeded over a 2-decade period. The initial report in 1994 [11] led to concrete experimental attempts in 2001 [27], which were followed by the generation of monovalent [28, 29], then bivalent [30], trivalent [31, 32], tetravalent [33], and finally hexavalent scaffoldins [34], thereby enabling the self-assembly of more complex designer cellulosomal systems.

Two types of chimaeric enzymes can be incorporated into designer cellulosomes. For cellulosomal enzymes, the sequence of the dockerin region has to be identified and then artificially replaced using an appropriate molecular cloning strategy by the dockerin expressing the desired specificity. For non-cellulosomal enzymes, which usually contain a CBM module attached to their catalytic

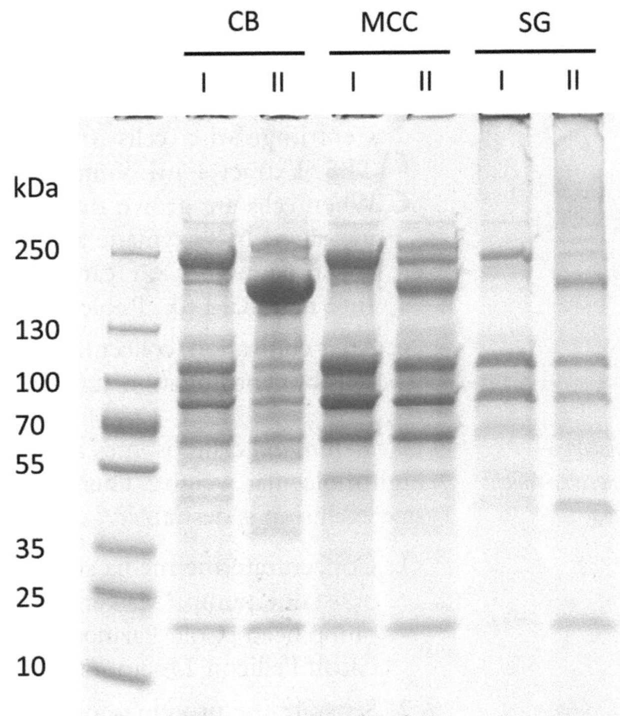

Fig. 3 SDS-PAGE analysis of *C. clariflavum* cellulosome fractions. Cellulosomes produced in cellobiose-, microcrystalline cellulose-, and switchgrass-containing growth media were concentrated and isolated by gel-filtration chromatography. For each carbon source, two peaks of cellulosome proteins were identified (*I* and *II*), representing very high- and high-molecular-weight fractions, respectively. The cellulosomes were loaded onto gradient SDS gels (4–15%) at an amount of 20 μg. *CB* cellobiose, *MCC* microcrystalline cellulose, *SG* switchgrass

module, the CBM module can be either replaced or left intact, depending on the enzyme type. In both cases a dockerin of the desired specificity needs to be incorporated, usually at the C-terminus of the engineered protein. The catalytic activity of the resultant chimaeric enzymes has to be subsequently compared to that of the wild-type form, in order to ensure that by conversion to the cellulosomal mode or by replacing the dockerin module, catalytic efficiency is not significantly affected.

Additional breakthroughs involved investigation of novel cellulosome geometries [35], inclusion of nonnative types of bacterial [28] and fungal cellulases [36] and oxidative enzymes [37] into designer cellulosome formats. Nevertheless, incorporation of additional cohesins to the hexavalent scaffoldin sequence eventually proved problematic, since the extended proteins were unstable, difficult to express using standard *E. coli* molecular biological tools and due to the limited number of distinct well-characterized cohesin–dockerin specificities available for cellulosome design. Increasing the number (and types) of enzymatic subunits integrated into designer cellulosome is, however, critical, in order to enhance degradation of plant cell wall residues and render future biotechnological applications cost-effective. Recently, an original strategy has been proposed for incorporating additional catalytic subunits into designer cellulosomes [13]. This approach involves the use of an adaptor scaffoldin: a chimaeric scaffoldin that is able to both incorporate enzymes and attach to an additional scaffoldin. Similar to the primary scaffoldin, the adaptor scaffoldin was found experimentally to require a CBM, in order to promote enhanced synergistic action on cellulosic substrates [13].

3.2.1 Construction of Interacting Chimaeric Scaffoldins

In order to create an extended designer cellulosome, two separate chimaeric scaffoldins have to be designed: a primary or platform scaffoldin and an adaptor scaffoldin.

1. Design of the primary scaffoldin. This type of scaffoldin is composed of several type I cohesins with different specificities for incorporation of the enzymes, a CBM, and a type II cohesin for interaction with the adaptor scaffoldin.
2. Design of the adaptor scaffoldin. An adaptor scaffoldin should comprise several type I cohesins, a CBM and a relevant type II dockerin to attach to the type II cohesin of the primary scaffoldin (*see* **Note 3**).
3. The chimaeric scaffoldins, i.e., primary (platform) and adaptor scaffoldins, used herein to exemplify this methodology, are represented in Fig. 4.
4. The adaptor scaffoldin **Scaf1** was obtained by integrating the following gene fragments—CBM3a (from CipA, *C. thermocellum*) and cohesins A (third cohesin of ScaC from *A. cellulolyticus*), B (third cohesin from ScaB, *B. cellulosolvens*), and T (third cohesin from CipA, *C. thermocellum*) culminating in the type II dockerin from the *C. thermocellum* CipA scaffoldin.
5. The primary (platform) scaffoldin **Scaf2** was obtained by fusing a *C. thermocellum* type II cohesin (from anchoring scaffoldin OlpB) to the following fragment: cohesins C (first cohesin of CipC, *Clostridium cellulolyticum*) and A (third cohesin of ScaC from *A. cellulolyticus*), CBM3a (from CipA, *C. thermocellum*), cohesins T (third cohesin from CipA, *C. thermocellum*), G (cohesin from Orf2375, *Archaeoglobus fulgidus*), and F (first cohesin of ScaB, *R. flavefaciens*).
6. The different modules were assembled using regular molecular biology cloning tools with restriction enzymes [38] in linearized pET28a plasmid to form the chimeric scaffoldin (*see* **Note 4**).

3.2.2 Expression of Chimaeric Scaffoldins

1. Transform BL21 cells and plate with antibiotic.
2. Add 4 mL LB medium to the plate, resuspend cells and add to 0.5 L LB.

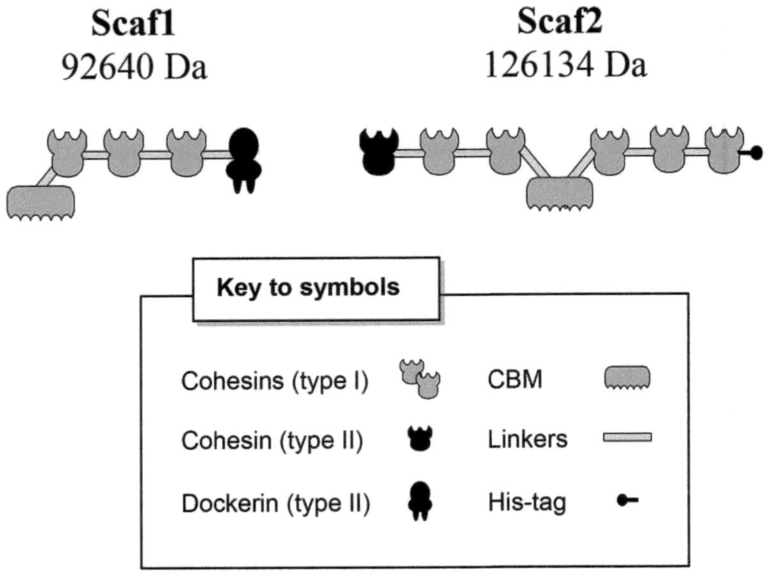

Fig. 4 Schematic representation of two interacting chimaeric scaffoldins. Scaf1 is the adaptor scaffoldin. The Scaf1 adaptor scaffoldin enables incorporation of three different chimaeric enzymes via its three divergent type I cohesins and interacts specifically via its type II dockerin with the type II cohesin of the major platform Scaf2, which incorporates five additional chimaeric enzymes via its five divergent type I cohesins. The resultant artificial cellulosome will thus be composed of a total of eight chimaeric enzymes

3. Grow cells to $A_{600\ nm}$ ~1.
4. Induction: Add 50 μL 1 M IPTG to a final concentration of 0.1 or 1 mM (*see* **Note 5**).
5. Grow cells at 37 °C for 3 h, at 16 °C for 17 h, at 30 °C for 5 h, or at 16 °C overnight (*see* **Note 5**).
6. Centrifuge cells for 15 min at 4500 × *g* at 4 °C.
7. Store pellet at −20 °C.

3.2.3 Purification of Chimaeric Scaffoldins

1. Prepare Wash, Elution, and Neutralization buffers in advance (*see* Subheading 2.2.3).
2. Resuspend pellet in 20 mL TBS + Protease Inhibitors.
3. Sonication: put sample in ice bath and cool down for 10 min. Sonicate the cell suspension at 40% amplitude with five burst of 30 s followed by intervals of 30 s for cooling (keep the suspension at all times on ice). Avoid foaming (*see* **Note 6**)
4. Centrifuge for 30 min at 20,000 × *g* at 4 °C.

Binding

1. Transfer supernatant to a new 50 mL tube and add 1 g of cellulose beads (IONTO-SORB, Czech Republic).
2. Centrifuge for 30 min at 20,000 × *g* at 4 °C.

3. Incubate for 1–2 h on the rotator at 4 °C.
4. Centrifuge for 5 min at 4000 × *g* at 4 °C (Save supernatant fluids to examine by SDS-PAGE as unbound fraction).

Washing

1. Wash pellet three times with 45 mL Wash Buffer #1 and three times with 45 mL TBS as follows.
   (a) Rotate 5 min at room temperature.
   (b) Centrifuge at 4000 × *g* for 5 min at 4 °C, and discard supernatant.

Elution

1. Elute protein from pellet two times with 5–10 mL Elution Buffer.
2. Rotate 5 min at room temperature.
3. Centrifuge at 4000 × *g* for 5 min at 4 °C and save supernatant.
4. Transfer supernatant to a new tube and neutralize with Neutralization Buffer.

3.2.4 SDS-PAGE

Purity of the proteins is examined by SDS-PAGE. Protein concentration is estimated optically by measuring absorbance at 280 nm, based on the known amino acid composition of the protein using the Protparam tool [39]. Proteins are stored in 50% (v/v) glycerol at −20 °C (*see* **Note 7**).

1. Prepare the different fractions containing the proteins of interest with SDS-containing Sample Buffer at a minimum amount of 5 μg of sample. When the concentration is unknown (following purification or isolation of the complex from cellulosomal bacteria), insert the maximal volume of sample (usually the wells of the gel can be loaded with 30–60 μL).
2. Prepare an SDS polyacrylamide gel at 6–20%, depending on the size of your protein/complex (10% for average sized proteins) by using the following two tables to prepare resolving and stacking gels.
   (a) Resolving gel (Table 1).
   (b) Stacking gel, 5 mL (Table 2).
3. Boil the samples for 5 min at 100 °C.
4. Perform electrophoresis for 45 min at 200 V.

3.3 Analysis of Cohesin–Dockerin Interactions

3.3.1 Modified Affinity-Based ELISA

The specificities of both dockerin- and cohesin-containing proteins are examined semiquantitatively using affinity-based ELISA [15]. Each module must bind selectively with its matching partner and exhibit no or very poor binding to non-matching counterparts. In addition, in order to ensure that the adaptor scaffoldin is able to bind simultaneously to recombinant enzymes via its type I cohesin modules and to the primary scaffoldin via its type II dockerin module, a modified ELISA-based assay should be performed (Fig. 5; *see* Subheading 2.3.1 for preparation of reagents.

Table 1
Volumes for preparing resolving gels from 6 to 20%

	6%	8%	10%	12%	15%	20%
H_2O (mL)	10.8	9.5	8.1	6.8	4.8	1.5
Acrylamide (mL)	4	5.3	6.7	8	10	13.3
Resolving buffer (mL)	5	5	5	5	5	5
10% APS (mL)	0.2	0.2	0.2	0.2	0.2	0.2
TEMED (mL)	0.016	0.012	0.008	0.008	0.008	0.08

APS ammonium persulfate, *TEMED* tetramethylethylenediamine

Table 2
Volumes for preparing stacking gel

H_2O	3.45
Acrylamide	0.83
Stacking buffer	0.63
10% APS	0.05
TEMED	0.005

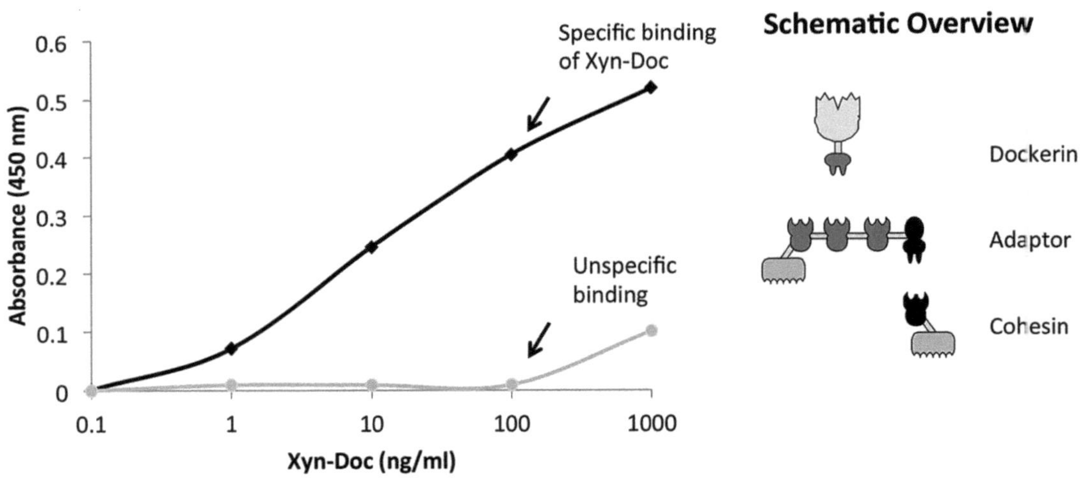

Fig. 5 Modified ELISA-based system for simultaneous binding of the adaptor scaffoldin. The monovalent type II cohesin-bearing scaffoldin is coated onto wells of the microtiter plate, followed by subsequent incubations of the adaptor scaffoldin and xylanase-fused dockerin (Xyn-Doc), the latter of which is recognized by anti-Xyn antibody. Only a relevant Xyn-Doc reveals appropriate binding signals, whereas a nonspecific one exhibits only poor interaction at very high concentrations. No signal was observed in the absence of adaptor scaffoldin

1. Coat wells of MaxiSorp ELISA plates (Nunc A/S, Roskilde, Denmark) with 100 μL/well (1 g/mL) of the relevant CBM-cohesin (type II) protein (matching-cohesin of the dockerin module of the adaptor) in Coating Solution, overnight at 4 °C.

2. Subsequent steps are performed at room temperature with all reagents at a volume of 100 μL/well.
3. Discard Coating Solution.
4. Add Blocking Buffer and incubate for 1 h.
5. Discard Blocking Buffer.
6. Interact with incremental concentrations (0.1–1000 ng/mL) of the adaptor scaffoldin in Blocking Buffer for 1 h.
7. Wash three times with Wash Buffer #2.
8. Samples are subjected to interaction with relevant Xyn-Docs at 100 ng/mL in Blocking Buffer for 1 h.
9. Plates are again washed three times.
10. Interact with primary antibody solution in Blocking Buffer for 1 h.
11. Wash three times with Wash Buffer.
12. Interact with secondary antibody solution in Blocking Buffer for 1 h.
13. Wash four times with Wash Buffer.
14. Detection: Add 100 μL/well of TMB substrate-chromogen.
15. Terminate reaction after 2–5 min with 50 μL/well 1 M H_2SO_4.
16. Measure absorbance at 450 nm using a tunable microplate reader. Plot absorbance as a function of adaptor concentration (usually results in a sigmoidal curve).

3.3.2 Non-denaturing PAGE

To check the extent of the interaction and determine exact equimolar ratios between matching pairs of cohesin and dockerin in the recombinant proteins, a differential mobility assay using non-denaturing PAGE is used. The optimal interaction may form at ratios somewhat different from the estimated 1:1 ratio, mainly due to impurities of the proteins that interfere with the real estimation of concentration. An example of two interacting chimaeric scaffoldins is provided in Fig. 6 (*see* **Note 8**; *see* Subheading 2.3.2).

1. Prepare solutions containing the two interacting proteins (i.e., chimaeric enzyme-scaffoldin or adaptor-primary scaffoldin) at different estimated molar ratios usually from 0.4:1 to 1.6:1 (1:1 ratio is about 4–8 μg) in Dilution Buffer in a total volume of 30 μL.
2. Incubate the tubes for 2 h at 37 °C.
3. Add 7.5 μL ND-Sample Buffer to 15 μL of the reaction mixture.
4. Load the samples onto non-denaturing gels. The stacking gel is prepared at 3.5% or 4.3%; the separating gel can be prepared at 6–15% (*see* **Note 9**).
5. Run the gel for a minimum of 2 h at 100 V to evaluate enzyme-scaffoldin combinations. To analyze interactions between the

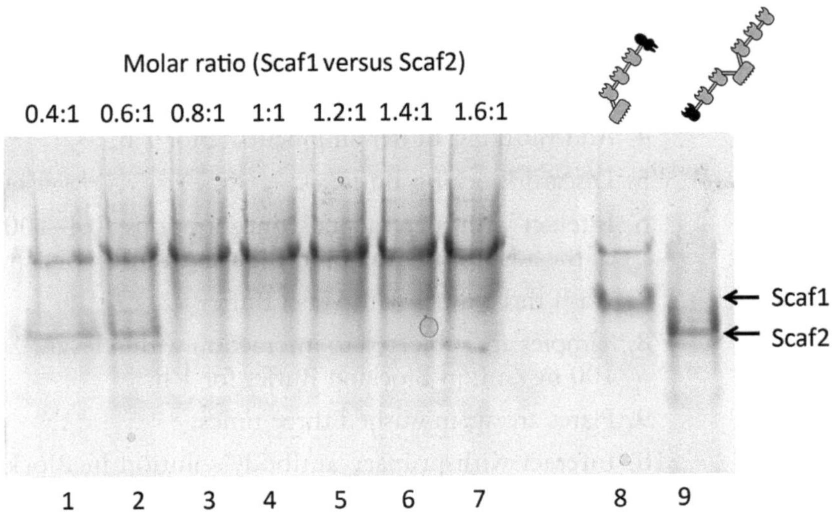

Fig. 6 Non-denaturing PAGE for determining stoichiometry of interaction between the adaptor scaffoldin Scaf1 and the primary platform scaffoldin Scaf2. *Lanes 1–7* correspond to complexes of the two scaffoldins at different ratios. *Lane 8* corresponds to Scaf1 alone, and *lane 9* corresponds to Scaf2 alone (100 pmol each). For all gels, the amount of Scaf2 was fixed at 100 pmol, and the amount of Scaf1 was adjusted according to the specified molar ratios. Here the exact equimolar ratio is reached in *lane 3*, equivalent to a 0.8:1 ratio according to initial estimates of concentration. At this ratio, both bands representing the two chimaeric scaffoldin have disappeared. Only the high heavy band representing the complex of the two can be observed. At lower ratios, Scaf2 is in excess, and at higher ratios, the excess Scaf1 band begins to appear

adaptor and primary/platform scaffoldins, electrophoresis should proceed for a minimum of 3 h but not exceeding that.

6. When both proteins are combined at the exact stoichiometric ratio, a single major band with altered mobility is usually observed.

3.3.3 Affinity Pull-Down Assay

In order to ensure that the designer cellulosome complex resulting from the binding of a chimaeric scaffoldin to several enzymes is indeed formed at exact equimolar ratios but also has the ability to strongly bind to the cellulosic substrate via the resident CBM, an affinity pull-down assay should be performed (Fig. 7). This approach will also show that the enzymes are indeed contained in the complex and not in the free form, i.e., released from the complex.

1. Prepare equimolar amounts of pure proteins (100 pmol each) in a final volume of 300 μL of Reaction Buffer (*see* Subheading 2.4.1).
2. Incubate samples for 2 h at 37 °C in a solution of 10% Cellobiose (Sigma-Aldrich) (*see* **Note 10**).
3. Gently mix the fractions with 30 mg of microcrystalline cellulose (Avicel, Sigma-Aldrich) for 1 h at 4 °C.
4. Centrifuge at 16,000 × *g* for 2 min.

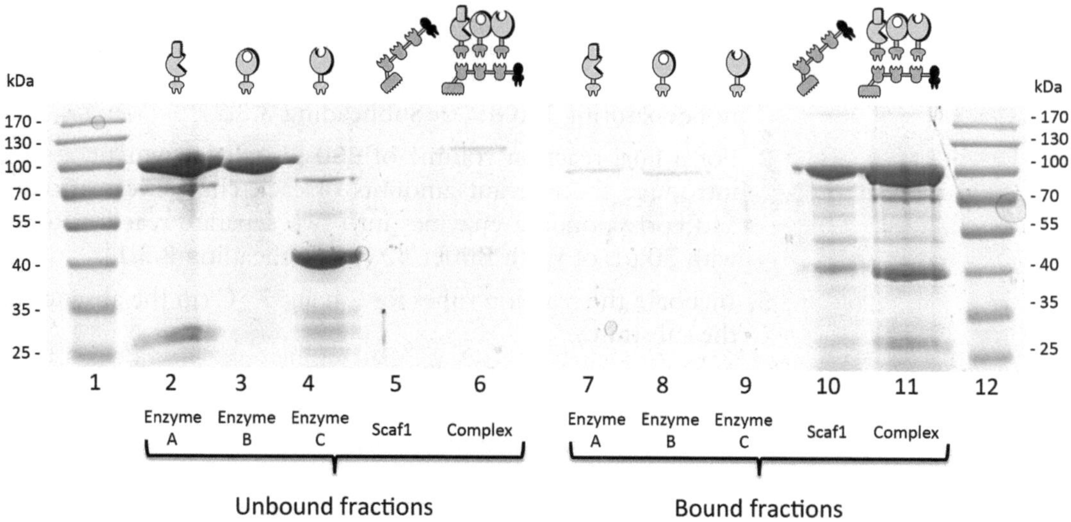

Fig. 7 Affinity pull-down assay. The three chimaeric enzymes and the adaptor scaffoldin Scaf1 were first assayed individually for binding to a cellulosic substrate. They were then mixed together at equimolar ratios and subsequently introduced to a cellulosic substrate (Avicel). The cellulose-binding ability of both individual proteins and the resultant complex was determined by examining the cellulose-unbound (*lanes 2–6*) versus cellulose-bound (*lanes 7–11*) fractions by SDS-PAGE. Samples include: *lane 1* and 12, molecular weight markers. *Lane 2–6*, unbound fractions with the following details: *lane 2–4*, three different chimaeric enzymes; *lane 5*, the adaptor scaffoldin Scaf1; *lane 6*, complex of the three enzymes and Scaf1. *Lanes 7–11* are bound fractions: *lanes 7–9*, three different chimaeric enzymes; *lane 10*, Scaf1; *lane 11*, complex of the three enzymes and Scaf1. In the presence of the chimaeric scaffoldin, the enzymatic components were associated with the cellulose-bound fraction, whereas in its absence they remained in the unbound fraction. In the example shown, Scaf1 and Enzymes A and B have similar molecular weights and are hard to distinguish individually in complex on the gel. Nevertheless, it is clear that the Sca1-complexed enzymes are all bound to the cellulosic matrix

5. Carefully remove supernatant fluids (containing unbound proteins) and add SDS-Sample Buffer (*see* Subheading 2.2.3) to a final volume of 60 μL.
6. Wash pellets (containing bound proteins) twice by resuspension in 200 μL of Wash Buffer #3 to eliminate nonspecific binding.
7. Centrifuge the washed pellets at 16,000 × *g* for 2 min, and suspend in 60 μL of SDS-Sample Buffer (*see* Subheading 2.2.3).
8. Boil both resultant unbound and bound fractions for 10 min and analyze by SDS-PAGE (*see* Subheading 3.2.4).

3.3.4 Assembly of Extended Designer Cellulosome

For assembly of designer cellulosome systems containing an adaptor scaffoldin, each scaffoldin is allowed to interact separately with the predetermined ratio of its respective enzymes. Once complexes are formed, the two scaffoldins (laden with enzymes) are mixed together, thereby initiating self-assembly, and subjected to additional interaction as described in the following

1. For each cellulosomal components, calculate the relevant amount of protein based on the absorbance measurement at OD_{280nm} and/or adjustment of equimolar ratios following nondenaturing PAGE (*see* Subheading 3.3.2).
2. For a final reaction volume of 200 μL (for enzymatic assay), introduce the relevant amounts of each chimaeric scaffoldin and corresponding enzymes into two separate reaction tubes with 30 μL of Wash Buffer #2 (*see* Subheading 2.3.1).
3. Incubate the reaction tubes for 2 h at 37 °C (in the absence of the substrate).
4. Mix the previously described reaction tubes into a new single reaction tube and incubate it for 2 additional hours at 37 °C (in the absence of the substrate).

3.4 Enzyme Activity Assays

The conditions for the enzyme assays of the complexed cellulosomal components have to be optimized, such as concentration, pH, and temperature. In addition, the linear range of activity exhibited by the native or chimaeric enzymatic cocktails also has to be calibrated [40]. The optimal substrates for cellulosomal activity are cellulosic substrates (as PASC and Avicel) or lignocellulosic ones (such as switchgrass or wheat straw) [32].

A typical example of proximity with or without targeting effect is illustrated in Fig. 8 [13]. The targeting effect can be deduced by the increased degradation profile of the CBM-restored enzymes in comparison with the free dockerin-bearing enzymes. While Adaptor1 provides an additional proximity effect by virtue of its resident CBM (i.e., higher degradation than CBM-restored enzymes), Adaptor2 (with an added external CBM) provides only a proximity effect among the enzymes, since the observed degradation is higher than that of the free enzymes but lower than the CBM-restored enzyme. In addition, the benefit of proximity between two chimaeric scaffoldins is also exemplified in Fig. 9: only the cellulosomal system with two attached chimaeric scaffoldins ultimately provides an advantage over the wild-type enzyme mixture. While these effects can be readily deduced using designer cellulosomes [40], precise interpretation of the data is problematic using natural cellulosomal extracts, since cellulosome composition is not controlled, even if mass spectrometric analysis is used for their analysis [10].

The enzymatic concentration in designer cellulosomes depends on the cellulosic substrate, and the temperature of the reaction depends on the origin of the enzyme(s) used (*see* **Note 11**). After termination of the reaction, samples are centrifuged at maximum speed to remove intact substrate, and the amount of reducing sugars can be determined using DNS assay or high-performance anion exchange chromatography (HPAEC), as previously described [40].

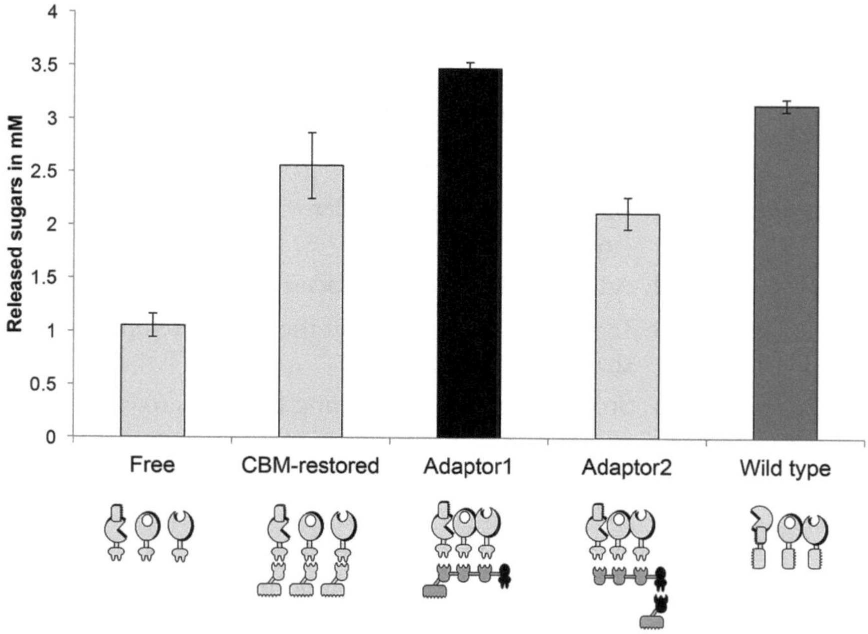

Fig. 8 Comparative degradation of Avicel by various artificial cellulosomal complexes and free enzymes. The composition of the complexes and free-enzyme systems is indicated as *pictograms*. The *black bar* indicates the optimal configuration shown by the scaffoldin-bound dockerin-bearing enzymes, and the *dark grey bar* represents the wild-type CBM-bearing free enzyme mixture. In this example, the use of a CBM-bearing scaffoldin (Adaptor1) is preferable to external inclusion of a CBM via an additional monovalent scaffoldin (Adaptor2) or to the performance of the mixture of free CBM-restored enzymes. Enzymatic activity is defined as mM total reducing sugars after 48 h using a glucose standard curve. Standard deviations are indicated

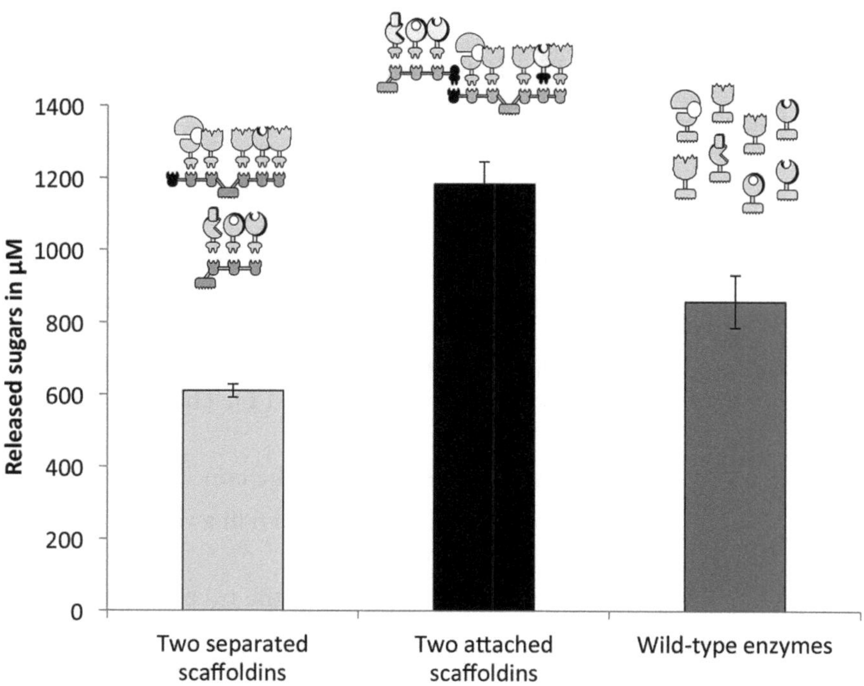

Fig. 9 Comparative degradation of wheat straw by two different chimaeric cellulosomal enzymatic complexes and wild-type enzymes after 48 h. The composition of the complexes and free enzyme system (each of which bears a CBM for substrate targeting) is indicated as pictograms. The *black bar* indicates the optimal configuration while the *dark grey bar* represents the wild-type enzyme mixture. Enzymatic activity is defined as μM total reducing sugars using a glucose standard curve. Standard deviations are as indicated

3.4.1 Substrate Degradation

See Subheading 2.4.1 for preparation of materials and reagents. All assays must be performed in triplicate.

Hydrolysis of Phosphoric Acid-Swollen Cellulose (PASC)

1. Mix 100 μL Reaction Buffer with the desired cellulosome sample (*see* **Note 11**).
2. Add 100 μL of PASC suspension (final concentration 7.5 g/L).
3. Incubate for 30–90 min at the desired temperature in a vertical shaking incubator.
4. Stop reaction by transferring the tubes to an ice–water bath.

Avicel Hydrolysis

1. Mix 120 μL Reaction Buffer with the desired cellulosome sample (*see* **Note 11**).
2. Add 80 μL of 10% Avicel suspension
3. Incubate for 18–72 h at the desired temperature in a vertical shaking incubator.
4. Stop reaction by transferring the tubes to an ice–water bath.

Xylan Hydrolysis

1. Mix 100 μL Reaction Buffer with the desired cellulosome sample (*see* **Note 11**). The final reaction volume is 200 μL.
2. Add 100 μL of 2% xylan solution.
3. Incubate for 20–60 min at the desired temperature.
4. Stop reaction by transferring the tubes to an ice–water bath.

Switchgrass Hydrolysis

1. Mix 50 μL Reaction Buffer with the desired cellulosomal composition sample (*see* **Note 11**). The final reaction volume is 200 μL.
2. Add acid-pretreated switchgrass [10] at a final concentration of 5 mg/mL
3. Incubate for 24 h at 60 °C in a vertical shaking incubator.
4. Stop reaction by transferring the tubes to an ice–water bath.

Wheat Straw Hydrolysis

1. Mix 100 μL Reaction Buffer the desired concentration of the cellulosome sample (*see* **Note 11**). The final reaction volume is 200 μL.
2. Add hatched wheat straw at the concentration of 3.5 g/L.
3. Incubate for a minimum of 16 h in a vertical shaking incubator at the desired temperature.
4. Stop reaction by transferring the tubes to an ice–water bath.

3.4.2 DNS Analysis of Released Sugars

1. Following enzymatic assay, centrifuge all the tubes for 2 min at 16,000 × *g*.
2. Take 100 μL of the supernatant of each reaction tube and transfer to a new one.

3. Add 150 μL of DNS solution to each tube.
4. Boil the resulting tubes during 10 min.
5. Measure absorbance at 540 nm using a tunable microplate reader (*see* **Note 12**).

4 Notes

1. Concentrated HCl (12 N) can initially be used, followed by reduced concentrations of lower strengths.
2. When working with cellulosomal components, a stock solution of 2 M $CaCl_2$ should always be available. The calcium solution will be used for numerous manipulations since the dockerin/cohesin binding event is calcium-dependent.
3. The arrangement of the different modules composing the two chimaeric scaffoldins must be considered [38] as well as the length of linkers between them [39].
4. Cloning of large chimaeric scaffoldins may be quite challenging in pET28a plasmids. Indeed, the insert almost doubles the size of the vector. Particularly, high concentrations of insert versus vector (about 10–1) should be used allowing the ligation reactions to last for longer periods (at least 1 day).
5. In order to examine the preferred conditions for expression of the desired protein, the final concentration of IPTG, the temperature and the period of incubation can be adjusted at the indicated values. An optimization step composed of low-scale growths with the different parameters can be performed in order to determine the optimal conditions.
6. Do not leave while the sonicator is in operation. It is possible that the beaker breaks or turns in the melting ice.
7. Determination of the scaffoldin concentration according to the absorbance at 280 nm should be taken with care. The effective stoichiometry of components must be readjusted using non-denaturing PAGE mobility assay, as the scaffoldin contains very few tryptophans, and hence has a particularly low extinction coefficient.
8. When performing non-denaturing gel electrophoresis in the presence of an adaptor scaffoldin, i.e., a recombinant cohesin-bearing protein that also contains a type II dockerin, the formation of homodimers via the type II dockerin may be observed [41] which may be confusing. Only at the optimal molecular ratio (1:1) between the type II dockerin and the matching type II cohesin will the homodimer not be observed.
9. As the assembled complex is more important than the component parts, the percentage of the gel should be lower to accommodate the high-molecular-weight material.

10. Cellobiose binds to the catalytic module of the enzymes, thus blocking potential weak binding to the cellulosic substrate.
11. The enzymatic concentration in designer cellulosomes for the degradation of recalcitrant substrate should be: for cellulosic materials, like PASC and Avicel, in the range of 0.1–2 μM of total enzymatic concentration, and for crude natural substrates, like switchgrass and wheat straw, in the range of 0.3–5 μM. In contrast, for degradation of simple purified substrates, such as CMC and xylan, the enzymatic concentration should be in a range of 0.1–1 nM. The required concentration of natural cellulosome extracts should be ~25 μg/mL. The temperature of incubation should correspond to the species from which the chimaeric enzymes originate (for designer cellulosomes) or from the species of the purified natural cellulosomes (i.e., from mesophilic or thermophilic bacteria).
12. In order to convert the OD measurements at 540 nm to amount of released sugars, a standard curve of known amount of glucose (0.5–5 mM) should be prepared and subjected to DNS analysis in the same time.

References

1. Himmel ME, Xu Q, Luo Y et al (2010) Microbial enzyme systems for biomass conversion: emerging paradigms. Biofuels 1:323–341. doi:10.4155/bfs.09.25
2. Lamed R, Setter E, Bayer EA (1983) The cellulosome: a discrete cell surface organelle of clostridium thermocellum which exhibits separate antigenic, cellulose-binding and various cellulolytic activities. Biotechnol Bioeng Symp 13:163–181
3. Bayer EA, Belaich J-P, Shoham Y, Lamed R (2004) The cellulosomes: multienzyme machines for degradation of plant cell wall polysaccharides. Annu Rev Microbiol 58:521–554. doi:10.1146/annurev.micro.57.030502.091022
4. Bayer EA, Kenig R, Lamed R (1983) Adherence of *Clostridium thermocellum* to cellulose. J Bacteriol 156:818–827
5. Dassa B, Borovok I, Lamed R et al (2012) Genome-wide analysis of *Acetivibrio cellulolyticus* provides a blueprint of an elaborate cellulosome system. BMC Genomics 13:210. doi:10.1186/1471-2164-13-210
6. Hamberg Y, Ruimy-Israeli V, Dassa B et al (2014) Elaborate cellulosome architecture of acetivibrio cellulolyticus revealed by selective screening of cohesin-dockerin interactions. Peer J 2:e636. doi:10.7717/peerj.636
7. Rincon MT, Dassa B, Flint HJ et al (2010) Abundance and diversity of dockerin-containing proteins in the fiber-degrading rumen bacterium, *Ruminococcus flavefaciens* FD-1. PLoS One 5:e12476. doi:10.1371/journal.pone.0012476
8. Izquierdo J a., Goodwin L, Davenport KW, et al (2012) Complete genome sequence of *Clostridium clariflavum* DSM 19732. Stand Genomic Sci 6:104–115. doi: 10.4056/sigs.2535732
9. Artzi L, Dassa B, Borovok I et al (2014) Cellulosomics of the cellulolytic thermophile *Clostridium clariflavum*. Biotechnol Biofuels 7:100. doi:10.1186/1754-6834-7-100
10. Artzi L, Morag E, Barak Y et al (2015) *Clostridium clariflavum*: key cellulosome players are revealed by proteomic analysis. MBio 6:1–12. doi:10.1128/mBio.00411-15
11. Bayer EA, Morag E, Lamed R (1994) The cellulosome-a treasure-trove for biotechnology. Trends Biotechnol 12:379–386. doi:10.1016/0167-7799(94)90039-6
12. Pagès S, Bélaïch A, Bélaïch JP et al (1997) Species-specificity of the cohesin-dockerin interaction between *Clostridium thermocellum* and *Clostridium cellulolyticum*: prediction of specificity determinants of the dockerin domain. Proteins 29:517–527
13. Stern J, Moraïs S, Lamed R, Bayer EA (2016) Adaptor scaffoldins: an original strategy for extended designer cellulosomes, inspired from

nature. MBio 7:e00083. doi:10.1128/mBio.00083-16.

14. Haimovitz R, Barak Y, Morag E et al (2008) Cohesin-dockerin microarray: diverse specificities between two complementary families of interacting protein modules. Proteomics 8:968–979. doi:10.1002/pmic.200700486
15. Barak Y, Handelsman T, Nakar D et al (2005) Matching fusion protein systems for affinity analysis of two interacting families of proteins: the cohesin-dockerin interaction. J Mol Recognit 18:491–501. doi:10.1002/jmr.749
16. Shiratori H, Sasaya K, Ohiwa H et al (2009) *Clostridium clariflavum* sp. nov. and *Clostridium caenicola* sp. nov., moderately thermophilic, cellulose–/cellobiose-digesting bacteria isolated from methanogenic sludge. Int J Syst Evol Microbiol 59:1764–1770. doi:10.1099/ijs.0.003483-0
17. Bayer EA, Shoham Y, Lamed R (2013) Lignocellulose-decomposing bacteria and their enzyme systems. In: Rosenberg E, DeLong EF, Lory S et al (eds) Prokaryotes-prokaryotic physiol biochem. pp 215–265. Springer: Berlin
18. Lamed R, Setter E, Bayer EA (1983) Characterization of a cellulose-binding, cellulase-containing complex in *Clostridium thermocellum*. J Bacteriol 156:828–836
19. Bayer EA, Setter E, Lamed R (1985) Organization and distribution of the cellulosome in *Clostridium thermocellum*. J Bacteriol 163:552–559
20. Bayer EA, Lamed R (1986) Ultrastructure of the cell surface cellulosome of *Clostridium thermocellum* and its interaction with cellulose. J Bacteriol 167:828–836
21. Lamed R, Naimark J, Morgenstern E, Bayer EA (1987) Specialized cell surface structures in cellulolytic bacteria. J Bacteriol 169:3792–3800
22. Xu Q, Resch MG, Podkaminer K et al (2016) Dramatic performance of *Clostridium thermocellum* explained by its wide range of cellulase modalities. Sci Adv 2:e1501254. doi:10.1126/sciadv.1501254
23. Raman B, Pan C, Hurst GB et al (2009) Impact of pretreated Switchgrass and biomass carbohydrates on *Clostridium thermocellum* ATCC 27405 cellulosome composition: a quantitative proteomic analysis. PLoS One 4:e5271
24. Morag E, Lapidot A, Govorko D et al (1995) Expression, purification and characterization of the cellulose-binding domain of the scaffoldin subunit from the cellulosome of *Clostridium thermocellum*. Appl Environ Microbiol 61:1980–1986
25. Fendri I, Tardif C, Fierobe H-P et al (2009) The cellulosomes from *Clostridium cellulolyticum*: identification of new components and synergies between complexes. FEBS J 276:3076–3086. doi:10.1111/j.1742-4658.2009.07025.x
26. Han SO, Yukawa H, Inui M, Doi RH (2005) Effect of carbon source on the cellulosomal subpopulations of *Clostridium cellulovorans*. Microbiology 151:1491–1497. doi:10.1099/mic.0.27605-0
27. Fierobe HP, Mechaly A, Tardif C et al (2001) Design and production of active cellulosome chimeras. Selective incorporation of dockerin-containing enzymes into defined functional complexes. J Biol Chem 276:21257–21261. doi:10.1074/jbc.M102082200
28. Caspi J, Irwin D, Lamed R et al (2006) *Thermobifida fusca* family-6 cellulases as potential designer cellulosome components. Biocatal Biotransform 24:3–12
29. Caspi J, Irwin D, Lamed R et al (2008) Conversion of *Thermobifida fusca* free exoglucanases into cellulosomal components: comparative impact on cellulose-degrading activity. J Biotechnol 135:351–357. doi:10.1016/j.jbiotec.2008.05.003
30. Caspi J, Barak Y, Haimovitz R et al (2009) Effect of linker length and dockerin position on conversion of a *Thermobifida fusca* endoglucanase to the cellulosomal mode. Appl Environ Microbiol 75:7335–7342. doi:10.1128/AEM.01241-09
31. Caspi J, Barak Y, Haimovitz R et al (2010) *Thermobifida fusca* exoglucanase Cel6B is incompatible with the cellulosomal mode in contrast to endoglucanase Cel6A. Syst Synth Biol 4:193–201. doi:10.1007/s11693-010-9056-1
32. Fierobe HP, Mingardon F, Mechaly A et al (2005) Action of designer cellulosomes on homogeneous versus complex substrates: controlled incorporation of three distinct enzymes into a defined trifunctional scaffoldin. J Biol Chem 280:16325–16334. doi:10.1074/jbc.M414449200
33. Moraïs S, Barak Y, Caspi J et al (2010) Cellulase-xylanase synergy in designer cellulosomes for enhanced degradation of a complex cellulosic substrate. MBio 1:3–10. doi:10.1128/mBio.00285-10.
34. Moraïs S, Morag E, Barak Y et al (2012) Deconstruction of lignocellulose into soluble sugars by native and designer cellulosomes. MBio. doi:10.1128/mBio.00508-12
35. Mingardon F, Chanal A, Tardif C et al (2007) Exploration of new geometries in cellulosome-like chimeras. Appl Environ Microbiol 73:7138–7149. doi:10.1128/AEM.01306-07
36. Mingardon F, Chanal A, López-Contreras AM et al (2007) Incorporation of fungal cellulases in bacterial minicellulosomes yields viable, synergistically acting cellulolytic com-

plexes. Appl Environ Microbiol 73:3822–3832. doi:10.1128/AEM.00398-07

37. Arfi Y, Shamshoum M, Rogachev I et al (2014) Integration of bacterial lytic polysaccharide monooxygenases into designer cellulosomes promotes enhanced cellulose degradation. Proc Natl Acad Sci U S A 111:9109–9114. doi:10.1073/pnas.1404148111
38. Eschbach S, Hofmann C, Maerz M, et al (1990) Molecular cloning. A laboratory manual. 2. Auflage. Hrsg. von J. Sambrook, E. F. Fritsch, T. Maniatis, Cold Spring Harbor Laboratory Press. ISBN 0-87969-309-6. Biol Unserer Zeit 20:285–285. doi: 10.1002/biuz.19900200607
39. Gasteiger E, Hoogland C, Gattiker A et al (2005) Protein identification and analysis tools on the ExPASy server. Proteomics Protoc Handb:571–607. doi:10.1385/1-59259-890-0:571
40. Vazana Y, Moraïs S, Barak Y et al (2012) Designer cellulosomes for enhanced hydrolysis of cellulosic substrates. Methods Enzymol 510:429–452. doi:10.1016/B978-0-12-415931-0.00023-9
41. Adams JJ, Webb BA, Spencer HL, Smith SP (2005) Structural characterization of type II dockerin module from the cellulosome of *Clostridium thermocellum*: Calcium-induced effects on conformation and target recognition. Biochemistry 44:2173–2182. doi: 10.1021/bi048039u

Part II

Analysis of Protein-Carbohydrate Interactions

Chapter 9

Affinity Electrophoresis for Analysis of Catalytic Module-Carbohydrate Interactions

Darrell Cockburn, Casper Wilkens, and Birte Svensson

Abstract

Affinity electrophoresis has long been used to study the interaction between proteins and large soluble ligands. The technique has been found to have great utility for the examination of polysaccharide binding by proteins, particularly carbohydrate binding modules (CBMs). In recent years, carbohydrate surface binding sites of proteins mostly enzymes have also been investigated by this method. Here, we describe a protocol for identifying binding interactions between enzyme catalytic modules and a variety of carbohydrate ligands.

Key words Affinity electrophoresis, Polyacrylamide gel electrophoresis, Polysaccharide, Surface binding site, Carbohydrate binding module, Dissociation constant

1 Introduction

Native polyacrylamide gel electrophoresis (PAGE) has been used for many years when electrophoretic analyses are required of proteins in their folded state, such as to monitor complex formation or enzyme activity. Affinity electrophoresis is the use of Native PAGE with a potential polysaccharide ligand embedded in the acrylamide matrix. Proteins that bind the ligand migrate more slowly than they do in native gel control. Interactions with ligands result in a gel shift, which is a similar effect to what is observed with DNA binding protein gel shift assays [1, 2]. Affinity gel electrophoresis (AGE) has seen use as a tool for detecting and visualizing the interactions between proteins and carbohydrates, often in the context of CBM binding [3–6]. Surface binding sites are ligand binding sites situated on the catalytic module of the enzyme, but distant from the active site, and have also been studied by affinity electrophoresis [7–11]. The mostly larger size of these non-truncated enzymes adds some additional challenges as compared to working with just CBMs; however, it is feasible with some optimization.

D. Wade Abbott and Alicia Lammerts van Bueren (eds.), *Protein-Carbohydrate Interactions: Methods and Protocols*, Methods in Molecular Biology, vol. 1588, DOI 10.1007/978-1-4939-6899-2_9, © Springer Science+Business Media LLC 2017

Here, we present a method for performing AGE with non-truncated enzymes, particularly those that lack CBMs, and a variety of potential polysaccharide ligands.

2 Materials

2.1 Polysaccharides

All polysaccharides are made as a 3× stock and the final concentration in the gel is given in parentheses.

1. *Amylopectin (0.1%)*

   Dissolve 150 mg in 3 mL DMSO (can stir in a 50 mL graduated cylinder). Once the amylopectin is fully dissolved in the DMSO, slowly add hot water (above 60 °C) to 50 mL while stirring vigorously. Make up to 50 mL with water in a volumetric flask.

2. *Pullulan (0.05%)*

   Dissolve 75 mg in 3 mL DMSO (can stir in a 50 mL graduated cylinder). Once the pullulan is fully dissolved in the DMSO, slowly add hot water (above 70 °C) to 50 mL while stirring vigorously. Make up to 50 mL with water in a volumetric flask.

3. *Amylose DP440 (0.05%)*

   Dissolve 75 mg in 10 mL DMSO (can stir in a 50 mL graduated cylinder). Once the amylose is fully dissolved in the DMSO (approximately 10 min), slowly add very hot water (above 80 °C) to 50 mL while stirring as vigorously as possible. Make up to 50 mL with water in a volumetric flask.

4. *Glycogen (0.1%)*

   Dissolve 150 mg in 40 mL water. Make up to 50 mL with water in a volumetric flask.

5. *Laminarin (Laminaria digitata) (0.1%)*

   Dissolve 150 mg in 40 mL water, may require some light heat. Make up to 50 mL with water in a volumetric flask.

6. *Curdlan (0.05%)*

   Dissolve 75 mg in 40 mL water at approximately 50 °C. Stir for 20 min. Make up to 50 mL with water in a volumetric flask. Note that it will not fully go into solution, only form a suspension. Shake prior to dispensing for casting gels.

7. *Konjac Glucomannan (0.05%)*

   Dissolve 75 mg in 40 mL water at approximately 50 °C. Make up to 50 mL with water in a volumetric flask.

8. *Galactomannan (0.05%)*

Dissolve 150 mg in 40 mL water, microwave on high until solution is clear, and stir until cooled. Adjust volume to 50 mL. It will not fully dissolve, hence shake prior to dispensing for gels.

9. *Galactan (0.05%)*

Dissolve 150 mg in 40 mL water, microwave on high until solution is clear, stir until cooled. Adjust volume to 50 mL.

10. *Sugar Beet Arabinan (0.1%)*

Dissolve 150 mg in 40 mL water at approximately 50 °C. Make up to 50 mL with water in a volumetric flask.

11. *Wheat Arabinoxylan (low viscoscity) (0.1%)*

Dissolve 150 mg in 40 mL water at approximately 50 °C. Make up to 50 mL with water in a volumetric flask.

12. *Birchwood and Oatspelt Xylan (0.1%)*

Dissolve 150 mg in 40 mL water, microwave until solution is clear, stir until cooled. Adjust volume to 50 mL. Oatspelt xylan will not go fully into solution, resuspend prior to dispensing for gels.

13. *Lignin (desulfonated) (0.01%)*

Dissolve 15 mg in 50 mL water. Also, double the concentration of TEMED and APS used for polymerization.

14. *Hydroxyethyl Cellulose (0.1%)*

Make 50 mL of 10 mM sodium phosphate pH 6.0. Slowly add 150 mg of HEC to a vigorously stirred solution. Stir for 5 min. Adjust the pH to 8.0 and stir vigorously for 20 min.

15. *Barley Beta Glucan (0.1%)*

Dissolve 150 mg of beta glucan in 95% ethanol (about 40 mL), stirring for 20 min. Add 50 mL of water and boil for 10 min, followed by stirring at 80 °C for 1 h. Add water every 15 min or so to prevent drying out. After cooling, adjust volume to 50 mL with water. It will not fully go into solution, shake prior to dispensing for gels.

16. *Xyloglucan (0.05%)*

Stir 50 mL of water vigorously and slowly add 75 mg of xyloglucan. Stir vigorously on medium heat for 20 min. Make up to 50 mL with water in a volumetric flask.

17. *Hyaluronic Acid (0.025%)*

Stir 50 mL of water vigorously and slowly add 37.5 mg of hyaluronic acid. Stir vigorously on medium (150 °C) heat for 20 min. Make up to 50 mL with water in a volumetric flask.

18. *Rhamnogalaturonan I (0.1%)*

Dissolve 150 mg in 40 mL water. Make up to 50 mL with water in a volumetric flask.

2.2 Native PAGE Gels

1. 5× Running buffer: Add 30.29 g Tris base and 7.73 g boric acid in a glass beaker. Make up to 1 L with water. Check pH, if significantly different than 8.7, remake, do not attempt to adjust pH with acid or base as this will alter the ionic strength. Store at room temperature.
2. 5× Loading buffer: Add 0.66 g Tris base, 0.15 g boric acid, 0.02 g bromphenol blue, 10 mL glycerol, and 10 mL water to a glass beaker. Stir until solution is clear. Check pH, if significantly different from 8.7, discard the solution and remake. Store at room temperature.
3. Coomassie stain: Add 0.1% Coomassie Brilliant Blue R-250 to 10% acetic acid, 50% methanol, 40% water. Mix until solution is a clear dark blue.
4. Destain: 10% acetic acid, 50% methanol, 40% water.
5. Ammonium persulfate: Make a 100 mg/mL solution fresh each day in a 1.5 mL Eppendorf tube.
6. Gel cassettes.
7. TEMED.
8. 40% Acrylamide/bis-acrylamide solution.
9. Isopropanol.

2.3 Enzyme Preparation

1. Prepare enzyme to a concentration of 0.5–1 mg/mL.
2. Add sufficient 5× loading dye to bring it to a 1× final concentration.

2.4 Gel Visualization

1. Camera, scanner, or gel doc station for imaging the gels.
2. ImageJ analysis software (download from: https://imagej.nih.gov/ij/).

3 Methods

3.1 Making Native Page Gels

1. Following Table 1, (*see* **Note 1**) make a set of polysaccharide containing gels and a set of control gels with desired percentage of acrylamide. These recipes are sufficient for making two gels. Add water, polysaccharide (replace with additional water for control gels), running buffer, and acrylamide to a 50 mL conical tube. Mix by gentle swirling or inversion.
2. Add TEMED and APS to initiate polymerization and then gently swirl or invert again. Using a transfer pipette, slowly add

Table 1
Gel recipes for affinity gel electrophoresis

Ingredient	6% gel	8% gel	10% gel	12% gel	14% gel
Water (mL)	4.75	4	3.25	2.5	1.75
Polysaccharide[a] (mL)	5	5	5	5	5
Buffer (5×) (mL)	3	3	3	3	3
Acrylamide 40% (mL)	2.25	3	3.75	4.5	5.25
TEMED (μL)	10	10	10	10	10
APS (μL)	75	75	75	75	75

[a]For control gels, replace the polysaccharide with water

the gel solution to the cassette, leaving a 2.5 cm gap to the top of the gel cassette. Add a thin layer of isopropanol to remove bubbles and smoothen the surface. Allow approximately 30 min to 1 h for polymerization to occur.

3. Pour off isopropanol layer and rinse with water.
4. Make stacking gel, by adding 3.5 mL water, 1 mL running buffer, and 0.5 mL acrylamide to a 50 mL conical tube. Mix by gentle swirling or inversion.
5. Add 10 μL TEMED and 75 μL APS. Mix by gentle swirling or inversion. Use a transfer pipette to fill the remainder of the gel cassette with the stacking gel solution. Be careful to avoid introduction of bubbles, add a gel comb with the desired number of wells. Allow polymerization to occur for approximately 45 min.
6. Gels can be kept for up to 1 week if covered with wet paper towels and plastic wrap, storing at 4 °C.

3.2 Running Gels

1. Place one polysaccharide gel and a control gel of the same acrylamide concentration into a gel chamber, removing the sticker at the bottom of the gel, with the gel combs facing toward the middle of the well formed by the two gels.
2. Fill this middle well with 1× running buffer and remove the well combs.
3. Add 5 μL of the control protein or ladder to the first well. Add experimental proteins to each of the additional wells. Different load levels can be tried as some proteins run in tighter bands than others. At 0.5–1 mg/mL, 5–10 μL is recommended.
4. Run at a constant voltage. Voltage and time may need to be optimized for the specific protein (*see* **Note 2**). Examples include 75 V for 6 h for 100 kDa protein, p*I* of 6.0 on a 6% gel, or 75 V for 2 h for a 30 kDa protein with p*I* 4.0 on a 12% gel.

3.3 Staining Gels

1. Open gel cassette and carefully remove the gel, discarding the stacking gel.
2. Place in a small container and cover with staining solution. Incubate 1–2 h at room temperature with gentle shaking.
3. Remove staining solution (dispose of in proper chemical waste, contains acid and methanol). Add destain solution. Incubate for 2 h—overnight with at least one additional change of solvent. Use gentle shaking.
4. Remove destaining solution (dispose of in proper chemical waste, contains acid and methanol).
5. Scan or photograph the gels.
6. Gel can be stored in 5% acetic acid.

3.4 Measure Protein Migration

1. Open the gel in ImageJ (download from: https://imagej.nih.gov/ij/).
2. Using the measurement function, measure the distance from the top of the separating gel to the band for each reference and sample protein for both the control and polysaccharide containing gels (*see* Fig. 1).
3. Divide the migration distance in the polysaccharide gel by the migration distance in the control gel for each protein to get the migration distance ratio (MDR).
4. For each sample protein divide its MDR by the MDR of the reference protein to get the Normalized MDR.
5. If the Normalized MDR is less than 1 (*see* **Note 3**) the protein is binding to the polysaccharide.

3.5 Quantitative Affinity Electrophoresis

Besides being used as a qualitative method to screen for potential ligands, AGE can also be used quantitatively to estimate binding constants for the enzymes by using a range of polysaccharide concentrations. The original equation (Eq. 1) developed by Takeo and Nakamura [12] assumes that there is a linear relationship between the concentration of the ligand and the mobility of the enzyme from which the apparent dissociation constants ($K_{D,app}$) can be determined:

$$\frac{1}{R_{mi}} = \frac{1}{R_{mo}}\left(1 + \frac{c}{K_{D,\,app}}\right) \tag{1}$$

R_{mi} and R_{mo} are migration distances of sample relative to reference protein in the presence and the absence of polysaccharide, respectively; c is the polysaccharide concentration. Thus, the inverse relative migration distances ($1/R_{mi}$) plotted as a function of c yield $1/K_{D,\,app}$ as the negative x-axis intercept of the straight line in a Takeo-Nakamura plot [12]. Later, as demonstrated by Tomme et al. [2] with CBMs the linearization is unnecessary to obtain quantitative binding data; however, this approach can require more

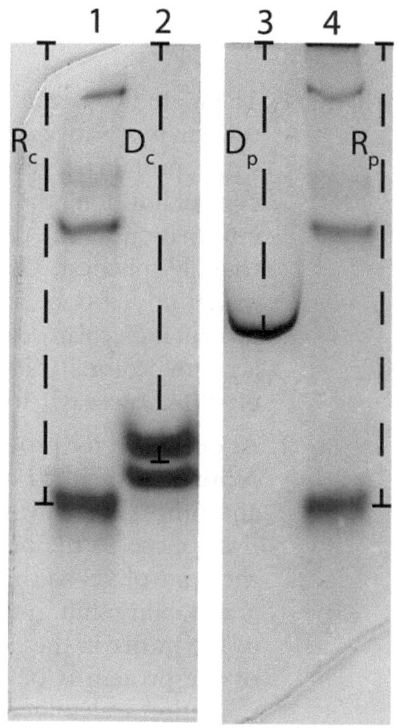

Fig. 1 Affinity electrophoresis of the barley α-amylase 1 (AMY1). The control gel (polysaccharide free) is on the *left*, while the polysaccharide gel is on the *right*. *Lanes 1* and *4* are the NativeMark protein ladder (Invitrogen), *lanes 2* and *3* are AMY1. Both gels are 12% polyacrylamide run at 100 V for 4.5 h. The ratio of Dp:Dc is the MDR, which is divided by Rp:Rc to give the normalized MDR. Note that AMY1 runs as two bands in the control gel, but only one in the polysaccharide gel, so the midpoint of the two bands is used for measurement purposes

data points than the linear Takeo-Nakamura plot, which can be problematic in studies where either or both the amount of enzyme and the ligand are limiting. Tomme et al. [5] showed that the data for the CBMs can be plotted with the relative migration (R_m) ($R_m = R_{mo} - R_{mi}$) of the enzyme as a function of c to the general rectangular hyperbolic form of the 1:1 binding isotherm (Eq. 2) [5].

$$R_m = \frac{R_{mo} + R_{mi} K_{A,app} c}{1 + K_{a,app} c} \tag{2}$$

where $K_{a\ app}$ is the apparent association constant. For convenience R_{mo} can be normalized to 1 and Eq. 2 then becomes:

$$R_{m,norm} = \frac{1 + R_{mi,\text{norm}} K_{a,app} c}{1 + K_{a,app} c} \tag{3}$$

where $R_{m,\text{norm}}$ is the R_m normalized to R_{mo} ($R_{m,\text{norm}} = R_m/R_{mo}$) and $R_{mi,\text{norm}}$ is R_{mi} normalized to R_{mo} ($R_{mi,\text{norm}} = R_{mi}/R_{mo}$). Eq. 3 can be used to regress K_a and $R_{mi,\text{norm}}$ from affinity data plotted as $R_{m,\text{norm}}$ of the enzyme against the c [5].

4 Notes

1. There are five factors that will determine how quickly a protein will run in a Native PAGE gel: protein molecular weight, protein shape, protein p*I*, gel pore size, and voltage. Additionally, we can control how long the gel is run for. Often we will not have much information about protein shape and so we assume that it is roughly spherical. Of the remaining factors we can most easily match up gel pore size (directly related to % acryclamide) and protein molecular weight. For proteins less than 20 kDa, use 14% acrylamide, for 20–40 kDa, use 12%, for 40–60 kDa use 10%, for 60–80 kDa use 8%, for >80 KDa use 6% kDa.
2. As described for protein molecular weight and gel pore size in **Note 1**, protein p*I* can best be adjusted for with the voltage and time. Generally, we shall want the protein to migrate relatively close to the bottom of the gel (though not run off) in the control gel so that we can easily see the difference if there is a mobility shift in the polysaccharide containing gel. The pH of the buffer in the described system is 8.7. The further the p*I* of the protein is below this, the quicker it will run. Protein migration scales linearly with both voltage and time, so we can use the measurement volt-hours to describe how to run the gel (e.g., 100 V for 1 h is 100 V-h). For $pI < 4.5$ use 150 V-h, for $4.5 < pI < 5.5$ use 300 V-h, for $5.5 < pI < 6.5$, use 450 V-h, for $6.5 < pI < 7.5$ use 600 V-h. For $pI > 7.5$ another buffer system will need to be used (*see* McLellan [13]). It is also possible to switch the electrodes on your gel system to allow migration of proteins with p*I* above the pH of the gel buffer. While protein migration is determined by V-h, there are certain voltage restrictions. Lower percentage gels are more sensitive to the heat generated at higher voltages, which may alter pore size or even result in gel failure. It is not recommended to exceed 100 V for gels 8% or less or 120 V for gels >8%.
3. Whether a value less than one is significant can be determined through replication. However, generally, the cutoff for accepting the shift as representative of binding needs to take into account the width of the band (can sometimes be large in Native PAGE). Thus, the difference in migration distance between the control and polysaccharide gel should be greater than the band thickness to be considered positive for binding.

Acknowledgments

This work was supported by a grant to B.S. from the Danish Council for Independent Research|Natural Sciences.

References

1. Fried M, Crothers DM (1981) Equilibria and kinetics of lac repressor-operator interactions by polyacrylamide gel electrophoresis. Nucleic Acids Res 9(23):6505–6525
2. Garner MM, Revzin A (1981) A gel electrophoresis method for quantifying the binding of proteins to specific DNA regions: application to components of the *Escherichia coli* lactose operon regulatory system. Nucleic Acids Res 9(13):3047–3060
3. Abbott DW, Boraston AB (2012) Quantitative approaches to the analysis of carbohydrate-binding module function. Methods Enzymol 510:211–231.doi:10.1016/B978-0-12-415931-0.00011-2
4. Tomme P, Creagh AL, Kilburn DG, Haynes CA (1996) Interaction of polysaccharides with the N-terminal cellulose-binding domain of Cellulomonas fimi CenC. 1. Binding specificity and calorimetric analysis. Biochemistry 35:13885–13894. doi:10.1021/bi961185i
5. Tomme P, Boraston A, Kormos J, Warren R, Kilburn D (2000) Affinity electrophoresis for the identification and characterization of soluble sugar binding by carbohydrate-binding modules. Enzyme Microb Technol 27:453–458
6. Moraïs S, Lamed R, Bayer E (2012) Affinity electrophoresis as a method for determining substrate-binding specificity of carbohydrate-active enzymes for soluble polysaccharides. Methods Enzymol 908:119–127. doi:10.1007/978-1-61779-956-3
7. Cockburn D, Nielsen MM, Christiansen C, Andersen JM, Rannes JB, Blennow A, Svensson B (2015) Surface binding sites in amylase have distinct roles in recognition of starch structure motifs and degradation. Int J Biol Macromol 75:338–345. doi:10.1016/j.ijbiomac.2015.01.054
8. Ludwiczek ML, Heller M, Kantner T, McIntosh LP (2007) A secondary xylan-binding site enhances the catalytic activity of a single-domain family 11 glycoside hydrolase. J Mol Biol 373:337–354. doi:10.1016/j.jmb.2007.07.057
9. Blennow A, Viksø-Nielsen A, Morell MK (1998) α-glucan binding of potato-tuber starch-branching enzyme I as determined by tryptophan fluorescence quenching, affinity electrophoresis and steady-state kinetics. Eur J Biochem 252:331–338
10. Wilkens C, Andersen S, Petersen BO, Li A, Busse-Wicher M, Birch J, Cockburn D, Nakai H, Christensen HE, Kragelund BB, Dupree P, McCleary B, Hindsgaul O, Hachem MA, Svensson B (2016) An efficient arabinoxylan-debranching α-L-arabinofuranosidase of family GH62 from *Aspergillus nidulans* contains a secondary carbohydrate binding site. Appl Microbiol Biotechnol 100(14):6265–6277. doi:10.1007/s00253-016-7417-8
11. Wilkens C, Auger KD, Anderson NT, Meekins DA, Raththagala M, Abou Hachem M, Payne CM, Gentry MS, Svensson B (2016) Plant α-glucan phosphatases SEX4 and LSF2 display different affinity for amylopectin and amylose. FEBS Lett 590(1):118–128. doi:10.1002/1873-3468.12027
12. Takeo K, Nakamura S (1972) Dissociation constants of glucan phosphorylases of rabbit tissues studied by polyacrylamide gel disc electrophoresis. Arch Biochem Biophys 153(1):1–7
13. McLellan T (1982) Electrophoresis buffers for polyacrylamide gels at various pH. Anal Biochem 126(1):94–99. doi:10.1016/0003-2697(82)90113-0

Chapter 10

Quantifying CBM Carbohydrate Interactions Using Microscale Thermophoresis

Haiyang Wu, Cédric Y. Montanier, and Claire Dumon

Abstract

MicroScale Thermophoresis (MST) is an emerging technology for studying a broad range of biomolecular interactions with high sensitivity. The affinity constant can be obtained for a wide range of molecules within minutes based on reactions in microliters. Here, we describe the application of MST in quantifying two CBM-carbohydrate interactions, a CBM3a toward cellulose nanocrystals and a CBM4 against xylohexaose.

Key words Microscale thermophoresis (MST), Binding studies, Fluorescence quenching, Fluorescent label, K_d (dissociation constant)

1 Introduction

Protein-carbohydrate interactions are involved in a myriad of important biological functions. Therefore, many techniques are used to quantify and characterize those interactions, such as Isothermal Titration Calorimetry (ITC) that allows characterization of thermodynamics during interaction process, Affinity Gel Electrophoresis (AGE) using soluble polysaccharides, solid-state depletion assay using insoluble polysaccharides, UV difference, and fluorescence spectroscopy [1].

Thermophoresis, or thermodiffusion/Soret effect, was first described by Carl Ludwig in 1856. This technique uncovered the different responses of particles in a mobile phase toward a gradient of temperature. The miniaturized thermophoresis, namely MST, was developed in the last decade to monitor the interactions and calculate the thermodynamic constants among biomolecules ranging in size from 10^1 to 10^7 Da in only few microliters. The description of thermophoresis in a frame of a local thermodynamic equilibrium allows one to quantify the thermophoretic behavior based on the charge, size, and hydration shell of biomolecules [2, 3]. Basically, the MST instrument contains an infra-red (IR) laser that

D. Wade Abbott and Alicia Lammerts van Bueren (eds.), *Protein-Carbohydrate Interactions: Methods and Protocols*, Methods in Molecular Biology, vol. 1588, DOI 10.1007/978-1-4939-6899-2_10, © Springer Science+Business Media LLC 2017

induces a microscopic temperature gradient (e.g., 1°K/10 μm) in a glass capillary, a high-power LED (light-emitting diode) as a source of light excitation, a fluorescence detector, and a sample tray of 16 capillaries [4, 5]. Briefly, the labeled molecule is mixed with a decreasing concentration of non-fluorescent ligand and each mixture is loaded into one of the 16 glass capillaries. Initial fluorescence is measured as a base line in each capillary and the IR laser is switched on. The IR laser beam is focused onto the mid-point of capillary and strongly absorbed by aqueous solution resulting in a temperature gradient from the laser spot to the outer region. Simultaneously, a dichroic mirror allows fluorescence from labeled molecule to be continuously monitored, while complexes migrate in the capillaries. The fluorescence in the laser spot area decreases or increases during thermophoresis, at a velocity related to the concentration of ligand, due to change in size, charge, or hydration shell of the complex. Each capillary is scanned in minutes one after another. The constant of dissociation K_d is deduced from the binding curve derived by plotting the normalized fluorescence (F_{Norm}) at a given time against each ligand concentration (L) [3, 5].

Depending on the molecule investigated, the source of fluorescence could be intrinsic (e.g., indole side chain of tryptophan), coming from attached dyes (e.g., Fluorescein, Alexa546) or from fluorescent protein (e.g., GFP, YFP). The fluorescence excitation/detection spectrum range would lead to choose appropriate apparatus. Several labeling strategies could be employed to target proteins by attaching, for instance, specific fluorophore molecule to the ε-amine group of lysine, thiol group of cysteine, fusion protein (such as GFP), or even unnatural amino acids.

Here, we present the use of the MST for the quantification of CBM/ligand interactions. The CBMs used in our protocol are the CBM3a from *Clostridium Thermocellum* [6] and a newly identified CBM4 from termite gut microbiota [7]. They display affinity for insoluble crystalline cellulose and soluble xylan respectively. Both CBMs were labeled to primary amine group of lysine using an adapted protocol from the protein labeling kit provided by NanoTemper and then MST experiments were performed using Monolith NT.115 instrument.

2 Materials

All solutions were prepared using deionized water dH_2O and analytical grade reagents. Buffers were prepared and stored at room temperature. Protein samples were kept on ice until use.

2.1 Chemicals and Kits

1. Deionized water (dH_2O).
2. Protein Labeling Kit RED-NHS (Cat# MO-L001 RED-NHS, NanoTemper Technologies GmbH).

3. DMSO.
4. Regents for SDS-PAGE (*see* **Note 1**).
5. Xylohexaose.
6. Cellulose nanocrystals.
7. Sodium dihydrogen phosphate dehydrate ($NaH_2PO_4{\cdot}2H_2O$).
8. Disodium hydrogen phosphate dodecahydrate ($Na_2HPO_4{\cdot}12H_2O$).
9. Imidazole.
10. 5% Pluronic® F-127 (Nanotemper).
11. Sodium dodecyl sulfate (SDS, optional for SD test).
12. Dithiothreitol (DTT, optional for SD test).

2.2 Buffers and Solutions

1. Labeling buffer: 50 mM sodium phosphate buffer pH 7.0. To prepare 0.5 M NaH_2PO_4 (monobasic), weigh 78 g of $NaH_2PO_4{\cdot}2H_2O$ and transfer to a 1 L glass beaker containing about 800 mL of dH_2O. Dissolve by magnetic stirring. Adjust the volume up to 1 L with dH_2O and filter at 0.22 μm. To prepare 0.5 M Na_2HPO_4 (dibasic), weigh 179 g of $Na_2HPO_4{\cdot}12H_2O$ in 1 L and prepare the solution as described above. Mix 250 mL of the monobasic stock solution and 700 mL of the dibasic stock solution to obtain 0.5 M sodium phosphate buffer pH 7 stock solution. Check the pH. Transfer 10 mL of the 0.5 M stock solution to 90 mL of dH_2O in a measuring cylinder to obtain 100 mL of 50 mM sodium phosphate buffer pH 7.0 (*see* **Note 2**).
2. Elution buffer: 50 mM sodium phosphate, 200 mM imidazole buffer pH 7.0. Dissolve 1.36 g of imidazole in 100 mL of labeling buffer. Check the pH and adjust with phosphoric acid, if needed.
3. Xylohexaose (300 mM). Weigh 0.015 g of xylohexaose in a 1.5 mL micro centrifuge tube. Add 62 μL of dH_2O and vortex to solubilize. Store at 4 °C (*see* **Note 3**).
4. Cellulose nanocrystals at 20 g/L [8].
5. SD mix (optional for SD test): 4% SDS, 40 mM DTT. Weigh 0.4 g of SDS, 0.0617 g of DTT and dissolve in 10 mL of dH_2O. Store at -20 °C

2.3 Biological Material

Proteins: CBM3a [6, 9] and CBM4 [7] were produced in house as His-tagged fusion proteins.

2.4 Sample Preparation Materials

1. Vivapure Metal Chelate Mini spin columns (Cat# VS-MC01MC12) (*see* **Note 4**).
2. TALON® Metal Affinity Resin (Cat# 635504, Clontech) (*see* **Note 5**).

3. MO-K005 Monolith™ NT.115 MST Premium Coated Capillaries (*see* **Note 6**).
4. 0.2 mL individual PCR tubes (*see* **Note 7**).
5. 1.5 mL micro centrifuge tubes.

2.5 Instruments

1. Monolith® NT.115 Blue/Red (Cat # G008, NanoTemper Technologies GmbH) (*see* **Note 8**).
2. Bench centrifuge (*see* **Note 9**).
3. Magnetic stirrer.

2.6 Computer Software

1. MO.Control software.
2. MO.Affinity Analysis software.

3 Methods

3.1 Labeling and Purifying Proteins

Protein Labeling Kit RED-NHS protocol was used to label our CBMs (http://www.helsinki.fi/biosciences/corefacilities/microscalethermophoresis/Protein_Labeling_Manual_V012-RED-NHS.pdf). Procedure for purification of labeled CBMs was modified as our CBMs displayed affinity for the resin of the desalting column provided in the kit (*see* **Note 4**).

1. As our CBMs were already in 50 mM phosphate buffer pH 7.0, the buffer exchange step was not necessary (*see* **Note 2**).
2. Protein concentration was adjusted to 20 μM in 100 μL in the labeling buffer (*see* **Note 10**). The original concentration of protein is determined by measuring the absorbance at 280 nm and calculated according to the Beer-Lambert Law using theoretical extinction coefficient determined by protparam tool on expasy website (http://web.expasy.org/protparam/) [10].
3. According to the protocol of the manufacturer, the solid fluorescent dye concentration was set to approx. 435 μM by adding 30 μL of DMSO into the solid dye, and vigorously vortex until complete dissolution (*see* **Note 11**).
4. 15 μL of dye was added to 85 μL of labeling buffer to reach 3.2-fold concentration of the protein (*see* **Note 12**).
5. Mix 100 μL of protein to 100 μL of dye and incubate 1 h at room temperature, protected from light by aluminum foil. Then, place labeled protein on ice.

   We changed protocol from Step C of Labeling Kit protocol.
6. Add 100 μL of TALON® Metal Affinity Resin into the Vivapure Metal Chelate Mini spin column, centrifuge at 1500 RCF 1 min at 10 °C to remove storage buffer.

7. Wash the resin with 500 µL of ice cold dH_2O, leave caps open (*see* **Note 13**), centrifuge as described in **step 6**, and discard the flow through.
8. Equilibrate the resin with 400 µL of ice cold labeling buffer, centrifuge as described in **step 6**, and discard the flow through. Repeat this step three times.
9. Load 200 µL of labeled protein (*see* **step 5**) to the center of the column. Gently mix the protein-dye mixture with resin by shaking and incubate on ice for 15 min. Centrifuge as described in **step 6** to remove excess of free dye.
10. Wash with 400 µL of ice cold labeling buffer, centrifuge as described in **step 6**, and discard the flow through. Repeat this step two times.
11. Add 200 µL of ice cold elution buffer, resuspend gently the resin by shaking and incubate on ice for 5 min. Centrifuge at 1500 RCF 3 min at 10 °C; collect the flow through as the labeled protein solution #1. Repeat one time and collect the flow through as labeled protein solution #2.
12. Assess the concentration of the labeled protein by SDS-PAGE (*see* **Note 1**).

3.2 Assay Optimization

Before setting up an assay, determining the optimized condition is required. Regarding this, the starting guide NT115 was followed: (http://www.helsinki.fi/biosciences/corefacilities/microscale-thermophoresis/StartingGuide_NT115.pdf). Briefly, the concentration of labeled protein is adjusted to reach an initial fluorescence count between 500 and 1500. Different types of capillary and additives such as BSA, Tween 20, or Pluronic® F-127 have to be tested to avoid sample aggregation or adsorption to the inner surface of capillary. Using Tween 20 will also result in an increasing of the sample fluorescence. Buffer composition may also influence fluorescence. Proper experimental condition will result in a symmetrical fluorescence peak (*see* **Note 14**).

1. The ligands, buffers, and all the additives used in the assay are scanned with LED power 60% to make sure there is no background fluorescence.
2. Premium capillaries were used for both CBM3a and CBM4, and Pluronic® F-127 was used at a final concentration of 0.05% in the dilution buffer to avoid severe adsorption to the surface of the capillaries (*see* **Note 14**).
3. To obtain an initial fluorescence count between 500 and 1500 while fixing LED power to 60%, MST power to 20%, CBM3a and CBM4 were diluted to 4.3×10^{-4} g/L and 0.057 µM respectively with dilution buffer containing Pluronic® F-127 (0.05% final) (*see* **Note 15**).

4. The highest concentration of the ligand in the reaction should be 20-fold above the expected K_d, but not below 100 nM (the lowest K_d that can be detected by NT.115 is 1 nM). For CBM3a, the highest concentration of cellulose nanocrystals was 10 g/L. For CBM4, the highest concentration of xylohexaose was 150 mM (*see* **Note 16**). Both ligands were diluted using the dilution buffer.

3.3 Assay Setup

1. A titration series of 16 dilutions of the ligand is prepared in 0.2 mL PCR tubes. Transfer 20 μL of the highest ligand concentration into vial 1 (*see* **Note 16**), then proceed to cascade dilution as follows: add 10 μL of dilution buffer into vial 2 to vial 16. Transfer 10 μL from vial 1 to vial 2, mix by pipetting up and down three times (*see* **Note 17**). Then transfer 10 μL from vial 2 to 3. Repeat this step successively for all vials. Remove 10 μL from vial 16 after mixing to end up with 10 μL final.
2. Prepare 2× stock solution of labeled protein (*see* **Note 18**). CBM3a and CBM4 were diluted to 8.6 × 10^{-4} g/L and 0.114 μM respectively with dilution buffer and Pluronic® F-127 (0.1% final). Add 10 μL of protein stock to each vial of the dilution series (*see* **Note 19**). Mix by pipetting up and down three times.
3. Fill capillaries by sticking the capillary horizontally into the reaction tube to aspirate the sample (*see* **Note 20**). The capillary containing the highest ligand concentration is placed in the front of the sample tray.
4. Place the sample tray in the instrument and perform the scan with the MO. Control software (*see* **Note 21**).

3.4 Data Setting

3.4.1 CBM3a

1. After performing the capillary scan and MST experiment, the initial fluorescence intensity was checked (Fig. 1a). The variation of fluorescence is less than ±10% and the MST signal could be analyzed (*see* **Note 22**).
2. Select the tab "MST Analysis Set" in the MO.Affinity Analysis software. The fluorescence ratio before and after switching on the laser is calculated and plotted against the ligand concentration. A sigmoidal dose-response curve is obtained (*see* **Note 23**). Choose K_d model, enter the concentration of CBM3a in the same unit used as for the ligand.
3. Choose the ΔF_{Norm} as *y*-axis (Fig. 1b). The sigmoidal curve needs to have (a) a clear baseline and a clear saturation (at least three points respectively), (b) an amplitude/noise >3 (*see* **Note 24**).
4. Then the K_d value is determined for CBM3a. MO.Affinity Analysis software could also evaluate the consistency among repeated experiments by simply including all the repeated

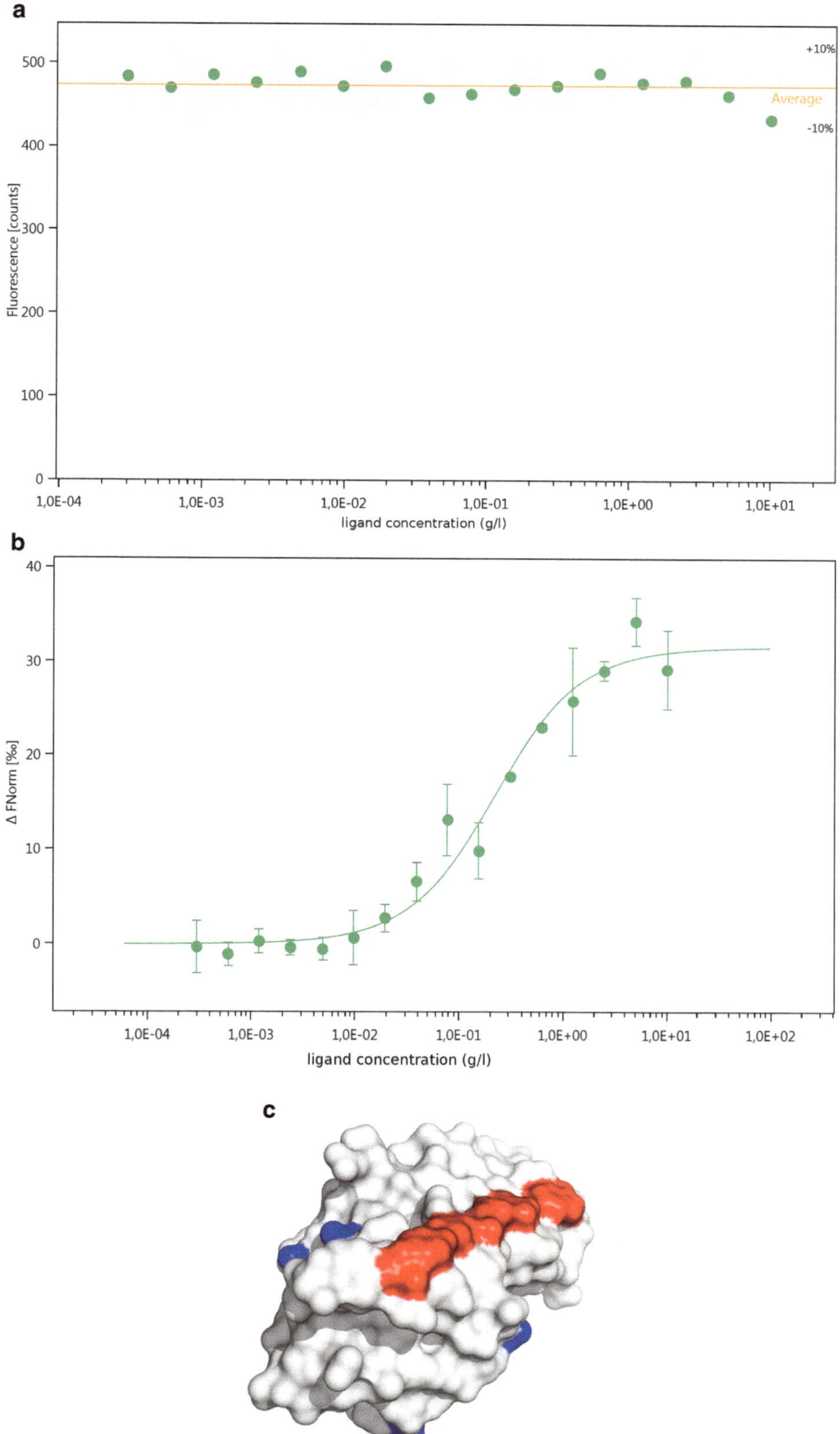

Fig. 1 CBM3a. (**a**) Initial fluorescence of CBM3a toward different concentrations of cellulose nanocrystals. The data showed a homogenous initial fluorescence, which indicated that MST experiment could be performed. (**b**) Measurement of CBM3a binding to cellulose nanocrystals. K_d apparent = 0.24 ± 0.05 g/L. (**c**) Surface representation of CBM3a (pdb 1NBC). In red residues involved in cellulose binding. In blue, lysine residues potentially labeled with NHS-dye. None of these residues are in closed vicinity to the flat surface of the binding site

experiments that have been done in the exact same conditions into the same analysis set, and errors bars will automatically appear in the dose-response curve.

5. The K_d of CBM3a toward cellulose nanocrystals deduced from the binding curve is 0.24 ± 0.05 g/L. Regarding the complexity of cellulose nanocrystals, K_d would be considered as an apparent K_d.

3.4.2 CBM4

1. The initial fluorescence shown in Fig. 2a displays a typical quenching sigmoidal curve not compatible with proper MST experiment. In such a case, it is recommended to perform a SDS-denaturation Test or SD test to investigate if the quenching comes from nonspecific interactions with the ligand and the fluorophore, nonspecific adsorption to tube walls, protein aggregation, or a binding event where the fluorophore is in close vicinity to the binding site (*see* **Note 25**). To perform a SD Test, the samples 1–3 and 14–16 from the 16 reaction series are centrifuged 10 min at maximum speed to avoid sample adsorption to the tube walls. An aliquot of 10 μL of SD mix is added to each sample and well mixed by pipetting up and down three times. The mixture is incubated for 5 min at 95 °C and centrifuge briefly to denature the protein. The fluorescence intensity of each sample is measured after filling capillaries (*see* **Note 26**). In our case, protein denaturation leads to restore uniform fluorescence intensity in all capillaries, meaning that the variation of fluorescence was induced by a binding event. Select the tab "Initial Fluorescence Analysis Set" in the MO.Affinity Analysis software, select the raw data and then move to "Dose Response Fit." Choose K_d Model, fix the concentration of CBM4, a sigmoid curve is derived and well fitted to the data.
2. Choose ΔRaw fluorescence [counts] as *y*-axis, and then the dose-response curve appears showing that the fluorescence intensity decreases with increasing concentrations of the ligand (Fig. 2b). The sigmoidal curve needs to have (a) a clear baseline and a clear saturation (at least three points respectively), (b) an amplitude/noise >3 (*see* **Note 24**).
3. The K_d of CBM4 toward xylohexaose derived from the binding curve was 2.5 ± 0.3 mM.

The two proteins were labeled on the lysine. However unlike CBM3a binding site (Fig. 1c) quenching fluorescence within CBM4 could be explained by the presence of a putative labeled lysine close to the binding site (Fig. 2c).

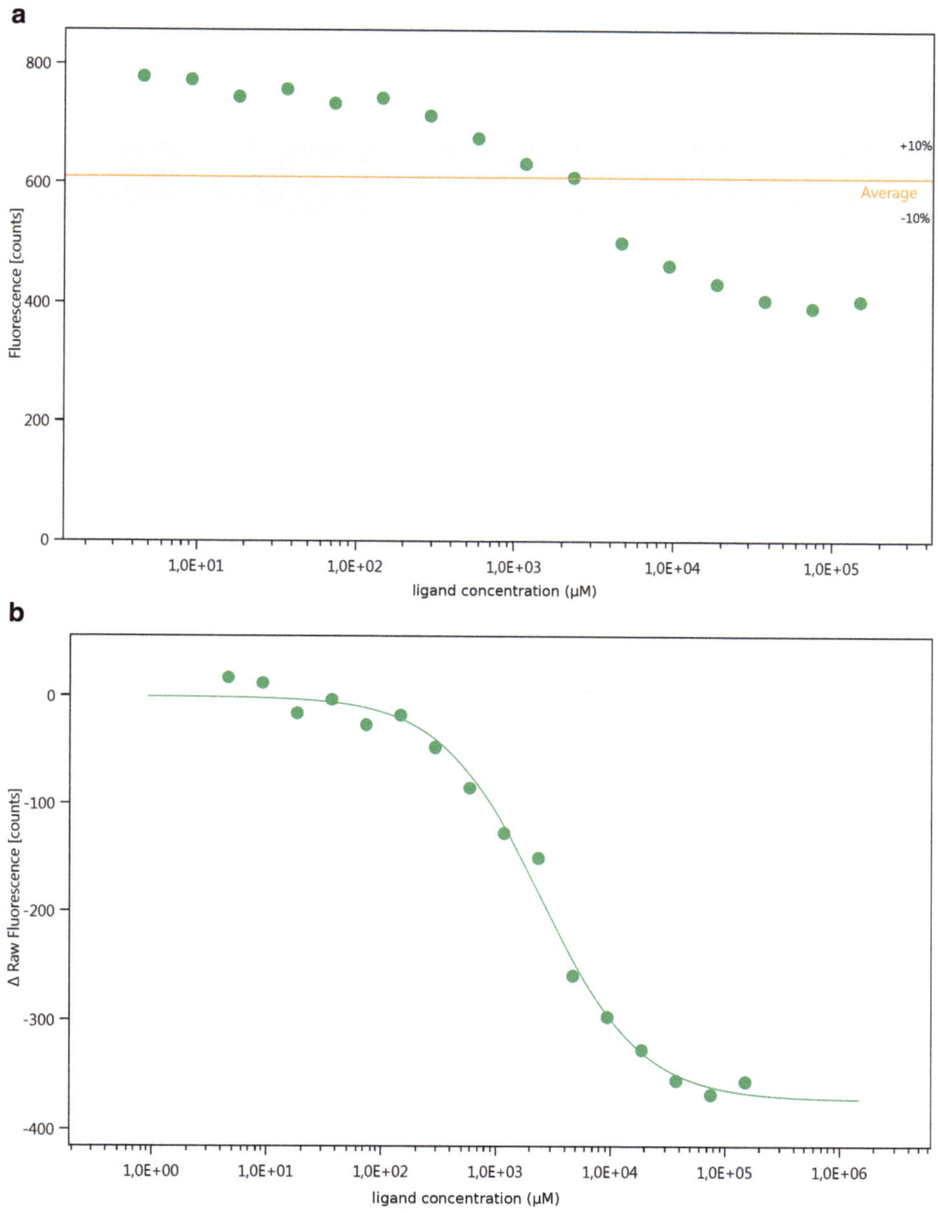

Fig. 2 CBM4. (**a**) Initial fluorescence of CBM4 toward different concentrations of xylohexaose. Quenching of fluorescence indicated that MST experiment could not be performed. (**b**) Measurement of CBM4 binding to xylohexaose. K_d = 2.5 ± 0.3 mM. (**c**) Multiple protein sequence alignment of CBM4 with family 4 CBMs of known crystallographic structure. In *red/green*, main residues involved in ligand recognition. In *purple*, CBM4 lysine residue potentially labeled with NHS-dye. Close proximity between Lys56 and Tyr57 in the binding site may explain the quenching of fluorescence during titration

c

```
2Y6G      MLV------------ANINGGFESTPAGVVTDLAEGVEGWDLNVGSSVTNPPVFEVLETS
CBM4      -LV-----PRGSHMASLVDGSFE---EGMG--------GWTPAFNA-----QNYTVV-DT
3K4Z      MLEDNSSTLPPYKNDLLYERTFD---EGLCY-------PWHTCEDSGGK--CSFDVV-DV
1GU3      -------------ASPIGEGTFD-----------DGPEGWVAYGTD--------GPL-DT
1GUI      ---------------SINNGTFD---EPIVNDQANNPDEWFIWQAGDYG--IS-GAR-VS
                            :  *:                *

2Y6G      DAPEGNKVLAVTVNGVGNNPFNIQATALPVNVRPGVTYTYTIRARAEQDGAVVSFTVG--
CBM4      EAVDGTQSLQ-VVVSGSTGKYDAQLNSPAFPIVAGHHYQLSFWIKGSVPGKVGIDFPNAE
3K4Z      PGQPGNKAFAVTVLDKGQNRWSVQMRHRGLTLEQGHTYRVRLKIWADASCKVYIKIGQ--
1GU3      ST----GALCVA-VPAGSAQYGVGVVLNGVAIEEGTTYTLRYTATASTDVTVRALVGQ--
1GUI      DYGVRDGYAYITIADPGTDTWHIQ-FNQWIGLYRGKTYTISFKAKADTPRPINVKILQ--
                     .    .   :        . :  *  *       ..    :

2Y6G      --NQ-SFDEYGRLHHQQITTEWQPFTFEFTVSDQ----------ETVIRAPIHFGY--AA
CBM4      LGNQYPWVNGAE-LADVGTSWTQVIYNTETVGD---PAMIATADNGAMTVRLLLGGVAE-
3K4Z      --MGEPYAEYWNNK------WSPYTLTAGKVLEIDETFVMDKPTDDTCEFTFHLGGELAA
1GU3      --NGAPYGTVLDTSPALTSEPRQVTETFTASATYP---ATPAADDPEGQIAFQLGGFSA-
1GUI      --NHDPWTNYFAQTVNLTADWQTFTFTY----------THPDDADEVVQISFELGEGTA-
                :                                     :      : :*

2Y6G      NVGNTIYIDGLAIVDL--------------------------------------------
CBM4      ---CTYLIDAVEITDL--------------------------------------------
3K4Z      TPPYTVYLDDVSLYDPEYTKPVEYILPQPDVRVNQVGYLPEGKKVATVVCNSTQPVKWQL
1GU3      -DAWTLCLDDVALDSE--------------------------------------------
1GUI      ---TTIYFDDVTVSPQ--------------------------------------------
              *  :* : :

2Y6G      ------------------------------------------------------------
CBM4      ------------------------------------------------------------
3K4Z      KNAAGVVVLEGYTEPKGLDKDSQDYVHWLDFSDFATEGIGYYFELPTVNSPTNYSHPFDI
1GU3      ------------------------------------------------------------
1GUI      ------------------------------------------------------------

2Y6G      ------------
CBM4      ------------
3K4Z      RKDILEHHHHHH
1GU3      ------------
1GUI      ------------
```

Fig. 2 (continued)

4 Notes

1. After purifying the protein by IMAC, determining the protein concentration was not possible by UV absorption because of imidazole. In this case, to avoid dialysis dilution, we performed sodium dodecyl sulfate polyacrylamide gel electrophoresis (SDS-PAGE) with labeled protein and a serial dilution of protein (CBM3a or CBM4) as standards, and then quantified the labeled protein using the quantifying tool in the software of Image Lab 5.2.1.

2. The buffer of the labeling kit is an amine-free buffer at pH 8. Our protein storage buffer was 50 mM phosphate buffer pH 7.0. It is also an amine-free buffer. Labeling efficiency using NHS-ester group is optimal at pH 8 but could also be performed at pH 7, not below. The $Na_2HPO_4{\cdot}12H_2O$ can be dissolved faster if the dH_2O is warmed up to about 50 °C. If the binding event is calcium dependent, HEPES buffer is recommended instead of the phosphate buffer.

3. Dealing with costly ligand such as xylohexaose, it is better to start with a rather high concentration (in our case 150 mM) for the very first trial before knowing the K_d to make sure the CBM can be saturated.
4. The Gravity Flow Column B for purification provided in Monolith NT™ Protein Labeling Kit is a dextran-based gel filtration column. CBM3a and CBM4 displayed affinity for dextran, and were retained by the desalting column. Therefore, vivapure metal chelate mini spin columns were selected because they are made with a cellulose-free-based membrane. Furthermore, it allows the use of small volume of IMAC resins in the purification step.
5. The TALON Metal Affinity Resin is made with Sepharose CL-6B, an agarose-based polysaccharide polymer material, which has no affinity with the CBMs studied.
6. There are three different types of capillaries for NT.115, standard treated capillaries, hydrophobic capillaries, and premium coated capillaries. The premium coated one is the best to tackle with samples that have significant adsorption to surfaces, although it is the most expensive. We recommend trying with the standard treated capillaries first using buffer supplemented with additives such as BSA or detergent if the protein sticks to the inner surfaces of the capillary. If the adsorption is still a problem, then try with other capillaries.
7. 20 μL of each sample is usually prepared for MST experiment; thus, a 0.2 mL individual tube is enough. Bigger reaction tubes are not recommended since the large surface area to volume ratio can increase adsorption to the tube wall and loss of protein. Besides, the 0.2 mL strip tubes are also recommended for their convenient manipulation, especially in the serial dilution step.
8. A red channel is always recommended as it allows measuring in complex bioliquids (cell lysate, serum) and no autofluorescence of a molecule in this channel has been observed.
9. A refrigerated and temperature-regulated centrifuge is recommended for the centrifuge of protein samples to maintain their stability.
10. We recommend preparing a little bit more than 100 μL (e.g., 120 μL) for accurate pipetting afterward.
11. It is possible to aliquot dye in DMSO and store at −20 °C.
12. In our case, as the pH of the labeling buffer is not optimal we used a higher concentration of dye to improve the labeling efficiency that was 3.2-fold of the protein concentration.
13. The maximum volume of the column is 600 μL, and adding 500 μL of dH_2O into the column does not leave space for the air.

Therefore, leaving the cap open is recommended to ensure that the wash buffer is completely removed.

14. Buffers containing detergent (0.05% pluronic acid, 0.05% tween-20), BSA (0.05–0.5 mg/mL), casein, and others can be tested. A modified pH or ionic strength can also improve the performance. In the case of binding-event induced quenching such as CBM4, BSA should be avoided because the quenching is resumed after adding BSA
15. Concentration of the CBM3a was expressed in g/L to be in accordance with the unit of its substrate concentration, cellulose nanocrystals.
16. The highest concentration of xylohexaose in the reaction is 150 mM; therefore, 300 mM stock solution is required as the ligand will be diluted twice when adding the same volume of protein.
17. Vortex will result in a possible adsorption of protein to the inner wall of the tube and considering the very low concentration of protein, even small adsorption would result in a significant decreasing of protein concentration and so to a non-optimal assay.
18. The protein and buffer additives will be diluted twice after mixing with the same volume of ligand; keep this in mind when designing the experiment.
19. Stock solution of both protein and ligand required to be centrifuged at full speed 5 min before mixing to avoid aggregates in capillaries.
20. Avoid touching the capillary in the middle where the optical measurement will be performed because it could modify the fluorescence.
21. Detailed introduction of the manipulation of the instrument and the software can be downloaded elsewhere http://www.helsinki.fi/biosciences/corefacilities/microscalethermophoresis/Manual_NT115.pdf.
22. It is not recommended to use the data to determine a binding affinity if the deviation is more than ±10%.
23. If a sigmoidal dose-response curve is not obtained with 20% nor 40% MST power, use 60% or 80% MST power.
24. The amplitude is the difference between bound and unbound states, the noise is the variations of fluorescence in the baseline. If a high noise level is observed, experimental conditions need to be improved by using detergents for instance.
25. The SD test allows verifying that the fluorescence quenching is due to the binding event. SD test cannot be performed with samples containing potassium (200 mM or more) because the SDS will precipitate.

26. The SD test is performed only with sample 1–3 and 14–16 which correspond to the highest and lowest ligand concentrations respectively. In the case of binding event, these contrasted samples should have no difference in fluorescence. See also (http://openwetware.org/images/5/59/FAQ_Fluorescence_Changes_V11-2.pdf).

Acknowledgment

We acknowledge the Fédération de Recherche Agrobiosciences, Interactions et Biodiversité (FR 3450), CNRS, Université de Toulouse, UPS, Castanet-Tolosan, France, and the IDEX "UNITI" Université de Toulouse (GO-AHEAD project) for the Nanotemper Monolith NT.115 facilities.

References

1. Abbott DW, Boraston AB (2012) Quantitative approaches to the analysis of carbohydrate-binding module function. Methods Enzymol 510:211–231.doi:10.1016/B978-0-12-415931-0.00011-2
2. Duhr S, Braun D (2006) Thermophoretic depletion follows boltzmann distribution. Phys Rev Lett 96:1–4. doi:10.1103/PhysRevLett.96.168301
3. Duhr S, Braun D (2006) Why molecules move along a temperature gradient. Proc Natl Acad Sci U S A 103:19678–19682. doi:10.1073/pnas.0603873103
4. Jerabek-Willemsen M, André T, Wanner R et al (2014) MicroScale thermophoresis: interaction analysis and beyond. J Mol Struct 1077:101–113. doi:10.1016/j.molstruc.2014.03.009
5. Baaske P, Wienken CJ, Reineck P et al (2010) Optical thermophoresis for quantifying the buffer dependence of aptamer binding. Angew Chem Int Ed Engl 49:2238–2241. doi:10.1002/anie.200903998
6. Tormo J, Lamed R, Chirino AJ et al (1996) Crystal structure of a bacterial family-III cellulose-binding domain: a general mechanism for attachment to cellulose. EMBO J 15: 5739–5751
7. Bastien GG, Arnal GG, Bozonnet S et al (2013) Mining for hemicellulases in the fungus-growing termite Pseudacanthotermes militaris using functional metagenomics. Biotechnol Biofuels 6:78. doi:10.1186/1754-6834-6-78
8. Martinez T, Texier H, Nahoum V et al (2015) Probing the functions of carbohydrate binding modules in the cbel protein from the oomycete phytophthora parasitica. PLoS One 10:1–14. doi:10.1371/journal.pone.0137481
9. Blake AW, Mccartney L, Flint JE et al (2006) Understanding the biological rationale for the diversity of cellulose-directed carbohydrate-binding modules in prokaryotic enzymes. J Biol Chem 281:29321–29329. doi:10.1074/jbc.M605903200
10. Gasteiger E, Hoogland C, Gattiker A et al (2005) Protein identification and analysis tools on the ExPASy Server. In: Proteomics protocols handbook. Humana Press, Totowa, NJ, pp 571–607

Chapter 11

Characterization of Protein-Carbohydrate Interactions by NMR Spectroscopy

Julie M. Grondin, David N. Langelaan, and Steven P. Smith

Abstract

Solution-state nuclear magnetic resonance (NMR) spectroscopy can be used to monitor protein-carbohydrate interactions. Two-dimensional ^{1}H-^{15}N heteronuclear single quantum coherence (HSQC)-based techniques described in this chapter can be used quickly and effectively to screen a set of possible carbohydrate binding partners, to quantify the dissociation constant (K_d) of any identified interactions, and to map the carbohydrate binding site on the structure of the protein. Here, we describe the titration of a family 32 carbohydrate binding module from *Clostridium perfringens* (*Cp*CBM32) with the monosaccharide *N*-acetylgalactosamine (GalNAc), in which we calculate the apparent dissociation of the interaction, and map the GalNAc binding site onto the structure of *Cp*CBM32.

Key words Chemical shift perturbation, Dissociation constant (K_d), Heteronuclear single quantum coherence (HSQC), Nuclear magnetic resonance (NMR), Protein-carbohydrate interactions

1 Introduction

Solution-state nuclear magnetic resonance (NMR) spectroscopy is a robust method for studying interactions between proteins and their respective ligands. Beyond determining the high-resolution structure of a protein-ligand complex, other NMR techniques such as the transferred-NOESY experiment and saturation transfer difference spectroscopy are used to characterize protein-ligand interactions (reviewed in [1–3]).

A particularly valuable NMR experiment for monitoring titrations of ligand to a protein solution is two-dimensional ^{1}H-^{15}N heteronuclear single quantum coherence (^{1}H-^{15}N HSQC) [4], which can provide information regarding the ligand binding site on the protein structure, and the affinity and stoichiometry of the interaction. As a prerequisite to collecting a ^{1}H-^{15}N HSQC spectrum, the protein of interest should be uniformly ^{15}N-labeled such that the chemical shifts of backbone amide nitrogen and proton pairs (NH) can be detected. For a folded and monodisperse protein

D. Wade Abbott and Alicia Lammerts van Bueren (eds.), *Protein-Carbohydrate Interactions: Methods and Protocols*, Methods in Molecular Biology, vol. 1588, DOI 10.1007/978-1-4939-6899-2_11, © Springer Science+Business Media LLC 2017

sample, the majority of resonances display good dispersion with each peak representing a backbone NH group for every non-proline amino acid residue in the protein sequence. The precise chemical shift of each NH resonance is determined by multiple factors, including secondary and tertiary structure, pH, temperature, and solvent. Importantly, the chemical shift of each resonance is sensitive to minute changes in its chemical environment, such as those induced by the binding of a ligand. To monitor a protein-ligand titration by NMR spectroscopy, a ^{1}H-^{15}N-HSQC spectrum of the protein of interest is collected after each incremental addition of ligand and the observed changes are quantified to characterize the interaction.

A typical 1:1 protein-ligand interaction may be modeled as:

$$\mathrm{P} + \mathrm{L} \underset{k_{on}}{\overset{k_{off}}{\rightleftarrows}} PL \qquad (1)$$

where P, L, and PL represent the protein, ligand, and the protein-ligand complex, respectively, and k_{on} and k_{off} represent the association and dissociation rates of the PL complex. In such a system, the dissociation constant (K_{d}) is defined by the ratio of k_{on} and k_{off} or by the equilibrium concentrations of P, L, and PL:

$$K_d = \frac{k_{off}}{k_{on}} = \frac{[\mathrm{P}][\mathrm{L}]}{[PL]} \qquad (2)$$

Thus, the value of K_{d} can be determined either by measuring the k_{off} and k_{on} rate constants or by quantifying the equilibrium concentrations of P, L, and PL. This chapter will focus on the use of NMR spectroscopy to quantitatively derive the K_{d} for protein-ligand interactions.

Assuming a 100 μM protein sample and that, as a rule of thumb for accurate K_{d} measurements the sample concentration should be <10-fold greater than the K_{d}, NMR spectroscopy is particularly well suited to measure the K_{d} of protein-ligand interactions that are of moderate affinity ($K_{\mathrm{d}} > 10$ μM) for proteins up to a molecular mass of approximately 25 kDa. Cryogenic probes and high-field NMR also allow the protein concentration to be decreased so that higher affinity interactions may be accurately measured.

When analyzing a series of ^{1}H-^{15}N HSQC spectra corresponding to a titration of a protein with a ligand, the kinetics of the PL complex will determine if slow, fast, or intermediate chemical exchange will be observed [5]. If the dissociation of the PL complex is slow (i.e., $k_{\mathrm{off}} << \Delta\omega$, where $\Delta\omega$ is the frequency difference between the free and bound signals) two distinct sets of resonances will be observed in the ^{1}H-^{15}N HSQC spectra during the course of the titration. Initially, only the set of resonances corresponding to the apo state of the protein is visible. Upon addition of ligand a second set of resonances corresponding to the bound state of the protein will begin to appear, while the intensity of resonances corresponding to the free state will accordingly decrease. The relative intensities of the free state and bound state resonances are a direct measurement of the proportion of protein in the free and bound states. As the protein becomes saturated with ligand, only resonances corresponding to the bound state are observed. This exchange regime is referred to as slow chemical exchange.

If instead the PL complex dissociation is fast (i.e., $k_{\text{off}} >> \Delta\omega$), a single set of resonances is observed during the titration. The position of each resonance in the ^{1}H-^{15}N HSQC spectrum is a fractional average of the free and bound chemical shift values. As a ligand is added to the protein, resonances corresponding to residues that are distal to the binding interface should not experience chemical shift changes, as their chemical environment remains constant. In contrast, resonances that correspond to residues in close proximity to the binding interface should transition from the free state chemical shift values to those of the bound state as ligand is added. The trajectory of this movement of resonances is informative; in a simple 1:1 binding model all peaks move linearly from the free to bound states at the same rate. If subsets of resonances shift from the free state to the bound state at different rates, or do not follow a linear trajectory, multiple binding events or confounding factors, such as changes in ionic strength or pH, may be occurring.

Finally, at intermediate rates of PL complex dissociation (i.e., $k_{\text{off}} \sim \Delta\omega$) resonances move in the same direction as for fast exchange, but increased linewidths and decreased intensity of resonances are observed midway through the titration. Of note, in some cases, resonances experience only minor line broadening during a titration, while others may disappear quickly and completely once ligand is added and only reappear when saturation has occurred. As long as the resonance can be observed in the ^{1}H-^{15}N HSQC spectra throughout a titration, the analysis and K_d estimation for a titration series is the same for both intermediate and fast exchange.

In NMR analysis software, such as NMRViewJ [6] (www.onemoonscienctific.com/nmrviewj) or CcpNmr Analysis [7] (www.ccpn.ac.uk/v2-software/software/analysis), the resonance trajectories in the fast exchange regime can be analyzed to extract an estimate of K_d for each resonance of a protein using Eq. 3, which is derived into a quadratic equation from Eq. 2, and depends on NMR observables that are easily determined from a series of ^{1}H-^{15}N HSQC spectra (http://structbio.vanderbilt.edu/chazin/wisdom/kdcalc.htm).

$$\delta_{obs} = \mathrm{A}\left(\mathrm{B} + x - \sqrt{(\mathrm{B} + x)^2 - 4x}\right) \quad (3)$$

where
$A = \Delta\delta_{bound}/2$; $B = 1 + K_d/a$; $x = b/a$; a is [P]; b is [L]
$\Delta\delta_{obs}$ and $\Delta\delta_{bound}$ are the differences between the free chemical shift and the observed or bound chemical shift, respectively, for each resonance experiencing a chemical shift change. The $\Delta\delta$ is a positive number that is the weighted sum of the ^{15}N and ^{1}H chemical shift changes for a particular resonance:

$$\delta = \sqrt{0.17(\ \delta_N)^2 + \ \delta_{HN}{}^2} \quad (4)$$

where $\Delta\delta_N$ and $\Delta\delta_{HN}$ are the chemical shift changes for a resonance in the ^{15}N and ^{1}H dimensions of the ^{1}H-^{15}N HSQC, respectively.

Generally, nonlinear, least-square fitting of the data is applied to determine the values of A and B in Eq. 3. From this fit the K_d is determined for individual residues. The estimated K_d of multiple residues in a protein should be similar and may be averaged to get a better approximation of the K_d of the protein-ligand interaction. If the backbone HN resonances of the protein have been sequentially assigned and the three-dimensional structure of the protein is known, the binding site may be mapped to the exact positions within the protein fold thereby identifying the ligand binding site.

When performing a NMR-based titration experiment, care should be taken to plan the experiment such that the volumes of ligand added are practical to pipette and that the protein sample will become saturated. In order to plan a titration experiment, increment volumes may be calculated using the following equation based on conservation of mass:

$$V_0 L_0 + V_s L_s = L_d (V_0 + V_s) \quad (5)$$

which can be rearranged to

$$V_s = V_0 \frac{L_d - L_0}{L_s - L_d} \quad (6)$$

where V_0 is the total volume of the sample before the subsequent addition of ligand (in μL); L_0 is the ligand concentration before the subsequent addition (in μM); Ls is the concentration of the ligand stock (in μM); V_s is the volume of ligand stock to be added (in μL); L_d is the desired ligand concentration after the addition (in μM).

NMR-based titrations employing the ^{1}H-^{15}N HSQC experiment have been particularly successful for monitoring the interactions of several carbohydrate-binding modules (CBMs), which tend to have molecular masses of 14–25 kDa, with their glycan ligands [8–11]. While the K_d for CBM-carbohydrate interactions vary based on the particular type of CBM and the degree of polymerization of the carbohydrate, most are generally weak (K_d in the mM to μM range) and display fast exchange on the NMR timescale [12–16].

In this chapter, we use the specific example of monitoring an interaction between a family 32 CBM from *Clostridium perfringens* (*Cp*CBM32) and *N*-acetylgalactosamine (GalNAc) using the ^{1}H-^{15}N HSQC experiment. Following the collection and processing of the full set of ^{1}H-^{15}N HSQC spectra over the course of the titration, the spectra are overlaid in NMR analysis software such as NMRViewJ [6] or CcpNmr Analysis [7], and a standard integrated titration analysis tool in each software program is used to derive a K_d value for the binding event. With the completed sequential backbone resonance assignments and a high-resolution structure, we will also map the GalNAc binding site on *Cp*CBM32.

2 Materials

Prepare all solutions using distilled deionized H_2O and analytical grade reagents. Prepare and store buffer solutions at room temperature unless otherwise stated. Unless otherwise stated, autoclave solutions to 121 °C for 45 min.

2.1 Preparation of M9 Minimal Media for Isotopic ^{15}N-Labeling of CpCBM32

1. Lysogeny broth (LB) media: Dissolve 1 g lysogeny broth solids in 50 mL H_2O.
2. 10× M9 salt solution: Dissolve 60 g $Na_2HPO_4{\cdot}7H_2O$, 30 g KH_2PO_4 and 5 g NaCl into 1 L H_2O. Adjust pH to 7.4.
3. 100× Trace element solution: Dissolve 0.6 g $FeSO_4{\cdot}7H_2O$, 0.6 g $CaCl_2{\cdot}2H_2O$, 0.12 g $MnCl_2{\cdot}4H_2O$, 0.08 g $CoCl_2{\cdot}2H_2O$, 0.07 g $ZnSO_4{\cdot}H_2O$, 0.03 g $CuCl_2{\cdot}2H_2O$, 0.002 g H_3BO_3, and 0.025 g $(NH_4)_6Mo_7O_{24}{\cdot}4H_2O$ into 70 mL H_2O. Once completely dissolved, add 0.5 g EDTA and stir until dissolved (*see* **Note 1**). Use H_2O to bring final volume up to 100 mL.
4. 100× NH_4Cl solution: Dissolve 100 g NH_4Cl in 1 L H_2O.
5. ^{15}N NH_4Cl (^{15}N, 99%) (Cambridge Isotope Laboratories, Inc.): Store over desiccant.
6. 20% (w/v) glucose solution: Dissolve 20 g glucose into 100 mL H_2O.
7. 10× ^{15}N BioExpress growth media: Obtain 10× concentrate BioExpress Cell Growth Media (U-15N, 98%) (Cambridge Isotope Laboratories, Inc.). Filter sterilize and prepare 5 mL aliquots. Store at 4 °C.
8. 0.1 M $CaCl_2$ solution: Dissolve 1.11 g $CaCl_2{\cdot}2H_2O$ in 100 mL H_2O.
9. 1 M $MgSO_4$ solution: Dissolve 12.04 g $MgSO_4$ in 100 mL H_2O.

10. 100 mg/ml thiamine solution: Dissolve 1 g of thiamine hydrocholoride into 10 mL H_2O. Filter sterilize and store at 4 °C.
11. 10 mg/mL biotin: Dissolve 100 mg d-biotin into 10 mL H_2O. Store at 4 °C.
12. Antibiotic solution: Prepare a 1000× stock of the required selection antibiotic. The type of expression plasmid used determines the concentration and type of selection antibiotic that is needed. Sterile filter and store at −20 °C.
13. Isopropyl β-d-1-thiogalactopyranoside (IPTG, 1 M): Dissolve 2.38 g IPTG into 10 mL H_2O. Filter sterilize and store in 1 mL aliquots at −20 °C.
14. 50 mL M9 minimal media: Add 5 mL 10× M9 salt solution to 43 mL H_2O.
15. 1 L M9 minimal media: Add 100 mL 10× M9 salt solution to 875 mL H_2O.

2.2 Solutions, Buffers, and Equipment for NMR Studies

1. 25 mM Tris–HCl pH 7.5, 50 mM NaCl (*see* **Note 2**): Add about 100 mL H_2O to a large beaker. Add 3.02 g Tris base and 2.92 g NaCl, and an additional 800 mL H_2O and stir until dissolved. Adjust the pH to 7.5, and then bring volume up to 1 L with H_2O.
2. Deuterium oxide (D_2O, D, 99.96%) (Cambridge Isotope Laboratories, Inc.): Store over desiccant.
3. 4,4-dimethyl-4-silapentane-1-sulfonic acid (DSS, 1 M): Dissolve 2.18 g DSS sodium salt into 10 mL H_2O.
4. 5 mm NMR tube with cap.
5. High-field NMR spectrometer (500 MHz or greater) configured for solution-state experiments.

2.3 Carbohydrate Solution Preparation

For the example CBM-glycan titration, 1 mL of 75 mM GalNAc (molar mass = 221.2 g/mol) was prepared. The method below is optimal for the accurate preparation of small volumes (1–2 mL) of carbohydrate solutions for which large volumes cannot be prepared for practical reasons.

1. Record the mass of an empty 1.5 mL microcentrifuge tube.
2. Weigh 16.6 mg GalNAc onto weigh paper. Transfer into the microcentrifuge tube (*see* **Note 3**). Record the mass of the microcentrifuge tube and GalNAc together.
3. Calculate the actual mass of GalNAc added to the microcentrifuge tube and add an appropriate amount of water so that the final solution is 75 mM. Store the carbohydrate solution at 4 °C.

3 Methods

3.1 Expression and Purification of ^{15}N-Labeled Protein (See Note 4)

1. On Day 1, inoculate 5 mL LB media supplemented with 5 μL of antibiotic stock solution from a glycerol stock or streaked plate of *Escherichia coli* strain BL(21) transformed with a pET28a(+) expression vector encoding a hexa-histidine tag and *Cp*CBM32. Grow culture overnight at 37 °C with shaking at 200 rpm.
2. On Day 2, finish preparing the autoclaved 50 mL M9 minimal media by adding 500 μL 100× NH_4Cl, 750 μL 20% (w/v) glucose solution, 50 μL each of $CaCl_2$, $MgSO_4$, thiamine, biotin, and antibiotic solutions. Add the 5 mL overnight culture and grow culture overnight at 37 °C with shaking at 200 rpm.
3. On Day 3, complete preparation of the autoclaved 1 L M9 minimal media by adding 10 mL 100× trace element solution, 1 g ^{15}N NH_4Cl, 20 mL 20% (w/v) glucose solution, 5 mL 10× ^{15}N BioExpress growth media, 1 mL each $CaCl_2$, $MgSO_4$, thiamine, biotin, and antibiotic stock solution. Add the contents of the 50 mL overnight culture and grow culture at 37 °C with shaking at 200 rpm to an OD_{600} of 0.6–0.8.
4. Induce protein expression by adding 1 mL of the IPTG stock solution (total concentration of 1 mM IPTG) and continue growth with shaking for 8 h.
5. Purify protein according to standard purification protocols (*see* **Note 5**).
6. Quantify the purified protein by UV absorbance using Beer's Law and the extinction coefficient of the protein construct. If an extinction coefficient for a protein has not been experimentally determined, a reasonable estimate can be calculated based on its primary sequence using the ProtParam tool of the ExPASy Bioinformatics Resource Portal [17].

3.2 NMR Sample Preparation

1. Prepare a 650 μL solution containing 100 μM *Cp*CBM32, 10% v/v D_2O, and 1 μM DSS. Use buffer that is matched to the protein sample to dilute the sample as necessary.
2. Using a 3 mm micro pH probe, verify the pH of the sample using dilute HCl or NaOH (*see* **Note 5**).
3. Transfer 600 μL of the *Cp*CBM32 protein sample to a 5 mm NMR tube (*see* **Note 6**).

3.3 NMR Acquisition

1. Using Eq. 6 calculate the volume additions required to complete the titration. For this example, we have calculated the required carbohydrate additions to titrate GalNAc to a final concentration of 5 mM (*see* **Notes** 7 and **8,** Table 1).
2. Place the sample into the NMR spectrometer and let the sample temperature equilibrate for 5–10 min. After equilibration, calibrate the tune, match, lock, and shims of the spectrometer

Table 1
Example calculations for the titration of GalNAc into a solution of *Cp*CBM32

Step no.	[L_d] (μM)	[L_s] (μM)	V_0 (μl)	[L_0] (μM)	V_s (μL)
1	0	75,000	600.0	0	0
2	250	75,000	600.0	0	2.0
3	500	75,000	602.0	250	2.0
4	1000	75,000	604.0	500	4.1
5	2000	75,000	608.1	1000	8.2
6	3000	75,000	616.3	2000	8.3
7	4000	75,000	624.6	3000	8.5
8	5000	75,000	633.1	4000	8.6

according to standard protocol. Calibrate the ^{1}H and ^{15}N offset frequencies, ^{1}H and ^{15}N pulse widths, determine a suitable ^{15}N spectral width, and collect a 2D ^{1}H-^{15}N HSQC spectrum of the *Cp*CBM32 protein sample (*see* **Note 9**, Fig. 1a).

3. Remove the NMR tube from the spectrometer and add the first increment of GalNAc (Table 1) to the upper inner side of the tube. To mix, cap the NMR tube and invert the sample several times. Firmly flick the NMR tube downward a final time to collect the sample in the bottom of the tube. Insert the NMR tube into the NMR spectrometer, let the sample equilibrate, calibrate the lock, and shim. Collect another 2D ^{1}H-^{15}N HSQC spectrum using identical experimental parameters.
4. Repeat **step 3** until the titration is complete. This will result in a set of 2D ^{1}H-^{15}N HSQC spectra each corresponding to a particular concentration of GalNAc.
5. Process the individual ^{1}H-^{15}N HSQC spectra using standard NMR processing software, such as NMRPipe [19], TopSpin, or VnmrJ.
6. Prior to quantitative titration analyses, the spectra can be overlaid and differentially colored to visualize site-specific chemical shift changes through the course of the titration (Fig. 1a).

Fig. 1 (continued) shown in *gray*. Residues whose ^{1}H-^{15}N resonances showed chemical shift changes greater than the mean but less than 1 SD above the mean are shown in white. (**c**) The binding curve for the titration of GalNAc into *Cp*CBM32 and calculated dissociation constant, as obtained using NMRViewJ [6]. (**d**) Residues displaying significantly changing chemical shifts were mapped onto the X-ray crystal structure of *Cp*CBM32 and are shown in *black*. The structure is displayed in cartoon format with perturbed residues denoted in stick format (*left*), as well as in surface format (*right*)

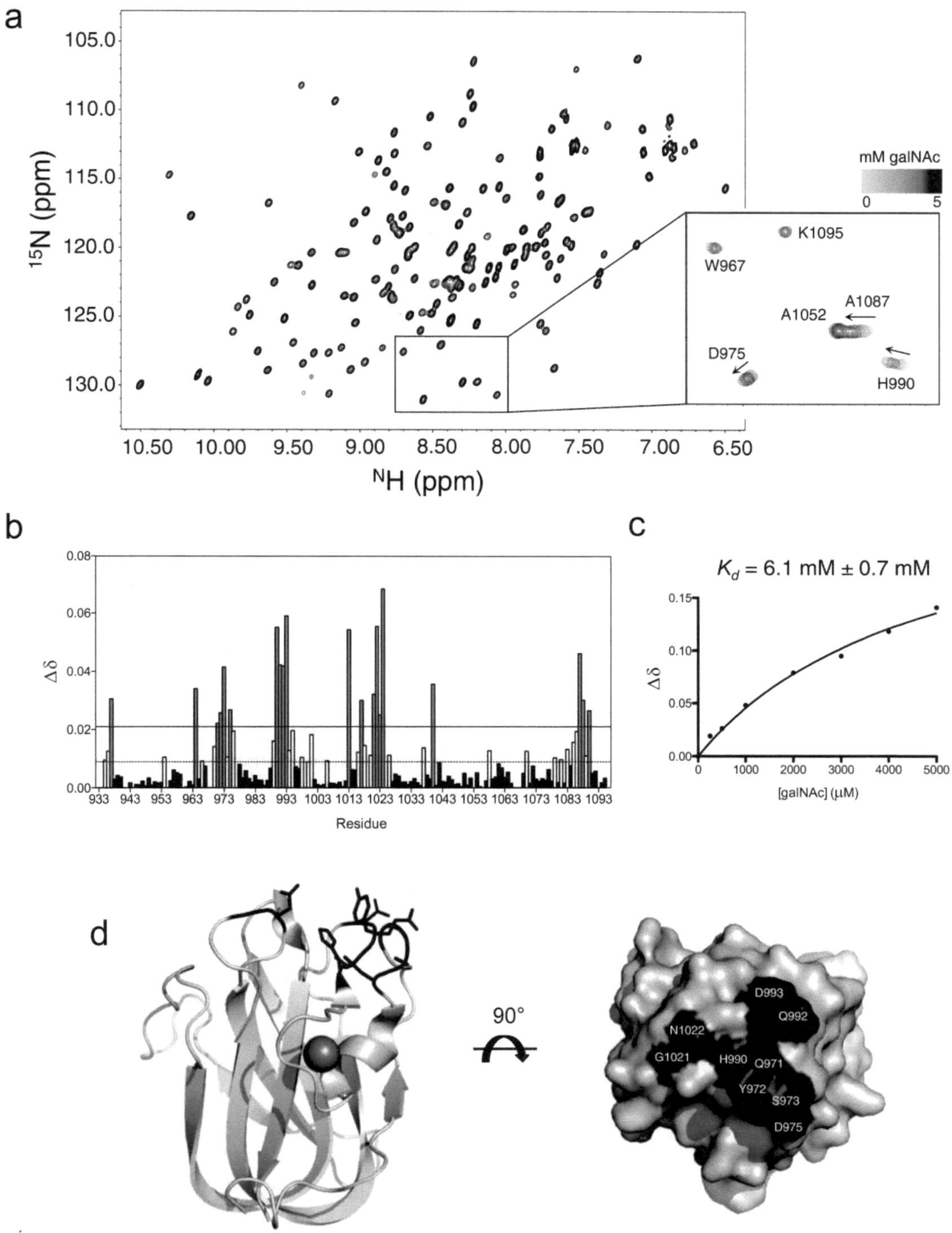

Fig. 1 (**a**) 2D ^{1}H-^{15}N HSQC spectrum of uniformly ^{13}C/^{15}N-labeled *Cp*CBM32 recorded at 600 MHz and 303 K. *Inset*: the titration of incremental additions of GalNAc to a 100 μM ^{15}N-labeled sample of *Cp*CBM32 was monitored using overlapping 2D ^{1}H-^{15}N HSQC spectra. (**b**) The resulting normalized residue-specific chemical shift perturbations were calculated using Eq. 4 and plotted as a function of residue. The calculated mean (*dotted line*) and one standard deviation (SD) [18] above the mean (*solid line*) are shown. Residues whose ^{1}H-^{15}N resonances showed chemical shift changes greater than 1 SD above the mean were considered significant and are

3.4 Data Analysis

In order to obtain the best estimate of K_d some precautions should be followed:

1. Adjust the data contours such that no resonances can be attributed to noise.
2. It is important to pick backbone HN resonances within a software package which are well resolved and not overlapped so that the K_d calculation will be as accurate as possible.
3. While theoretically any HN resonance that experiences a chemical shift change could provide a measurement of K_d, one should select those HN resonances that undergo the largest chemical shift change ($\Delta\delta$). Generally, we consider resonances with a $\Delta\delta$ greater than 0.5–1 standard deviations above the mean $\Delta\delta$ of all residues in a protein to be significant and directly associated with ligand binding (Fig. 1b). A K_d value for each of these significantly affected backbone resonances can be calculated individually, and then averaged into a single apparent K_d representing the entire binding event.

3.4.1 Titration Analysis with NMRViewJ

The Titration Analysis tool within NMRViewJ allows the user to track $\Delta\delta$ across a series of spectra. For a detailed discussion concerning the use of NMRViewJ, please consult the excellent documentation for the software [6].

1. Export the chemical shift list (.xpk file) from the apo and fully bound spectra (i.e., 0 mM GalNAc and 5 mM GalNAc). Using the amide proton and nitrogen perturbations from the apo and bound chemical shift list, calculate total weighted $\Delta\delta$ for each backbone amide resonance using Eq. 4.
2. Proceed with titration analysis using the standard Titration Analysis tool protocol in NMRViewJ. Briefly, K_d values derived from those significantly perturbed backbone amide resonances are interpolated from plots of the chemical shift change as a function of carbohydrate concentration, and a nonlinear fit is individually applied to significantly perturbed backbone amide resonances (Fig. 1c). An overall average K_d value and standard deviation can be calculated from these individual dissociation constants.

3.4.2 Titration Analysis with CcpNmr Analysis

CcpNmr Analysis has a straightforward interface that allows the user to assess chemical shift changes between spectra and to determine a K_d for individual residues of a protein. For a more detailed discussion, please consult the excellent CcpNmr Analysis documentation [6].

1. Load the ^1H-^{15}N HSQC spectra into CcpNmr Analysis and pick the peaks in each spectrum. The peaks do not need to be assigned.

2. To fit K_d values in Analysis, the titration data must be grouped together into a series. Use the "NMR Series" panel to do this. Create a new series and set the varied parameter to be ligand concentration. Then use the "Add Series Point" button to load in the set of ^1H-^{15}N HSQC spectra. It is convenient to name the spectra according to the concentration of ligand present for that data. Set the appropriate ligand concentration value and units for each spectrum.
3. Use the "Follow Shift Changes" tool to fit and analyze the titration data. Set the reference peak list to be the spectrum corresponding to free protein. Set the fitting function to Eq. 3 (*see* **Note 10**), and set the binding site concentration to what was used in the experiment (in this example, 100 μM). Finally, click "Group & Fit Peaks."
4. The fitted data will appear in the "Peak Groups & Analysis" window. In the resulting table the fit parameters including the K_d are listed for each peak. A plot of the experimental vs. fit data is available for each resonance, which can be used to find and correct outliers in the data and to identify the most reliable measurements. Finally, the data can be exported as a text file for import into other software.

3.4.3 Mapping the Binding Site

Titration analysis will result in a subset of significantly perturbed backbone amide residues that can be mapped onto the surface of the high-resolution three-dimensional structure of the protein using molecular graphics software such as PyMOL [20]. These residues should be differentially colored with respect to the rest of the structure to quickly and easily map the binding site, as affected residues generally cluster to form a contiguous binding site on the surface of the protein (Fig. 1d).

4 Notes

1. Ensure that all trace elements are dissolved in water before adding the EDTA. Mixing all reagents at once will result in a precipitate that is extremely difficult to redissolve.
2. For NMR studies, ionic strength of buffer solutions should considered since high salt concentrations can affect the spectrometer's shimming process. In our experience working with salt concentrations <150 mM is routine. The pH of the solution should be kept below 7.5 to avoid chemical exchange of surface backbone amide protons, which are used to monitor ligand binding.
3. The measurement and accurate transfer of mg amounts of carbohydrate powders for the preparation of small stock volumes may be required due to the cost of the carbohydrate, but can

lead to weighing inaccuracies. Small amounts of carbohydrate powders may be electrostatic and difficult to fully transfer into a microcentrifuge tube. Fold the weigh paper diagonally, ensuring that the paper is well creased. This will make the paper easier to handle and the carbohydrate easier to tip into the small opening of the microcentrifuge tube.

4. Due to the minimal composition of the M9 minimal media protein expression and purification yields from growth in this media are typically lower than those obtained from growth in enriched media. As such, depending on the expression level of a protein of interest, a larger volume of media may be required to ensure sufficient protein is produced for this experiment. Assuming a minimum 500 MHz spectrometer with a room temperature probe, a 100 μM protein sample is sufficient to obtain strong NMR signal. Each carbohydrate to be tested will require the preparation of a separate protein sample.
5. Briefly, harvest the cells by centrifugation at 6500 × *g* for 10 min and resuspend the pellet in 20 mM Tris–HCl pH 8.0, 0.5 M NaCl. Lyse the cells on ice by sonication using 1 s medium intensity sonic pulses for 2 min. Submit the cell lysate to high-speed centrifugation at 17,500 × *g* for 45 min to remove cell debris and insoluble proteins. Purify the hexa-histidine tagged protein of interest via Ni^{2+} affinity chromatography, using a step-wise imidazole gradient (0–500 mM) to elute the protein from the column. Assess the protein contents of each elution fraction by SDS-PAGE. Pool fractions containing the eluted protein and dialyze against a solution of 25 mM Tris–HCl pH 7.5, 50 mM NaCl to remove the imidazole. Concentrate the dialyzed protein to a volume of approximately 1–2 mL. Further purify the protein of interest by size exclusion chromatography using a Hi-Prep 16/60 S200 Sephacryl column. Pool and concentrate the fractions containing the purified protein, and assess the purity by SDS-PAGE.
6. We recommend the addition of 1–2 μL of 1 M HCl or NaOH at a time. After each addition, remove the pH probe, cap the microcentrifuge tube, invert gently to mix, and gently centrifuge to collect the sample at the bottom of the tube. Replace the probe and repeat until the desired pH is achieved.
7. The optimal sample volume for a 5 mm NMR tube is approximately 600 μL. Smaller volumes may be used but may be more difficult to shim to obtain good quality data. Transfer the solution into the NMR tube by pipetting onto the side of the tube. Should the solution get stuck at the top of the tube, cap the tube and gently flick the tube in a downward motion until the solution reaches the bottom. Use a thin strip of Parafilm to seal the seam between the cap and the tube to ensure that the sample does not dry out before use.

8. To prevent excessive dilution of the protein, the total volume of the sample at the end of the titration should not exceed 10% more than the initial starting sample volume. As such, the stock solutions of each carbohydrate must be sufficiently concentrated that this condition can be met, but dilute enough so that volumes that are accurate to pipette are used in each addition.
9. The addition of a single saturating dose of carbohydrate to a protein sample and the subsequent observation of the absence or presence of site-specific chemical shift changes can be used to qualitatively assess binding between the carbohydrate and protein prior to undertaking a full NMR-based titration. This is particularly helpful if the carbohydrate ligands of the protein are unknown. A separate protein sample should be prepared for each carbohydrate to be tested, as well as for each titration to be performed.
10. The collection parameters of the spectrum will vary depending on, among other things, the probe, field strength of the instrument, and the sample quality. Some effort should be made to refine the data collection parameters to optimize data quality and experimental time. These refined parameters should be consistently used over the course of the titration.
11. The fitting algorithm of CcpNmr analysis assumes that the concentration of protein does not change during a titration [7]. In order to accommodate this condition it is prudent to dissolve the carbohydrate stock solution in a small volume of NMR buffer containing free CBM. In this way, protein, salt, and pH changes are minimized during the titration and larger volumes of titrant may be added to achieve saturation.

References

1. Zuiderweg ER (2002) Mapping protein-protein interactions in solution by NMR spectroscopy. Biochemistry 41(1):1–7
2. Haselhorst T, Lamerz AC, Itzstein M (2009) Saturation transfer difference NMR spectroscopy as a technique to investigate protein-carbohydrate interactions in solution. Methods Mol Biol 534:375–386
3. Johnson MA, Pinto BM (2004) NMR spectroscopic and molecular modeling studies of protein-carbohydrate and protein-peptide interactions. Carbohydr Res 339(5):907–928
4. Kay LE, Keifer P, Saarinen T (1992) Pure absorption gradient enhanced heteronuclear single quantum correlation spectroscopy with improved sensitivity. J Am Chem Soc 114:10663–10665
5. Lian LY et al (1994) Protein-ligand interactions: exchange processes and determination of ligand conformation and protein-ligand contacts. Methods Enzymol 239:657–700
6. Johnson BA (2004) Using NMRView to visualize and analyze the NMR spectra of macromolecules. Methods Mol Biol 278:313–352
7. Vranken WF et al (2005) The CCPN data model for NMR spectroscopy: development of a software pipeline. Proteins 59(4):687–696
8. Grondin JM et al (2014) An unusual mode of galactose recognition by a family 32 carbohydrate-binding module. J Mol Biol 426(4):869–880
9. Viegas A et al (2013) Solution structure, dynamics and binding studies of a family 11 carbohydrate-binding module from *Clostridium thermocellum* (CtCBM11). Biochem J 451(2):289–300

10. Koay A et al (2007) Oligosaccharide recognition and binding to the carbohydrate binding module of AMP-activated protein kinase. FEBS Lett 581(26):5055–5059
11. Ohnuma T et al (2008) LysM domains from Pteris ryukyuensis chitinase-A: a stability study and characterization of the chitin-binding site. J Biol Chem 283(8):5178–5187
12. Boraston AB et al (2004) Carbohydrate-binding modules: fine-tuning polysaccharide recognition. Biochem J 382(Pt 3): 769–781
13. Boraston AB, Ficko-Blean E, Healey M (2007) Carbohydrate recognition by a large sialidase toxin from *Clostridium perfringens*. Biochemistry 46(40):11352–11360
14. Ficko-Blean E, Boraston AB (2009) N-acetylglucosamine recognition by a family 32 carbohydrate-binding module from *Clostridium perfringens* NagH. J Mol Biol 390(2):208–220
15. Ficko-Blean E et al (2012) Carbohydrate recognition by an architecturally complex alpha-N-acetylglucosaminidase from *Clostridium perfringens*. PLoS One 7(3):e33524
16. Ficko-Blean E, Boraston AB (2006) The interaction of a carbohydrate-binding module from a *Clostridium perfringens* N-acetyl-beta-hexosaminidase with its carbohydrate receptor. J Biol Chem 281(49):37748–37757
17. Wilkins MR et al (1999) Protein identification and analysis tools in the ExPASy server. Methods Mol Biol 112:531–552
18. Shekels LL et al (1998) Cloning and characterization of mouse intestinal MUC3 mucin: 3' sequence contains epidermal-growth-factor-like domains. Biochem J 330(Pt 3):1301–1308
19. Delaglio F et al (1995) NMRPipe: a multidimensional spectral processing system based on UNIX pipes. J Biomol NMR 6(3):277–293
20. DeLano WL (2002) The PyMOL molecular graphics system, Version 1.8 Schrödinger, LLC

Chapter 12

Measuring the Biomechanical Loosening Action of Bacterial Expansins on Paper and Plant Cell Walls

Daniel J. Cosgrove, Nathan K. Hepler, Edward R. Wagner, and Daniel M. Durachko

Abstract

Expansins are proteins that loosen plant cell walls but lack enzymatic activity. Here, we describe two protocols tailored to measure the biomechanical activity of bacterial expansin. The first assay relies on weakening of filter paper by expansin. The second assay is based on induction of creep (long-term, irreversible extension) of plant cell wall samples.

Key words Cellulose, Creep, Expansin, Filter paper, Plant cell wall, Wall loosening proteins

1 Introduction

Expansins are unique cell-wall loosening proteins that characteristically induce stress relaxation and creep of plant cell walls, yet no bona fide enzymatic activity has been identified for these proteins [1, 2]. Consequently, assessments of expansin activity are based on biomechanical assays of cell walls or similar cellulosic materials [3–7]. The walls of growing plant cells consist of a polylamellate scaffold of cellulose microfibrils noncovalently bonded to a matrix of complex polysaccharides (predominantly pectins and hemicelluloses) and small amounts of glycoproteins [8, 9]. The growing plant cell wall is mechanically strong and able to withstand large tensile forces created by cell turgor pressure, yet it is also able to expand irreversibly in surface area in a pH-dependent process mediated by expansins. Expansin activity is traditionally measured by induction of creep (slow, irreversible extension) of cell walls or similar cellulosic materials that are clamped in tension [10].

To briefly summarize cell wall biomechanics: the stress-strain properties of growing plant cell walls are complex; without expansins they behave to a first approximation like a hydrated *solid* material that combines visco-elastic and visco-plastic properties [11–13].

D. Wade Abbott and Alicia Lammerts van Bueren (eds.), *Protein-Carbohydrate Interactions: Methods and Protocols*, Methods in Molecular Biology, vol. 1588, DOI 10.1007/978-1-4939-6899-2_12, © Springer Science+Business Media LLC 2017

Treatment with expansin induces a visco-elastic *fluid* behavior, resulting in slow irreversible deformation (creep) when the walls are held in constant tensile stress above a yield threshold [2, 14]. When expansin is removed, the walls stop creeping and revert to their solid-like behavior, but the deformation (strain) remains.

Expansin is hypothesized to loosen selective sites ("biomechanical hotspots") where cellulose microfibrils are laterally bound to each other via noncovalent interactions involving hemicellulose such as xyloglucan [15–17]. Expansin action facilitates microfibril slippage and separation in response to mechanical tension in the cell wall, resulting in cell wall creep. Cell walls isolated from some growing plant tissues can extend irreversibly up to two-fold under the action of expansins, at rates that may equal or exceed the normal growth rate of the living plant tissues.

We currently recognize three major groups of expansins, differentiated by protein sequence and by activity. The first two groups, synthesized by plants, named α-expansins and β-expansins, are involved in plant growth, fruit softening, and many other events during plant development [10]. The third group is a heterogeneous collection of proteins from bacteria and other microbes [18, 19]. Phylogenetic studies indicate that microbial expansins arose by lateral gene transfer, probably multiple times, from plants to bacteria and fungi [20]. These xenologs facilitate colonization of plant surfaces and plant vascular systems by some (not all) commensal or pathogenic bacteria [18].

Assays for the three groups of expansins differ, reflecting differences in targets and/or cell wall structure. The assay of α-expansin generally relies on measurement of creep of isolated dicot cell walls, e.g., from hypocotyl walls, in a constant-force extensometer. The procedure was described in detail previously [21]. The assay of β-expansin is similar but makes use of isolated cell walls from the grass family (Poaceae, e.g., young coleoptile walls from maize or wheat seedlings) [22–24]. The use of different walls as substrates reflects the fact that the dominant hemicellulose of grass cell walls is arabinoxylan, whereas in dicots it is xyloglucan [25]. The two groups of plant expansins are believed to target different load-bearing junctions within the growing walls. Dicot walls and grass walls likely include both types of junctions, but their relative importance for wall mechanics varies in the two groups of plants. We have measured the activity of bacterial expansins by creep of grass cell walls [7] and cellulosic composites synthesized by *Gluconacetobacter xylinus* [4, 26, 27], and by breaking-strength measurements that make use of cellulosic filter paper as the substrate [27].

Here, we describe two protocols tailored for measuring bacterial expansin activity, by breaking strength of paper and by creep of wheat coleoptile cell walls. In the first method (Subheading 3.1), a strip of wet filter paper is clamped in a device that extends the paper at a constant rate while monitoring the force applied to the paper. The maximum force attained by the paper strip is taken as a measure of the breaking strength. In the second method (Subheading 3.2), a wall sample prepared from the growing coleoptile of a wheat seedling is held in a constant-force extensometer; after the initial visco-elastic transient has decayed, expansin is added to the sample and the resulting increase in extension rate is taken as a measure of expansin activity. These methods have been used to quantify expansin activity of several microbial expansins [26] and of site-directed mutants of EXLX1 from *Bacillus subtilis* [27].

2 Materials

Prepare solutions using reagent grade chemicals and deionized water and store at room temperature unless indicated otherwise.

2.1 Breaking Strength of Paper

1. Whatman filter paper No. 3 cut into 10 × 1.5-mm strips (*see* **Note 1**). Ten strips are used per treatment.
2. HEPES buffer: 25 mM HEPES (dissolve 0.596 g of 4-(2-hydroxyethyl)-1-piperazineethanesulfonic acid in 90 mL of water, adjust to pH 7.5 with 1 M KOH, then bring volume to 100 mL.
3. Expansin Solution: purified native or recombinant protein, e.g., EXLX1 from *Bacillus subtilis* [27], dissolved at a concentration of 50–200 μg/mL in HEPES buffer.
4. Commercial or custom-made extensometer (*see* Fig. 1): Suitable commercial instruments are often to be found in polymer material testing laboratories (e.g., in the past we have used an Instron Universal Testing System).

2.2 Preparation of Coleoptile Cell Walls

1. Wheat seeds (*Triticum spp*) (we use Seedway SW-50, but other cultivars and other grass species may also be used).
2. Seed germination paper such as Versa-Pak paper from www.seedburo.com.
3. Seed germination trays, 30.5 × 22 × 6 cm.
4. Light-tight box or dark room for growing the plant material.
5. Carborundum (320 grit, Fisher Scientific, C192-500).

2.3 Creep of Coleoptile Cell Walls

1. HEPES buffer: 25 mM HEPES (dissolve 0.596 g of (4-(2-hydroxyethyl)-1-piperazineethanesulfonic acid) in 90 mL of water, adjust to pH 7.5 with 1 M KOH, and then bring volume to 100 mL.
2. Expansin Solution (same as Subheading 2.1 **item** 3 above): purified native or recombinant protein, e.g., EXLX1 from

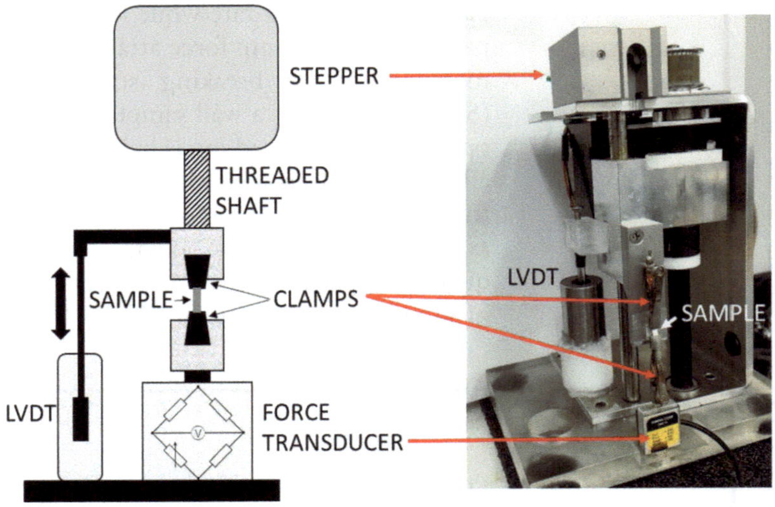

Fig. 1 Diagram of custom-built extensometer for stress-strain experiments. This is a mechanical testing device used to clamp the paper strips, extend the distance between the clamps at a constant rate (we use 1.5 mm/min, sometimes 3 mm/min), and monitor the increasing force until breakage occurs. The instrument consists of a finely threaded sliding stage scavenged from old equipment. Tissue samples are held in place with two smooth-jaw micro-alligator clips (Mueller BU-34C). Clips are stiffened at the ends with steel epoxy (J-B Weld). An additional piece of copper sheeting is soldered to clip ends to enlarge their gripping surface. Clamping surfaces have a piece of fine (~220 grit) waterproof automotive sandpaper glued to them to prevent sample slippage. A stepper motor (LIN Engineering 4218L-01D-02) controlled through a computer RS-232 serial port and driver (Automation Direct STP-DRV-4850) allows precisely controlled movement of the *upper clamp* using two gears and a toothed drive belt (also scavenged; 1:2 gear ratio) via the sliding stage. The *lower clip* is affixed directly to a force transducer (Futek LSB200) allowing measurement of stress applied to the sample. A linear displacement transducer, or LVDT, (Schaevitz 050HR) coupled to the *upper stage* allows measurement of strain imposed upon the sample

Bacillus subtilis [21], dissolved at a concentration of 50–200 μg/mL in HEPES buffer.

3. Constant force extensometer, custom-made (*see* Fig. 2 and **Note 2**).

3 Methods

3.1 Breaking Strength of Paper (See Note 3)

1. Soak ten paper strips in 1 mL of HEPES buffer or in Expansin Solution for 4 h at 25 °C with gentle shaking or inversion to equilibrate the solution with the strips (*see* **Note 4**).
2. Using fine forceps, clamp the paper strip between the two clamps of the extensometer, with 5 mm distance between the clamps.

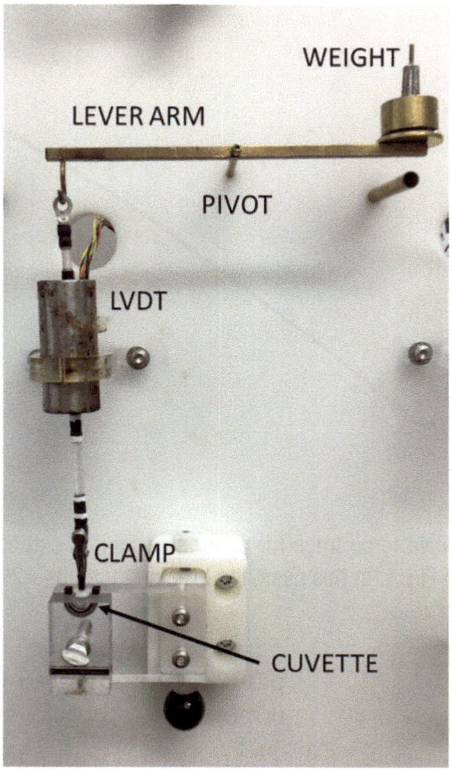

Fig. 2 Photograph of custom-built constant-force extensometer for cell wall creep experiments. The base of the wall specimen is clamped into the base of a 200 μL sample holder (*cuvette*) which is filled with buffer to keep the specimen wet throughout the experiment. The top of the specimen is clamped to a rod that bears the magnetic cylinder of the LVDT (linear variable differential transformer, the position sensor). An upward force is applied to the sample via a lever arm and counter weight. Counter weights are adjustable and are normally constant throughout an experiment, but can be added or removed during the course of an experiment if desired. For different cell wall materials, preliminary trial experiments are needed to establish an appropriate force (*see* **Note 9**). The LVDT is connected to a signal processor and the position is digitized and recorded by a microcomputer

3. Extend the strip at 1.5 mm/min while recording the clamping force (*see* Fig. 3 for an example). The force will increase, level off and then decrease, sometimes precipitously, as the paper fibers separate. The maximum force attained is taken as the breaking force (*see* **Note 5**).

3.2 Preparation of Coleoptile Cell Walls

1. Sow ~1000 wheat seeds (other grass species may also be used) on absorbent germination paper (e.g., Versa-Pak paper from), enclosed in covered germination trays.
2. Grow in a dark room or a light-tight box for 3–4 days at 28 °C.
3. Coleoptiles (*see* **Note 6**) from healthy seedlings are cut and gently abraded between thumb and forefinger coated with a

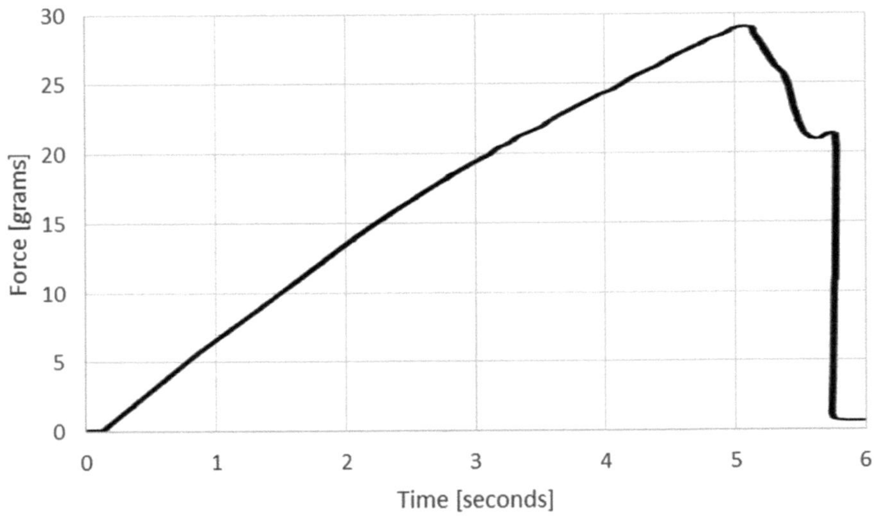

Fig. 3 Representative force trace for paper strip as it is extended and breaks, using the stress-strain extensometer shown in Fig. 1. In this case, the breaking force is ~29 gram force (~0.28 N)

paste of carborundum-water, then rinsed in distilled water to remove the carborundum (*see* **Note 7**).

4. The rolled leaf within the coleoptile is removed by gently snapping the base of the coleoptile with your fingers and carefully withdrawing the leaf from the coleoptile sheath.
5. Freeze and store coleoptiles in groups of ten at −80 °C in 1.5 mL microcentrifuge tubes.
6. Heat inactivate groups of ten coleoptile walls to remove endogenous expansin activity by dipping in boiling water for 15 s, then plunging into cold water (*see* **Note 8**).

3.3 Creep of Coleoptile Cell Walls

1. With fine forceps mount the wall specimen between the extensometer clamps, with 5 mm between the two clamps. This results in a tensile force of 20 g on the specimen (*see* **Note 9**).
2. Pipette 200 μL HEPES buffer into the sample cuvette to keep the wall sample wet.
3. (Optional) Replace the HEPES buffer in the cuvette with 200 μL 70 mM NaOH for 20 min, then wash extensively with deionized water and finally restore HEPES buffer to the cuvette (*see* **Note 10**).
4. The extension rate should stabilize to a low value (<1 μm/min) in ~10–15 min. (*see* **Note 11**).
5. Carefully remove the HEPES buffer by a pipette attached to a suction and replace with 200 μL Expansin Solution. The wall should begin extending within 1–2 min, although it may take a few more minutes for this to become obvious. Figure 4 shows a representative creep response to EXLX1 from *Bacillus subtilis*.

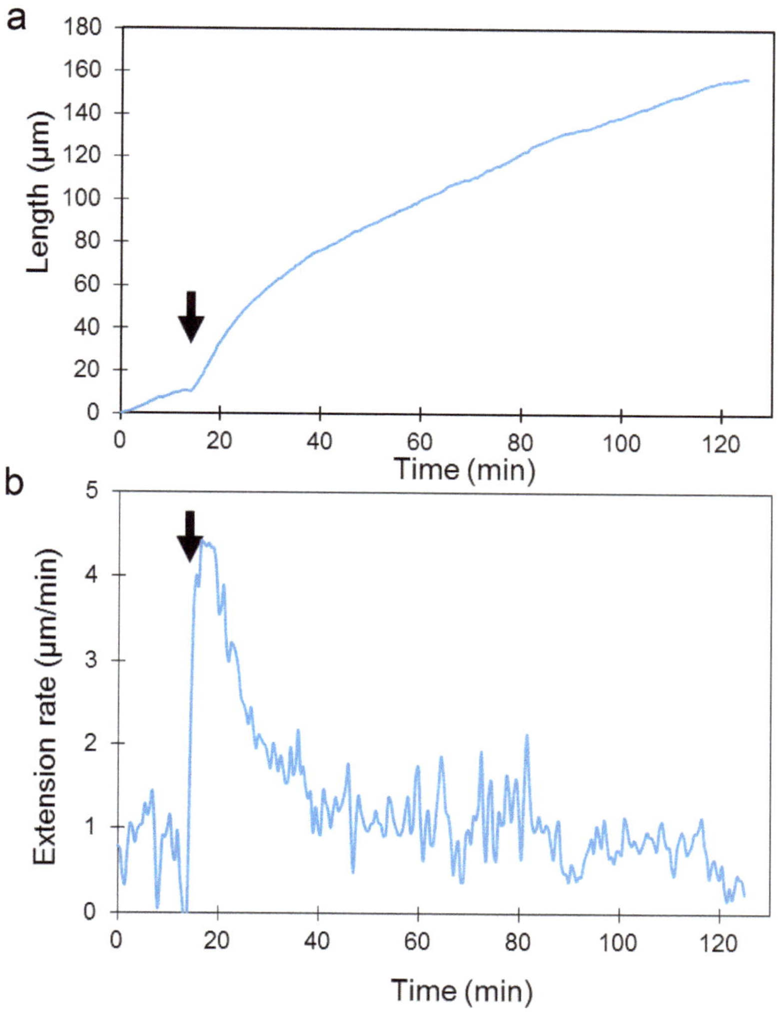

Fig. 4 Representative trace for (**a**) length and (**b**) extension rate of a wheat coleoptile wall upon treatment with bacterial expansin (200 μg/mL of BsEXLX1, applied at the time indicated by the *arrows*)

4 Notes

1. Whatman filter paper discs can be taped to 10 mm × 1.5 mm grid graph paper and cut using a fabric rotary cutter. Using the graph paper helps ensure consistent strip sizes. The instrument used to cut the filter paper needs to be sharp to avoid tearing or shearing of the paper's edge, which can result in reduced breaking strength. Scissors, unless brand new, are often too dull; a new single-edge razor blade can be substituted for a rotary cutter.
2. We have not found a commercial instrument suitable for these measurements, although they might be on the market. The measuring device is very sensitive to mechanical vibrations, blowing air, and temperature fluctuations. We locate our

instrument on a vibration-dampening table in a room with a proportional temperature controller (not on/off, which results in unacceptable temperature oscillations).

3. This assay is good for qualitative assessment of activity. It has not been tested for linearity of response, which would be needed for use as a quantitative assay.
4. Shorter or longer times may be used, but keep the times constant as the paper gradually weakens in water.
5. Examine the force-extension curve and the paper strip for slippage within the clamp and for anomalous tearing, e.g., as a result of the clamp tearing the paper strip. Indications of such technical problems include an erratic (saw tooth) force curve and regular breakage at the mouth of the clamp.
6. The coleoptile is a modified leaf sheath that covers and protects the delicate primary leaves. The coleoptile was used in classical studies on growth hormones such as auxin.
7. Abrasion with carborundum disrupts the cuticle and provides buffer and protein access to the underlying cell wall.
8. The following step is generally not used for wheat coleoptiles but may be needed for thicker materials such as maize coleoptiles and cucumber hypocotyls: Ten samples are pressed for 10 min between two glass microscope slides (22 × 75 mm) lined with Kimwipe paper, using a weight of ~400 g. This step is needed to express cell sap and to facilitate clamping of the sample, which otherwise tends to split and tear in the clamp if excessive cell fluids remain within the sample.
9. Different weights are needed for different cell wall samples. This is determined by trial and error. If the weight is too large, then breakage of the wall specimen occurs too soon or too frequently. If insufficient weight is applied, then the wall may not undergo creep.
10. Treatment with NaOH causes the breaking of ester bonds, including diferulate crosslinks between arabinoxylans, resulting in solubilization of arabinoxylan. Compared to untreated samples, coleoptiles incubated in NaOH display reduced variability and a ~5-fold increase in response to EXLX1 [27]. This step is useful for grass cell walls, but is generally not helpful for dicot walls or other samples.
11. High residual creep rate likely indicates mechanical damage to the sample or inadequate removal of NaOH.

Acknowledgment

This work was supported by United States Department of Energy Grant DE-FG02-84ER13179 from the Office of Basic Energy Sciences.

References

1. Cosgrove DJ (2016) Catalysts of plant cell wall loosening. F1000Research 5
2. Cosgrove DJ (2000) Loosening of plant cell walls by expansins. Nature 407:321–326
3. McQueen-Mason S, Durachko DM, Cosgrove DJ (1992) Two endogenous proteins that induce cell wall extension in plants. Plant Cell 4:1425–1433
4. Whitney SEC, Gidley MJ, McQueen-Mason SJ (2000) Probing expansin action using cellulose/hemicellulose composites. Plant J 22:327–334
5. Qin L, Kudla U, Roze EH, Goverse A, Popeijus H, Nieuwland J, Overmars H, Jones JT, Schots A, Smant G, Bakker J, Helder J (2004) Plant degradation: a nematode expansin acting on plants. Nature 427:30
6. Cho HT, Kende H (1997) Expansins in deepwater rice internodes. Plant Physiol 113: 1137–1143
7. Kerff F, Amoroso A, Herman R, Sauvage E, Petrella S, Filee P, Charlier P, Joris B, Tabuchi A, Nikolaidis N, Cosgrove DJ (2008) Crystal structure and activity of *Bacillus subtilis* yoaj (exlx1), a bacterial expansin that promotes root colonization. Proc Natl Acad Sci U S A 105:16876–16881
8. Cosgrove DJ (2005) Growth of the plant cell wall. Nat Rev Mol Cell Biol 6:850–861
9. Zhang T, Zheng Y, Cosgrove DJ (2016) Spatial organization of cellulose microfibrils and matrix polysaccharides in primary plant cell walls as imaged by multichannel atomic force microscopy. Plant J 85:179–192
10. Cosgrove DJ (2015) Plant expansins: diversity and interactions with plant cell walls. Curr Opin Plant Biol 25:162–172
11. Burgert I, Keplinger T (2013) Plant micro- and nanomechanics: experimental techniques for plant cell-wall analysis. J Exp Bot 64:4635–4649
12. Nolte T, Schopfer P (1997) Viscoelastic versus plastic cell wall extensibility in growing seedling organs: a contribution to avoid some misconceptions. J Exp Bot 48:2103–2107
13. Cleland RE (1984) The instron technique as a measure of immediate-past wall extensibility. Planta 160:514–520
14. Takahashi K, Hirata S, Kido N, Katou K (2006) Wall-yielding properties of cell walls from elongating cucumber hypocotyls in relation to the action of expansin. Plant Cell Physiol 47:1520–1529
15. Cosgrove DJ (2016) Plant cell wall extensibility: connecting plant cell growth with cell wall structure, mechanics, and the action of wall-modifying enzymes. J Exp Bot 67:463–476
16. Wang T, Park YB, Caporini MA, Rosay M, Zhong L, Cosgrove DJ, Hong M (2013) Sensitivity-enhanced solid-state nmr detection of expansin's target in plant cell walls. Proc Natl Acad Sci U S A 110:16444–16449
17. Cosgrove DJ (2014) Re-constructing our models of cellulose and primary cell wall assembly. Curr Opin Plant Biol 22C:122–131
18. Georgelis N, Nikolaidis N, Cosgrove DJ (2015) Bacterial expansins and related proteins from the world of microbes. Appl Microbiol Biotechnol 99:3807–3823
19. Pastor N, Davila S, Perez-Rueda E, Segovia L, Martinez-Anaya C (2014) Electrostatic analysis of bacterial expansins. Proteins 83:215–223
20. Nikolaidis N, Doran N, Cosgrove DJ (2014) Plant expansins in bacteria and fungi: evolution by horizontal gene transfer and independent domain fusion. Mol Biol Evol 31:376–386
21. Durachko DM, Cosgrove DJ (2009) Measuring plant cell wall extension (creep) induced by acidic ph and by alpha-expansin. J Vis Exp: JoVE 25:1263
22. Sampedro J, Guttman M, Li LC, Cosgrove DJ (2015) Evolutionary divergence of beta-expansin structure and function in grasses parallels emergence of distinctive primary cell wall traits. Plant J 81:108–120
23. Cosgrove DJ, Bedinger P, Durachko DM (1997) Group I allergens of grass pollen as cell wall-loosening agents. Proc Natl Acad Sci U S A 94:6559–6564
24. Li LC, Bedinger PA, Volk C, Jones AD, Cosgrove DJ (2003) Purification and characterization of four beta-expansins (zea m 1 isoforms) from maize pollen. Plant Physiol 132:2073–2085
25. Carpita NC (1996) Structure and biogenesis of the cell walls of grasses. Annu Rev Plant Physiol Plant Mol Biol 47:445–476
26. Georgelis N, Nikolaidis N, Cosgrove DJ (2014) Biochemical analysis of expansin-like proteins from microbes. Carbohydr Polym 100:17–23
27. Georgelis N, Tabuchi A, Nikolaidis N, Cosgrove DJ (2011) Structure-function analysis of the bacterial expansin exlx1. J Biol Chem 286:16814–16823

Part III

Visualization of Carbohydrates and Protein-Carbohydrate Complexes

Chapter 13

Bioinspired Assemblies of Plant Cell Walls for Measuring Protein-Carbohydrate Interactions by FRAP

Gabriel Paës

Abstract

The interactions of proteins involved in plant cell wall hydrolysis, such as enzymes and CBMs, significantly determine their role and efficiency. In order to go beyond the characterization of interactions with simple ligands, bioinspired assemblies combined with the measurement of diffusion and interaction by FRAP offer a relevant alternative for highlighting the importance of different parameters related to the protein affinity and to the assembly.

Key words Bioinspired, Diffusion, Interaction, Ligand, Protein, Enzyme, CBM, FRAP

1 Introduction

In the emerging bioeconomy era, many enzymes and noncatalytic proteins are involved in biorefineries for hydrolyzing plant cell wall polysaccharides such as cellulose and hemicellulose, in order to produce bio-based fuels, materials, and chemicals [1, 2]. Optimization of these processes to reduce operating costs and to maximize catalytic efficiency requires in-depth knowledge of the protein mode of interactions. Most of the time, measuring interactions of proteins with polysaccharides is performed with pure oligomer/polymer ligands or synthetic molecules, using different readily accessible techniques such as gravimetric approach [3], isothermal calorimetry [4], fluorescence titration [5], fluorescence spectroscopy [6], or more sophisticated techniques such as atomic force microscopy [7]. But once in the complex plant polymer network, proteins often do not behave in the same way as observed with simple ligands. Of course, this discrepancy originates from the much higher chemical and structural complexity of plant cell walls compared to simple ligands.

To address this issue, we have designed model assemblies that contain some of the polymers and chemical motifs found in secondary plant cell walls of grasses. These bioinspired assemblies are based

D. Wade Abbott and Alicia Lammerts van Bueren (eds.), *Protein-Carbohydrate Interactions: Methods and Protocols*, Methods in Molecular Biology, vol. 1588, DOI 10.1007/978-1-4939-6899-2_13, © Springer Science+Business Media LLC 2017

on a hemicellulose network whose polymers (feruloylated arabinoxylans, FAXs) are cross-linked through ferulic acid moieties to make gels under the action of a peroxidase [8, 9]. This network can be complexified by adding cellulose nano-crystals (CNCs) [10, 11]. Type and concentration of both polymers can be easily varied. Diffusion of a macromolecule such as a protein can be measured in these assemblies by using the fluorescence recovery after photobleaching (FRAP) technique implemented in a fluorescence confocal microscope [12–14]. The protein must be labeled with a fluorophore before it is embedded into the assembly, then its diffusion coefficient can be calculated. The possibility of measuring interactions of the labeled protein is based on the assumption that if its diffusion is slower than expected on the basis of its size and polymer entanglements, then the protein is interacting with the polymer assembly. Overall, in order to discriminate the role of the different parameters related to the protein and/or to the structure of the assembly, interactions must be calculated in different assemblies with varied polymer concentrations, so that statistical analysis eventually determines the relative influence of each parameter on the interaction, as previously demonstrated [15, 16].

The method presented emphasizes preparation of bioinspired assemblies and of labeled proteins. Statistical analysis of interaction data obtained from FRAP measurements is also discussed, while FRAP is briefly described since many relevant articles and reviews have already been published for performing accurate quantitative FRAP analysis.

2 Materials

Prepare all buffers using ultrapure water and analytical grade reagents, at room temperature.

2.1 Fluorescent Protein

1. Protein: as pure as possible and of known concentration, molecular weight (MW) and molar extinction coefficient; placed in a stable buffer, stored at 4 °C.
2. Fluorophore: many fluorophores are commercially available, varying in their emission spectrum, stability, etc. Here, fluorescein isothiocyanate (FITC, MW 389 g/mol) in powder is used, stored in the dark at 4 °C (*see* **Note 1**).
3. Conjugation buffer: 1 M sodium carbonate-bicarbonate buffer at pH 10.0. Weigh 10.6 g of sodium carbonate (MW 106 g/mol) and dissolve in 100 mL water; weigh 8.4 g of sodium bicarbonate (MW 84 g/mol) and dissolve in 100 mL water. Mix 27.5 mL of the sodium carbonate solution and 22.5 mL of the sodium bicarbonate solution, adjust the final volume with water to 200 mL. Using a sensitive pH-meter, check that the pH is close to 10.0, then store the assembly buffer at 4 °C.

4. Elution buffer: 50 mM phosphate buffer pH 6.2. Weigh 1.38 g of sodium phosphate monobasic (MW 138 g/mol) and dissolve in 100 mL water; weigh 2.68 g of sodium phosphate dibasic (MW 268 g/mol) and dissolve in 100 mL water. Mix 81.5 mL of the sodium phosphate monobasic solution and 18.5 mL of the sodium phosphate dibasic solution, adjust the final volume with water to 200 mL. Using a sensitive pH-meter, check that the pH is close to 6.2, then store the elution buffer at 4 °C.
5. Desalting/gel filtration columns: PD-10 type disposable columns prepacked with Sephadex G-25 for sample volume in the range 0.5–1.5 mL.
6. Precision balance (at least able to weigh in the range 0.1–1.0 mg).
7. One glass vial (1.0–1.5 mL capacity) with cap and stirring bar.
8. Aluminum foil.
9. Plastic tubes with cap (1.5–2.0 mL capacity).
10. UV-visible spectrophotometer and cuvette (<0.5 mL capacity).

2.2 Bioinspired Assemblies

1. Chambered 8-well coverglass (0.2–1.0 mL capacity for each well) (Fig. 1).
2. Hemicellulose: feruloylated arabinoxylan (FAX) extracted from maize (reference AXF purchased from Cambridge Biopolymers Ltd., UK).
3. Cellulose nano-crystals: stock solution can be prepared [17] or purchased from different companies (CelluForce NCC™ from CelluForce, Montréal, Canada).
4. Assembly buffer: 50 mM citrate-phosphate buffer pH 5.4. Weigh 1.92 g of citric acid (MW 192.1 g/mol) and dissolve in

Fig. 1 Chambered 8-well coverglass used for preparing bioinspired assemblies. Each well can contain a volume of 500 μL. On the photo, the first well contains a 300 μL FAX gel and fluorescently labeled protein

100 mL water; weigh 3.56 g of dibasic sodium phosphate (MW 178 g/mol) and dissolve in 100 mL water. Mix 22.2 mL of the citric acid solution and 27.8 mL of the sodium phosphate solution, adjust the final volume with water to 100 mL. Using a sensitive pH-meter, check that the pH is close to 5.4 and then store the assembly buffer at 4 °C.

5. Glass vial (5–10 mL capacity) with cap and stirring bar.
6. Plastic tubes with cap (1.5–2.0 mL capacity).
7. Peroxidase: weigh 0.2 mg of peroxidase and dissolve in 0.4 mL of assembly buffer to make a solution at 0.5 mg/mL, store at 4 °C.
8. Hydrogen peroxide: take 50 μL of the commercial stock solution at 350 mg/mL and mix with 450 μL of assembly buffer to make a 35 mg/mL solution, store at 4 °C.

2.3 Diffusion Measurement

FRAP measurements can be performed on most confocal microscopes. Generally, minimal equipment consists in a fluorescence inverted confocal microscope equipped with a laser wavelength corresponding to that of the fluorophore (488 nm for fluorescein). A FRAP analysis module embedded in the microscope software is also necessary. In addition, image analysis software such as the freeware ImageJ/Fiji and statistical analysis software are particularly helpful.

3 Methods

3.1 Desalting Protein

If the protein is available in powder, begin at Subheading 3.2.

1. Support the desalting column over a discarding beaker. Prepare the desalting column following the provider instruction. Usually, wash the column with 4–5 column-equivalent volumes of water.
2. Apply 1 mg of the protein sample, in a volume of 2.5 mL. Use water if a dilution of the protein stock is necessary. Let the sample enter the column completely.
3. Elute with 3.5 mL of water and collect seven fractions of 0.5 mL into plastic tubes.
4. Measure the absorbance at 280 nm of each fraction (*see* **Note 2**). Calculate the protein concentration in each fraction. Pool together the highest concentrated protein fractions and keep them at 4 °C (*see* **Note 3**). Calculate the quantity of desalted protein, which should be closed to 1 mg.

3.2 Labeling of Protein with a Fluorophore

1. Weigh FITC in the glass vial so that FITC molar quantity is ten times higher than that of the desalted protein (*see* **Note 4**). Add 100 μL of the conjugation buffer (*see* **Note 5**).

2. Add the desalted protein sample into the glass vial. Complete with water to reach a final volume of 1 mL.
3. Cover the vial with aluminum foil and let the reaction for 2 h with moderate stirring (60–90 rpm).
4. Support the column over a discarding beaker. Prepare the gel filtration column following the provider instruction (usually wash the column with 4–5 column-equivalent volumes of elution buffer).
5. Apply the reaction mixture. Let the sample enter the column completely. Discard the flow through, then add 1.5 mL of elution buffer. Discard the flow through.
6. Elute with 3.5 mL of elution buffer (*see* **Note 6**). A fast displacing yellow-orange band will appear, while a second band will stay at the top of the column. Collect only the first band (*see* **Note 7**).
7. Measure the absorbance at 280 and 495 nm of the collected fraction (*see* **Note 2**). Calculate the fluorescent protein concentration and the molar fluorophore: protein ratio using Eqs. 1 and 2, respectively:

$$\text{Fluorescent protein concentration} = \frac{A_{280} - 0.35 \times A_{495}}{\varepsilon_{\text{prot}}} \times MW_{\text{prot}} \qquad (1)$$

$$\text{Molar ratio} = \frac{A_{495}}{\varepsilon_{\text{fluo}} \times \dfrac{\text{Fluorescent protein concentration}}{MW_{\text{prot}}}} \qquad (2)$$

A_{280} and A_{495} are absorbances of the labeled protein at 280 and 495 nm, respectively.

$\varepsilon_{\text{prot}}$ is the molar extinction coefficient of the protein (in L/mol cm).

MW_{prot} is the protein molecular weight (in g/mol).

$\varepsilon_{\text{fluo}}$ is the molar extinction coefficient of the fluorophore (in L/mol cm).

8. Make sure that the molar ratio is in the range 0.5–2.0 (*see* **Notes 4** and **8**). If it is below, prepare another labeling reaction by increasing the amount of FITC; if it is higher, reduce the amount of FITC. Fluorescent protein concentration should be at least 1 g/L for subsequent use. If not, it should be concentrated (by using, for example, Vivaspin ultrafiltration unit with a molecular weight cutoff adapted to the protein size).
9. Store the fluorescent protein at 4 °C in the dark.

3.3 Bioinspired Assemblies

The bioinspired assemblies presented in this protocol contain hemicellulose (FAX) and cellulose nano-crystals (CNCs) found in secondary plant cell walls. But other polymers of interest can be used such as pectins, other hemicellulose, and cellulose-type polymers.

1. Prepare 2 mL of a 30 g/L FAX solution by dissolving 60 mg of FAX in 2 mL of the assembly buffer under moderate stirring in a glass vial at room temperature, overnight (at least 16 h).
2. Prepare 2 mL of 30 g/L CNC solution by dissolving the stock solution in the assembly buffer (*see* **Note 9**).
3. Total volume of the assembly is set at 300 μL, since it can be easily prepared without consuming too much material. Mix together in a plastic tube 100 μL of the FAX solution at 30 g/L and 100 μL of the CNC solution at 30 g/L to have final concentrations of 10 g/L for each. Add the fluorescent protein to get a final concentration of ca. 0.2–0.3 g/L (*see* **Note 10**). Add 7 μL of the hydrogen peroxide solution. Complete if necessary with the assembly buffer at 290 μL. Mix carefully by gently pipetting up and down without adding any bubble (*see* **Note 11**).
4. Transfer this mix into the 8-well chamber. Add 10 μL of the peroxidase solution and gently mix by pipetting (*see* **Note 11**). Gelation will start; let the reaction for at least 2 h in the dark. Diffusion measurements in the assemblies should be done on the same day or the day after the assemblies have been prepared. If not analyzed immediately they should be stored at 4 °C in the dark.
5. Prepare as many assemblies containing the labeled protein as necessary, depending on the number and values of parameters modulated (e.g., FAX and CNC concentration).

3.4 Measuring Diffusion and Interaction of Fluorescent Protein by FRAP

Excellent and comprehensive protocols regarding the measurement of diffusion by FRAP have already been published [18, 19]; therefore, only the main steps are presented below. FRAP measures the fluorescence recovery into a sample in which a region of interest (ROI) has been bleached irreversibly using a high intensity laser light. Depending on the characteristics of the labeled protein and of the chemical and physical properties of assembly, diffusion can vary notably (Fig. 2) [14]. Image and data processing gives access to the diffusion coefficient D that can be turned into an interaction coefficient. For FRAP data to be relevant, image scanning rate must be as fast as possible to acquire as much data as possible. Indeed, the fluorescence recovery is very fast in the first seconds, then slows down, to finally reach (theoretically) a plateau (Fig. 2).

1. Place the chamber containing the fluorescent protein in the assembly on the microscope objective, using a magnification of 63×. Adjust the laser intensity and detectors sensitivity to image the sample so that the image rate is below one frame/s. Since the assembly has no structure at such a magnification, the image should be homogeneously fluorescent.
2. Draw a circular ROI of 5 μm diameter. The ROI surface must not exceed 10% of the total surface of the image to leave enough unbleached labeled protein around the ROI.

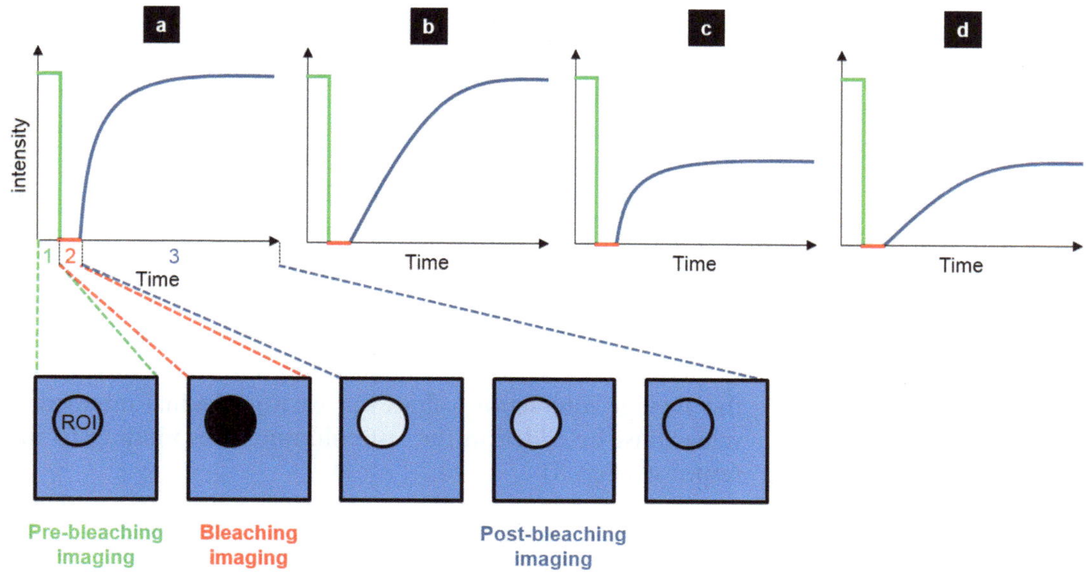

Fig. 2 Typical recovery curves obtained after FRAP experiments. The fluorescence intensity is measured before (step 1, in *green*), during (step 2, in *red*) and after photobleaching (step 3, in *blue*), while equilibrium takes place. Several cases can be observed depending on the velocity and the amplitude of the recovery. Recovery can be fast with a maximal recovery (**a**), or slow with a maximal recovery (**b**). It can also be fast with a limited recovery (**c**) and even slow with a limited recovery (**d**) (figure from [14])

3. Collect the FRAP data in at least five different ROIs in the sample and in at least three different samples. Processing of the data should take into account reference ROIs for the unintentional photobleaching. Overall, diffusion coefficient D should be obtained for the fluorescent protein in the different tested assemblies.
4. In order to get interactions values, calculate the pseudo-affinity constant $k^*_{\text{on}}/k_{\text{off}}$ of the labeled protein in each assembly by using Eq. 3:

$$\frac{k^*_{on}}{k_{off}} = \frac{D_{\text{free}} - D}{D} \tag{3}$$

D_{free} is the diffusion of the labeled protein in the assembly buffer (free diffusion) and D is the diffusion of the labeled protein in the considered assembly, also called an effective diffusion coefficient [18, 19].

3.5 Statistical Analysis to Highlight Meaningful Parameters

The goal of this section is to evaluate quantitatively the weight of each varied parameter (related to the fluorescent protein and/or to the assembly) to rank the parameters according to their influence on the affinity, called the response. Analysis is based on full-factorial experiments and statistical analysis, using for example, dedicated software like Design Expert 8.0, but other software can be used.

1. Define the parameters that have been varied and give them an alias for clarity. For each parameter, identify the number of levels and the values of each level. For example, if FAX concentration in assemblies has been varied at 0.5%, 1.0%, and 2.0% while the CNC concentration has been tested at 0% and 1.0%, one can call the first parameter *FAX*, having three numerical levels that are 0.5/1.0/2.0 and so on.
2. Prepare a table in which the first column corresponds to the first parameter, second column to the second parameter, and so on. Add the level values for each parameter so that all parameter combinations are presented, taking into account that not the average interaction values but each individual interaction value must be added in the last column corresponding to the response (Fig. 3).
3. Perform the different steps of the statistical analysis in Design Expert. Following the ANOVA analysis, *p*-values are calculated for each parameter, so that some parameters are considered to be statistically significant or not. Among the former ones, the F-values are also indicated: the higher these values are, the more impact the considered parameter has on the response. Overall, the different parameters can be ranked according to their F-values, highlighting those having a real importance

Experiment nb	Parameter 1	Parameter 2	Response values
1	Level 1	Level 1	
2	Level 1	Level 1	
3	Level 1	Level 1	
4	Level 2	Level 1	
5	Level 2	Level 1	
6	Level 2	Level 1	
7	Level 3	Level 1	
8	Level 3	Level 1	
9	Level 3	Level 1	
10	Level 1	Level 2	
11	Level 1	Level 2	
12	Level 1	Level 2	
13	Level 2	Level 2	
14	Level 2	Level 2	
15	Level 2	Level 2	
16	Level 3	Level 2	
17	Level 3	Level 2	
18	Level 3	Level 2	

Fig. 3 Typical example of the dataset organized as a full factorial experiment. Parameter 1 has three levels while parameter 2 has two levels. Data are from experiments performed in triplicate. Overall, 18 different experimental data values (responses) must be filled in

toward the response. Mathematically speaking, the response is equal to a sum of parameters and their combination, each having a coefficient related to their importance on the response. So if the parameters computed have numerical levels, the equation can be used to predict the response in conditions not tested experimentally [10]. The values for each parameter must remain between the highest and lowest levels used; otherwise, the prediction is not statistically acceptable.

4 Notes

1. Since the fluorophore will be used for FRAP experiments (i.e., photobleached), select a fluorophore that has not been engineered to be highly photostable.
2. Absorbance values should be below 1.0 to respect Beer-Lambert proportionality law between absorbance and concentration. Otherwise, dilute the sample.
3. In general, the highest protein concentration is collected in fractions two and three, even it can slightly vary depending on protein size.
4. FITC reacts with amine functions. If the protein is not circular, the N-terminal residue presents an amine function. Surface protein residues like Lys have an amine function. Knowing the number of Lys residues, or even better, the accessible Lys residues through the analysis of the protein 3D structure can help to adjust the fluorophore:protein ratio. This molar ratio can be varied from 1:1 to 100:1. Under-labeling reaction results in a non-fluorescently detectable protein. Over-labeling reaction results in altered biochemical properties (modified interactions and catalytic activity), physicochemical properties (stability), and spectral properties (loss in fluorescence due to fluorophore self-quenching).
5. Depending on the type of labeling, different pH and/or buffer conditions can be used [20]. Caution must be taken to check whether the protein is stable under the labeling conditions used, in particular regarding pH value.
6. Elution buffer should be selected so that the protein is stable in this buffer.
7. The first band contains labeled proteins and non-labeled proteins; it moves fast because the protein is excluded from the gel network. The second band contains only unreacted FITC and salts that are much smaller and thus move very slowly. If only one band appears after applying elution buffer, labeling has not worked properly: protein is eluted without having been conjugated. It can mean that no accessible amine function is

present on the surface of the protein. Try to increase the amount of protein to be labeled. If no improvement is noted, try another labeling protocol involving other functions (for example hydroxyls that are present on serine and threonine residues). To check the labeling protocol, a positive control can be a commercial bovine serum albumin protein that is well characterized and displays several amine functions on its surface.

8. Determination of labeled protein concentration can be tricky. Due to the protein labeling, fluorescence properties of both the protein and the fluorophore can be slightly altered. The best way to determine this parameter is to measure a fluorescence map of the sample by measuring fluorescence intensity when acquiring excitation on the range 250–600 nm and emission on the range 260–600 nm. Three spots should be visible: one for the protein for an excitation ca. 280 nm and an emission ca. 320 nm; two other ones for the fluorophore, for an emission ca. 520 nm (in the case of fluorescein) and an excitation ca. 280 nm and 490 nm. Taking as a reference the emission of the fluorophore alone and comparing this value to the emission of the fluorophore bound to the protein gives a good estimation of the molar concentration of the fluorophore. The same can be done by taking the un-labeled protein as a reference to calculate the molar concentration of the protein, so that the labeling ratio can be determined.
9. CNCs are only partially soluble in water buffer, so there is no need for stirring.
10. Concentration should be adapted depending on the fluorophore brightness and stability so that the labeled protein is easily imaged under a confocal microscope without unintentional photobleaching. Also, fluorescence can be dramatically modified by pH depending on the fluorophore considered. In the case of fluorescein, we have found that a pH ca. 5.5–6.0 was appropriate. In the case another fluorophore may be appropriate, pH should be adapted accordingly.
11. FAX solution is highly viscous. When pipetting, mixing, and transferring FAX solution, bubbles can appear. To get rid of them, use a tip to guide them on the side of the tube or of the tray.

Acknowledgment

This work was supported by INRA and the French National Research Agency (LIGNOPROG project ANR-14-CE05-0026).

References

1. Isikgor FH, Becer CR (2015) Lignocellulosic biomass: a sustainable platform for the production of bio-based chemicals and polymers. Polym Chem 6(25):4497–4559
2. Viikari L, Vehmaanperä J, Koivula A (2012) Lignocellulosic ethanol: from science to industry. Biomass Bioenergy 46:13–24
3. Guo FF, Shi WJ, Sun W, Li XZ, Wang FF, Zhao J, Qu YB (2014) Differences in the adsorption of enzymes onto lignins from diverse types of lignocellulosic biomass and the underlying mechanism. Biotechnol Biofuels 7
4. Guo J, Catchmark JM (2013) Binding specificity and thermodynamics of cellulose-binding modules from *Trichoderma reesei* Cel7A and Cel6A. Biomacromolecules 14(5):1268–1277
5. Paës G, Tran V, Takahashi M, Boukari I, O'Donohue MJ (2007) New insights into the role of the thumb-like loop in GH-11 xylanases. Protein Eng Des Sel 20(1):15–23
6. Wang LQ, Wang YQ, Ragauskas AJ (2010) A novel FRET approach for in situ investigation of cellulase-cellulose interaction. Anal Bioanal Chem 398(3):1257–1262
7. King JR, Bowers CM, Toone EJ (2015) Specific binding at the cellulose binding module-cellulose interface observed by force spectroscopy. Langmuir 31(11):3431–3440
8. Carvajal-Millan E, Guilbert S, Morel MH, Micard V (2005) Impact of the structure of arabinoxylan gels on their rheological and protein transport properties. Carbohydr Polym 60(4):431–438
9. Carvajal-Millan E, Landillon V, Morel MH, Rouau X, Doublier JL, Micard V (2005) Arabinoxylan gels: impact of the feruloylation degree on their structure and properties. Biomacromolecules 6(1):309–317
10. Paës G, Burr S, Saab M-B, Molinari M, Aguié-Béghin V, Chabbert B (2013) Modeling progression of fluorescent probes in bioinspired lignocellulosic assemblies. Biomacromolecules 14(7):2196–2205
11. Paës G, Chabbert B (2012) Characterization of arabinoxylan/cellulose nanocrystals gels to investigate fluorescent probes mobility in bioinspired models of plant secondary cell wall. Biomacromolecules 13:206–214
12. Ishikawa-Ankerhold HC, Ankerhold R, Drummen GPC (2012) Advanced fluorescence microscopy techniques-FRAP, FLIP, FLAP, FRET and FLIM. Molecules 17(4):4047–4132
13. Lippincott-Schwartz J, Snapp E, Kenworthy A (2001) Studying protein dynamics in living cells. Nat Rev Mol Cell Biol 2(6):444–456
14. Paës G (2014) Fluorescent probes for exploring plant cell wall deconstruction: a review. Molecules 19(7):9380–9402
15. Paës G, von Schantz L, Ohlin M (2015) Bioinspired assemblies of plant cell wall polymers unravel affinity properties of carbohydrate-binding modules. Soft Matter 11:6586–6594
16. Fong M, Berrin JG, Paës G (2016) Investigation of the binding properties of a multi-modular GH45 cellulase using bioinspired model assemblies. Biotechnol Biofuels 9:12
17. Aguié-Béghin V, Molinari M, Hambarzymyan A, Foulon L, Habibi Y, Heim T, Blossey R, Douillard R (2009) Preparation of ordered films from cellulose nanocrystals. In: Roman M (ed) Model cellulosic surfaces, vol 1019. ACS symposium series 1019: American Chemical Society, Washington DC, USA, pp 115–136
18. Mueller F, Karpova TS, Mazza D, McNally JG (2012) Monitoring dynamic binding of chromatin proteins in vivo by fluorescence recovery after photobleaching chromatin remodeling. In: RHH M (ed) Methods in molecular biology, vol vol 833. Humana Press, New York, pp 153–176
19. Sprague BL, McNally JG (2005) FRAP analysis of binding: proper and fitting. Trends Cell Biol 15(2):84–91
20. Hermanson GT (2008) Bioconjugate techniques, 2nd edn. Elsevier, London

Chapter 14

CBMs as Probes to Explore Plant Cell Wall Heterogeneity Using Immunocytochemistry

Louise Badruna, Vincent Burlat, and Cédric Y. Montanier

Abstract

Immunocytochemistry is a widely used technique to localize antigen within intact tissues. Plant cell walls are complex matrixes of highly decorated polysaccharides and the large number of CBM families displaying specific substrate recognition reflects this complexity. The accessibility of large proteins, such as antibodies, to their cell wall epitopes may be sometimes difficult due to steric hindrance problems. Due to their smaller size, CBMs are interesting alternative probes. The aim of this chapter is to describe the use of CBM as probes to explore complex polysaccharide topochemistry *in muro* and to quantify enzymatic deconstruction.

Key words Wheat straw section, Immunocytochemistry, Enzymatic hydrolysis, Fluorescence, Microscopy

1 Introduction

The plant cell wall organization is highly complex and continuous investigation needs to be performed to explore this complexity. Indeed, biomass is a combination of cellulose, hemicelluloses, pectins, proteins, and other polymers such as lignins. During 475 million years, terrestrial plants, and more particularly plant cell wall evolved up to the point making them particularly recalcitrant to chemical or biological attack [1]. Furthermore, its composition is highly specific among species, organs, cell types, and cell wall microdomains [2, 3]. In this context, immunocytochemistry displays many advantages to allow researchers to obtain more evidence about how all these macromolecules are spatially organized and how they interact with each other. For example, Knox and coworkers used this approach to localize precisely many polymer-specific epitopes *in muro* and describe at which developmental stage pectic (1→4)-β-D-galactan appears in pea [4].

Briefly, plant tissues sections have to be prepared in a way preventing or minimizing structural changes within plant cell wall [5]. Series of monoclonal antibodies specific to plant cell wall polymers

D. Wade Abbott and Alicia Lammerts van Bueren (eds.), *Protein-Carbohydrate Interactions: Methods and Protocols*, Methods in Molecular Biology, vol. 1588, DOI 10.1007/978-1-4939-6899-2_14, © Springer Science+Business Media LLC 2017

have been developed and are available (*e.g.*, http://www.plant-probes.net/index.php; https://www.ccrc.uga.edu/~carbosource/CSS_home.html; http://www.biosupplies.com.au/price.htm). These antibodies are directed against various epitopes included within polysaccharidic polymers such as mannans, xyloglucans, xylans, rhamnogalacturonans, homogalacturonans, callose, or proteins such as extensins or arabinogalactan proteins. For instance, LM10 and LM11 that target xylans could be used to discriminate (1→4)-β-D-xylan from (1→4)-β-D-arabinoxylan, respectively [6]. An anti-rat or anti-mouse IgG labeled secondary antibody is then used against the first antibody to enhance the signal and to enable detection by microscopy [4]. The secondary antibody may display an alkaline phosphatase activity or a peroxidase activity for use with chromogenic substrates. It may also be labeled with a fluorochrome or gold particles, for fluorescent or transmission electron microscopy (TEM), respectively.

CBMs are domains which are found independently or next to an enzyme that hydrolyzes polysaccharides from plant cell walls [7]. They can be produced alone with correct folding and functions. A lot of CBMs have been found in Nature with different ligand specificities, *e.g.*, xylans [8], crystalline cellulose [9], amorphous cellulose [10], mannans [11], or starch [12].

Pioneer studies in the 1980s used the carbohydrate binding capacities of CBM contained within carbohydrate hydrolyzing enzymes by producing carbohydrate hydrolyzing enzyme-colloidal gold complexes for labeling at the TEM level [13, 14]. Then, Johansson and coworkers introduced for the first time the use of CBM-fluorochrome to probe cellulose in lignocellulose material such as birchwood or pulp fibers [15]. The family 1 CBM from *Phanerochaete chrysosporium* Cel7D was covalently labeled to fluorescein isothiocyanate (FITC) through α-amino group. Alternatively, Knox and coworkers used Histidine-Tagged CBMs as molecular probes, which introduced experimental flexibility as the CBM could be revealed with anti-HistidineTag antibodies and then with a tertiary labeled antibody. This latter approach also presents the advantage of preventing inactivation of the CBM due to the FITC labeling [16].

CBMs have advantageous properties over immunocytochemistry, being easy to produce and purify, facilitating implementation and reproducibility of their use [17]. They show affinity for their ligand and relatively high selectivity for their target in a specific context [18]. Indeed, different CBMs displaying affinity to the same ligand in vitro may target different polysaccharides in plant cell walls sections [19]. CBMs could also be used to monitor enzymatic glycoside hydrolysis activity. In the situation where CBM ligands were masked within the dense plant cell wall matrix, their accessibility could be recovered using pretreatment with enzymes such as pectate lyase Pel10A from *Cellvibrio japonicus* decloaking cellulose microfibrils on tobacco [20]. Another parameter that can be considered is the three-dimensional orientation of the CBM within the protein that may reflect the ligand context through protein/carbohydrate interactions.

Other classical cytochemical labeling methods may also be useful in combination to CBMs or antibodies or by themselves. These include carmine [21] or calcofluor white [22] that recognizes β-glucans [23], Pontamine Fast Scarlet 4B that target cellulose [24]. Toluidine Blue O that block cell wall autofluorescence may also be used for better visualization of the exogenous fluorescent labeling [23].

Sample preparation is a crucial step and the way samples are prepared may affect the way CBMs or enzymes act on plant cell wall polymers. Sample preparation should also be performed in accordance with the microscopic technique to be used. The thickness of the plant cell wall section is important and it may be adjusted in direct link to the stiffness of the sample and to the resolution required. For instance, resin is more resolutive than paraffin/wax since allowing thinner sections, but this irreversible embedding technique may hide or restrict CBM/antibody accessibility, since the labeling is basically restricted to the surface of the section. In our protocol, we used wheat straw, a particularly rigid sample difficult to cut, but we choose a smooth weak fixation associated with a reversible paraffin embedding approach, minimizing as much as possible further chemical modification of the cell wall polymers and accessibility of the probes to their targets.

Otherwise, a sample predigestion may highlight polymer recognition by antibodies or CBMs as illustrated with a xylanase (hydrolyzing β-1,4 linkages of xylans main chain) and/or a pectate lyase (hydrolyzing homogalacturonans) used on tobacco [18]. Some works are also based on an esterase (splits off acetyl groups) [25] or a mannanase (breaking down mannans) [26].

Below, we will use the family 3 CBM3a from *Clostridium thermocellum* as a probe to illustrate our protocol. This CBM is part of the multicomponent multienzyme complex, called the cellulosome [27]. The cellulosome is a discrete cell surface organelle of *C. thermocellum* which exhibits separate cellulose binding and various cellulolytic activities [28]. CBM3a is part of the scaffold protein CipA and displays the highest affinity for crystalline cellulose. This CBM is widely used in immunocytochemistry [29–31], in glycan binding assay [32] or even as a probe using atomic force microscopy [33]. However, this protocol should be applicable to other CBMs or antibodies.

2 Material

2.1 Chemicals

1. Acetic acid.
2. Absolute ethanol.

3. Tert-butanol.
4. Distilled water (dH_2O).
5. Paraplast Plus® (Sigma-Aldrich).
6. Xylene.
7. 3-Aminopropyl-triethoxysilane.
8. Agarose.
9. Nonfat dry milk (Régilait).
10. Triton X-100.
11. Tris base.
12. Sodium chloride.
13. HCl.
14. Sodium dihydrogen phosphate dehydrate.
15. Disodium hydrogen phosphate dodecahydrate.
16. Prolong antifade mounting reagent (Thermofisher Scientific).

2.2 Fixatives and Buffers

1. 1 M Tris–HCl pH 7.5: weigh 121 g of Tris and dissolve in 900 mL of dH_2O. Bring to a pH of 7.5 with 1 M HCl. Bring the solution to a final volume of 1 L.
2. 5 M NaCl: weigh 292 g of NaCl and dissolve in 1 L of dH_2O.
3. 2 M $CaCl_2$: weigh 2.22 g of $CaCl_2$ and dissolve in 10 mL of dH_2O.
4. $TTBS_{500}$ buffer: weigh 6 g of Triton X-100, add 20 mL of 1 M Tris pH 7.5, 200 mL of 5 M NaCl, 2 mL of 2 M $CaCl_2$ and bring the solution to a final volume of 2 L.
5. $TTBS_{500}$-milk: weight 10 g of nonfat dry milk (Régilait) and dissolve in 200 mL of $TTBS_{500}$.
6. AA fixative: use 500 mL of ethyl alcohol, add 50 mL of acetic acid, and bring the solution to a final volume of 1 L (*see* **Note 1**).
7. 50 mM Sodium phosphate buffer pH 7: weigh 31.21 g of $NaH_2PO_4{\cdot}2H_2O$ (sodium dihydrogen phosphate dihydrate) and dissolve in 1 L of dH_2O to prepare a 0.2 M monobasic sodium phosphate stock solution. Weigh 71.64 g of $Na_2HPO_4{\cdot}12H_2O$ (disodium hydrogen phosphate dodecahydrate) and dissolve in 1 L of dH_2O to prepare a 0.2 M dibasic sodium phosphate stock solution. Mix 97.5 mL of the monobasic sodium phosphate with 152.5 mL of the dibasic sodium phosphate and check the pH. Add 750 mL of dH_2O.
8. Solution with 1% agarose: weigh 0.1 g of agarose and dissolve in 10 mL of dH_2O.

2.3 Biological Material

1. Wheat straw internodes [34].
2. Commercial enzymatic cocktail Ctec2 (Novozymes).
3. CBM3a (homemade production with an Histidine-Tag (*see* **Note 2**)).
4. Mouse monoclonal anti-Histidine Tag antibody (A00186-100, GenScript)
5. Goat anti-mouseIgG-FITC antibodies (Sigma, F2012).

2.4 Sample Preparation Materials

1. Dry oven.
2. Vacuum chamber.
3. Metallic embedding base molds (38 × 25 × 12 mm, Tissue-Tek 4133; Delta Microscopies, Mauressac, France).
4. White embedding ring adaptable to the above metallic molds.
5. Rotary microtome (Seit 1872 Jung AG Microtome).
6. Tissue Floating Bath (Premiere XH-1003, Delta Microscopies, Mauressac, France).
7. Large slide warmer (slide warmer, Labscientific XH-2001, Delta Microscopies, Mauressac, France).
8. Starfrort microscopy slides.
9. Polypropylene staining racks for 20 slides (Kartell polypropylene staining racks 354; Ref 391058 Dutscher SAS, Brumath, France).
10. Coverslips (24 × 50 mm).
11. Humid chamber (*see* **Note 3**).
12. Microincubation chamber (22 × 40 × 0.2 mm deep; (200 μL), #70324-20, Electron Microscopy Science).
13. Nanozoomer RS slide scanner (Hamamatsu).

2.5 Computer Software

1. ImageJ [35].
2. NDP scan (Hamamatsu, Japan).
3. NDP.view2 (Hamamatsu, Japan).

3 Methods

3.1 Plant Material Fixation

1. Submerge samples (about 1-cm-long wheat straw sections) in 20–30 mL of AA (*see* **Note 1**).
2. Do five cycles of vacuum infiltration (1 min vacuum and vacuum release in a vacuum chamber).
3. Incubate samples for 8 or 16 h at 4 °C in AA.
4. Wash four times in a solution containing 500 mL of absolute ethanol and 500 mL of water for 10 min each (*see* **Note 4**).

5. Dehydrate progressively in a gradient series of dH_2O/absolute ethanol/tert-butanol for 1 h each, at room temperature (adapted from [36]). Solutions are respectively composed of:
   (a) 50 mL of dH_2O, 40 mL of absolute ethanol, and 10 mL of tert-butanol.
   (b) 30 mL of dH_2O, 50 mL of absolute ethanol, and 20 mL of tert-butanol.
   (c) 15 mL of dH_2O, 50 mL of absolute ethanol, and 35 mL of tert-butanol.
   (d) 45 mL of absolute ethanol and 55 mL of tert-butanol.
   (e) 25 mL of absolute ethanol and 75 mL of tert-butanol.
6. Incubate in pure tert-butanol overnight at room temperature above 25 °C.
7. Change the tert-butanol solution and incubate for 1 h.
8. Replace half of the tert-butanol solution by melted Paraplast Plus and incubate in a dry oven at 60 °C for 5 h.
9. Renew with fresh solution containing 50 mL of tert-butanol and 50 mL of Paraplast Plus and incubate overnight at 60 °C.
10. Change the Paraplast Plus solution, once or twice per day, for 2 or 3 days (*see* **Notes 5** and **6**).

3.2 ParaplastPlus Embedding

1. Set the slide warmer at 62 °C and protect it with aluminum foil.
2. Place melted Paraplast Plus into metallic embedding base molds at 62 °C.
3. In parallel, fill Petri dishes with samples in melted Paraplast Plus at 62 °C.
4. Use preheated metallic spatula and tweezers to dispose the sample in the molds allowing further desired section orientation (*e.g.*, cross-section).
5. Cover each mold by a white embedding ring enabling further positioning of the paraffin bloc to the microtome harm.
6. Write the reference on each embedding ring using a pencil to avoid mark disappearing in **step 9**.
7. Incubate for at least 1 h at 4 °C to obtain the Paraplast Plus solidification.
8. Separate samples and molds (*see* **Notes 7** and **10**).
9. Immerse for 24–48 h in a solution containing 100 mL of acetic acid and 900 mL of absolute ethanol at 4 °C [37]

3.3 Microtomy

3.3.1 Microscopy Slides Coating Pretreatment

1. Place Starfrost microscopy slides in polypropylene staining racks for 20 slides.
2. Dip the rack for 5 min in 200 mL of freshly prepared solution containing 4 mL of 3-aminopropyltriethoxysilane and 196 mL of acetone within a glass-staining jar disposed under a fume hood.

3. Air-dry for 15 min under the fume hood.
4. Rinse the slides sequentially in three baths of dH_2O (200 mL) in three glass staining jars disposed under a fume hood.
5. Dry overnight at 37 °C (*see* **Note 8**).

3.3.2 Microtomy

1. Annotate slides (*see* **Note 9**).
2. Set the Tissue Floating Bath at 36 °C.
3. Cut serial sections of 10–20 μm with the rotary microtome (*see* **Notes 10–12**).
4. Align section ribbon on a kimwipe and carefully individualize the sections with a needle and a paintbrush.
5. Place a section in the dH_2O bath using a spatula (*see* **Note 13**).
6. Wait for about 2 min to flatten the sections (*see* **Note 14**).
7. Immerse a slide in the dH_2O with an angle of 45°.
8. Use a paintbrush to position and maintain the floating section in front of the slide and when there is a contact between the slide and the section, maintain the section position with the paintbrush while removing the slide from the dH_2O.
9. Place each section on the same orientation and on the same position on each slide (*see* **Notes 15** and **16**).
10. Dry slides overnight at 35 °C on the slide warmer to allow full adherence of the section to the slide.

3.4 Deparaffinization

Slides are handled in polypropylene staining racks for 20 slides fitting to glass-staining jars each containing around 200 mL of solution. Figure 1 shows a slide with a cross-section of wheat straw internodes ready for a labeling following paraffin solubilization.

1. Immerse slides in pure xylene for 10 min to solubilize paraffin.
2. Repeat this step in the second bath with the same conditions.
3. Sequentially rehydrate slides for 5 min in a solution with
   (a) 200 mL of pure ethanol (twice).
   (b) 200 mL of ethanol 95%.
   (c) 140 mL of ethanol and 60 mL of dH_2O.
   (d) 100 mL of ethanol and 100 mL of dH_2O.
4. Rinse slides in dH_2O, two times 5 min.
5. Dry slides on the slide warmer at 40 °C.

3.5 Digestion

1. Digest section with commercial enzymatic cocktail, Ctec2 diluted at 1/20 (200 μL) or sodium phosphate buffer 50 mM at pH 7 as a control (*see* **Notes 17–19**).
2. Incubate in a dH_2O saturated humid chamber placed in an oven at 50 °C for 1 h.
3. Wash in fresh $TTBS_{500}$ for 5 min, two times.

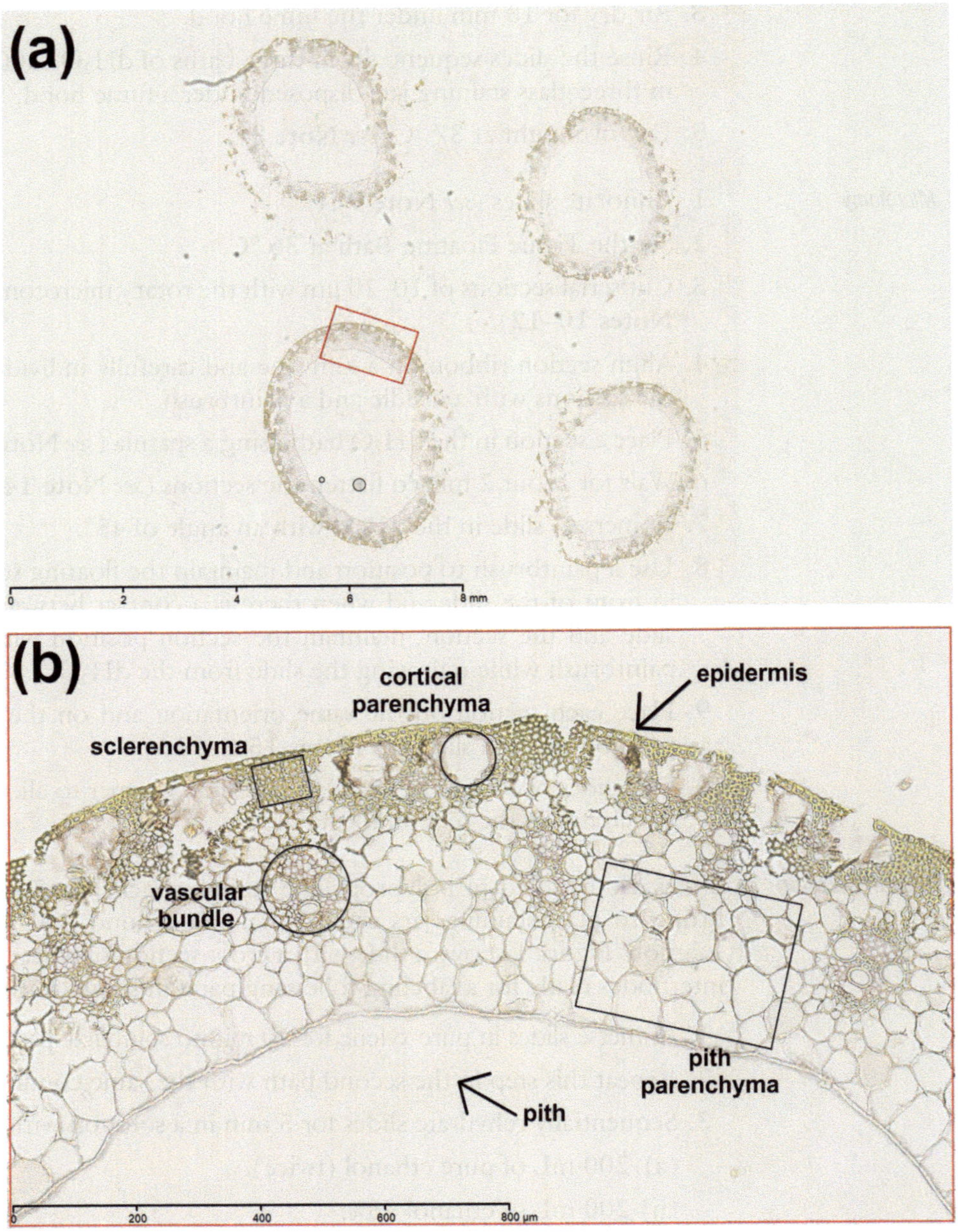

Fig. 1 Wheat straw internodes cross-section morphology, taken with Bright-field microscopy. (**a**) Overview of the tissue-array comprising 4 wheat straw internodes cross-sections. The red frame represents the zoom used for following image. (**b**) Main anatomical tissues from a wheat straw cross-section

3.6 Section Labeling and Visualization

1. Block nonspecific interactions with $TTBS_{500}$-milk for 30 min at room temperature (*see* **Note 20**).
2. Incubate the sample with CBM3a with an Histidine-Tag diluted at 3 μg/mL in $TTBS_{500}$-milk for 1–2 h at room temperature (100 μL) (*see* **Notes 21–23**).
3. Carefully remove coverslips (*see* **Note 24**).

4. Wash with 200 mL of $TTBS_{500}$ for 5 min, six times.
5. Incubate with a secondary antibody Anti-Histidine diluted at 1/100 in $TTBS_{500}$-milk for 1 h at room temperature (100 μL).
6. Process as in **step 3**.
7. Process as in **step 4**.
8. Add the tertiary antibody, Anti-mouse IgG-FITC diluted at 1/100 in $TTBS_{500}$-milk for 1 h at room temperature protected from light (100 μL).
9. Process as in **step 3**.
10. Wash with 200 mL of $TTBS_{500}$ for 5 min, three times.
11. Wash with 200 mL of dH_2O, quickly, six times to remove the detergent.
12. Dry on the slide warmer at 40 °C.
13. Mount in Prolong Antifade (add one drop per slide and add a coverslip).
14. Allow the mounting media to spread under the coverslip on the slide warmer at 40 °C.

    FITC detection can be done by epifluorescence or confocal microscopy. A slide scanner allows a higher throughput.
15. Place each slide on a Nanozoomer RS (slides scanner).
16. Set each parameter (FITC detection: 473–491 nm excitation filter set; 488 nm dichroic mirror; 510–540 nm emission filter set, DAPI detection: 381–393 nm excitation filter set; 405 nm dichroic mirror; 420–460 nm emission filter set, bright field, color balance) by testing with positive and negative control slides.
17. Scan and save images.

3.7 Section Analysis

The following analyses are possible on the images using the NDP. view2 freeware. A qualitative protocol consists in a comparison between series of images taken from serial sections synchronized on the same zone on the computer screen: with/without the first probe to estimate the background level and with/without an enzymatic treatment to estimate the enzymatic action. A quantitative protocol can be to count cells before and after an enzymatic treatment. The next step is an analysis to quantify the relative fluorescence intensity with ImageJ on selected areas, *e.g.*, pith parenchyma and sclerenchyma/cortical parenchyma/epidermis (*see* **Note 25**). The protocol is as follows.

1. Open images to be compared with NDP.view2.
2. Determine and apply the same exact parameters for the images to be analyzed (color balance, brightness, contrast, gamma, and so on).
3. Zoom on an area of interest.

4. Align, synchronize, and frame each image on the same area for serial sections (*see* **Notes 26** and **27**).
5. Save the images of this area.
6. Open first an image with ImageJ.
7. Split channels (red, green, blue).
8. Choose a reference channel.
9. Set a threshold with triangle method (*see* **Note 28**).
10. Measure the intensity.
11. Repeat **steps 6–10** for all images.
12. Save and plot data (Microsoft Excel file).
13. Compare data.

As an example, three wheat straw serial cross-sections individually treated for three conditions are illustrated in Fig. 2 and the quantified results of mean for at least two repeats of similar areas are represented in Fig. 3. Maximum fluorescence was obtained for Fig. 2b and the lowest fluorescence was for Fig. 2c allowing estimating the nonspecific background due to autofluorescence. The relative action of the enzymatic cocktail can therefore be selectively quantified for various tissues on the same sections.

This method and the presented examples clearly illustrate the possibilities to measure relative fluorescence intensities for different tissues found on the same section and to compare these with the same zone differentially treated on serial sections. This example shows that the enzymatic action is more pronounced in the sclerenchyma/cortical parenchyma/epidermis zone as compared to the pith parenchyma.

4 Notes

1. This fixative is adapted from the classical FAA fixative deprived from the 3.7% (v/v) cross-linking formaldehyde. FAA (100 mL of Formalin (37 mL of formaldehyde Sigma-Aldrich F1635 and 63 mL of dH_2O), 500 mL of ethyl alcohol, 50 mL of acetic acid, 350 mL of dH_2O could be used instead of AA. However, formaldehyde may prevent enzymatic pretreatment degradation and/or CBM recognition as it is a covalent cross-linker and could disrupt polymers.
2. CBM3a homemade production was done by the authors as in [20] with the Histidine-Tag at the *C*-terminal end of the construct. This CBM is also commercially available (Plant Probes or CZ00411 on NZytech).

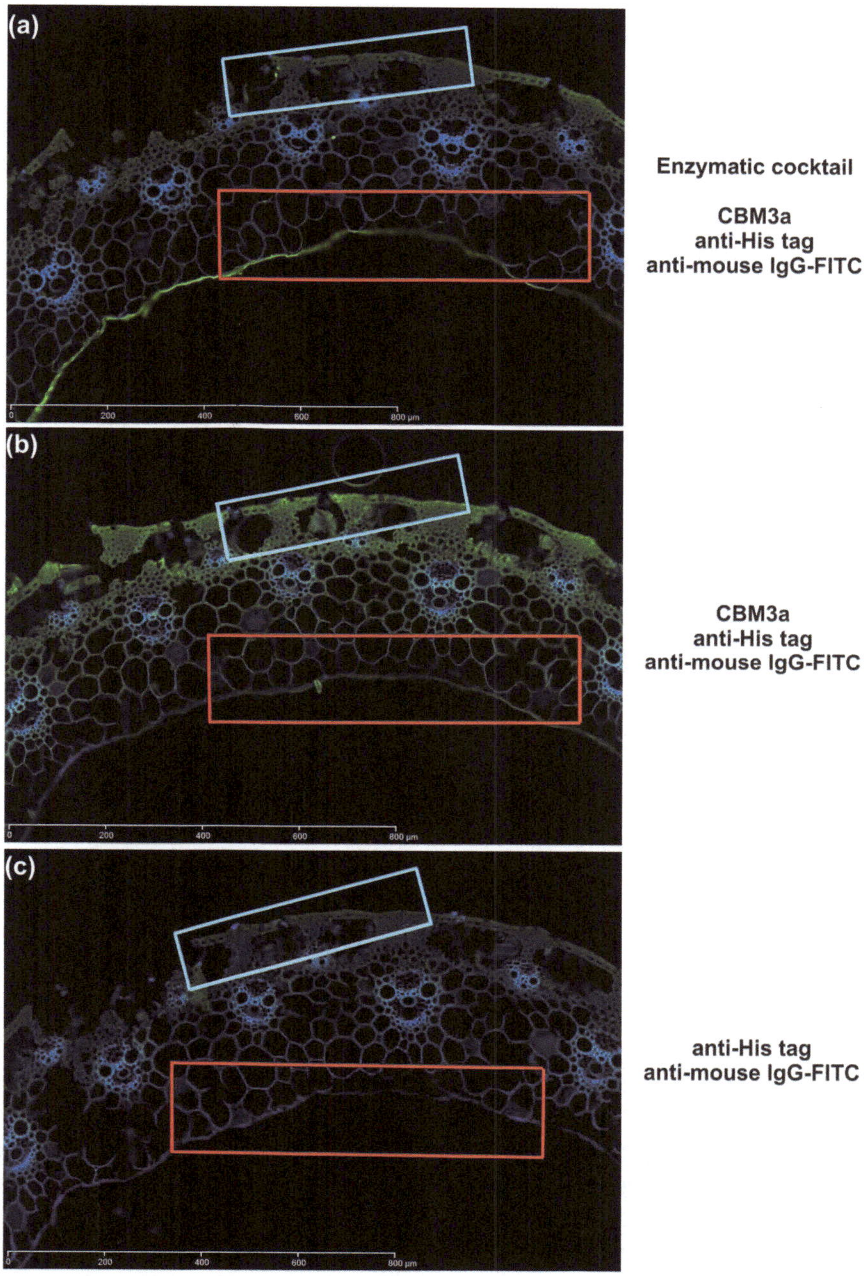

Fig. 2 Zoom on the same exact zone of three serial sections treated as described in the right panel. The frames correspond to examples of the zones used for image analysis quantified in Fig. 3: red for pith parenchyma and blue for a zone encompassing sclerenchyma/cortical parenchyma/epidermis. The crystalline cellulose detected by CBM3a is displayed in green (FITC channel). The blue channel shows the lignin autofluorescence used as a control (DAPI channel) to check the structure preservation

3. Humid chamber can be bought or homemade chamber can be assembled with large Petri dishes and two lines of broken Pasteur pipets taped at the bottom to accommodate the slides, preventing them from being flooded by the dH_2O placed at the bottom of the Petri dish.

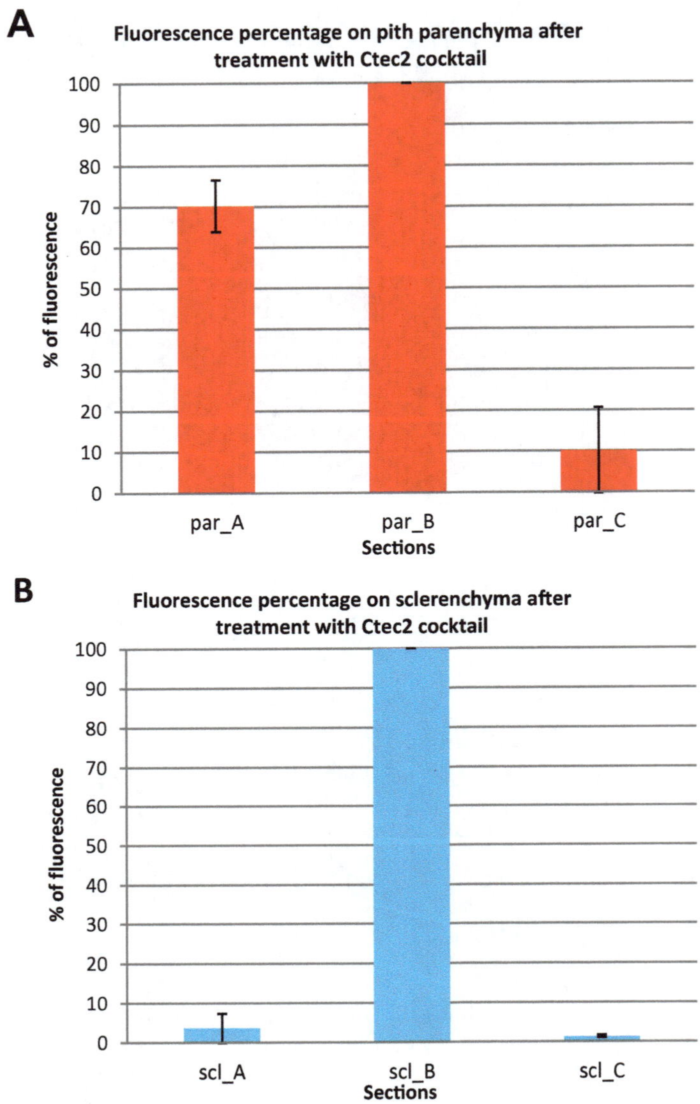

Fig. 3 Relative fluorescence quantification for two anatomical regions from images exemplified in Fig. 2. (**a**) The mean of two frames of similar anatomical views taken on one wheat straw cross-sections for pith parenchyma area ± standard deviation is shown.par_A, par_B and par_C refer to the zones exemplified in Fig. 2a–c, respectively. (**b**) The mean of two frames of similar anatomical views taken on two wheat straw cross-sections for sclerenchyma, cortical parenchyma and epidermis area ± standard deviation is shown. scl_A, scl_B and scl_C refer to the zone exemplified in Fig. 2a–c, respectively

4. Samples can be treated directly or stored for months at 4 °C in a solution containing 10 mL of ethanol and 10 mL of dH_2O.
5. Take off the cap from the tubes for these incubations to allow the evaporation of the residual tert-butanol.

6. The tissue embedding can be realized directly or Paraplast infiltrated tissues may be stored with cap at 4 °C for years. In this case, on restart, samples need to be incubated for 1 day at 60 °C before performing the tissue embedding.
7. The separation between metallic embedding base molds and samples could be done directly or after storage at 4 °C for years.
8. Slides after treatment can be stored at room temperature for months.
9. Use a pencil because pen marks would disappear with solvent.
10. The soaking of the Paraplast embedded bloc in a solution of acetic acid and ethanol (step A-2-9) for few weeks may improve sectioning efficiency for hard tissues.
11. Sample sectioning is easier when performed directly with cooled samples directly following taking out these from 4 °C.
12. Keep the same thickness for all sections to allow a comparison of labeling intensities between serial sections. 10 μm was chosen for wheat straw internodes because it is a relatively hard material.
13. Use this dH_2O bath for only four sections in parallel to keep the order of serial sections.
14. The incubation is material dependent, for some samples 20 s could be enough.
15. Be careful about the serial sections positioning on the sequential slides to be able to further compare the same area on serial sections treated with different conditions.
16. The use of tissue floating bath is useful, especially for breakable tissues since it avoids excessive volumes of dH_2O between the section and the slide. However, another method without such a bath exists: degas dH_2O in a vacuum chamber for at least 1 h to avoid air bubble formation. Annotate slides. Place on a slide warmer at 35 °C. Add 2–3 mL of degassed dH_2O. Cut serial sections of 10–20 μm with the rotary microtome. Separate sections. Place them on the same orientation and on the same position on each slide. Remove dH_2O in excess with a pipette. Dry overnight at 35 °C on the slide warmer.
17. To find reaction conditions, do preliminary tests with a range of concentrations for example. The aim is to find an intermediate concentration enabling sufficient digestion to observe changes and avoiding excessive digestion that could result in section detachment from the slide.
18. Use a micro-incubation chamber with a silicone seal to avoid the solution evaporation.
19. To improve micro-incubator use, cut a thin opening (1 mm wide) in the silicone. Firmly press this 200 μL micro-incubation

chamber on the top of the section area and distribute 200 μL through this opening. Finally, close it by adding 10–20 μL of 1% agarose in fusion (around 40 °C). After the incubation, remove agarose with the use of a razor blade. Such micro-incubation chambers may be washed and reused.

20. Sections should not dry during the labeling process, since this could result in excessive unspecific background. Insure to add the next solution at each step just after the removal of the previous one. For blocking and washing steps, slides are commonly handled in polypropylene staining racks for 20 slides fitting to glass staining jars each containing around 200 mL of solution. For incubation steps, slides are handled individually within a humid chamber with smaller volumes (100–200 μL with a coverslip on each slide to avoid the evaporation).
21. Samples may be labeled in several ways. For example, the first step can be done using antibody directed against cell wall polymer (e.g., rat monoclonal LM11 antibody from Plant Probes, which binds to unsubstituted β-1,4 xylans or arabinoxylans displaying a low degree of arabinose substitutions). The second step consists of using a secondary antibody. This one could be the Anti-RatIgG-FITC (Sigma) or the Anti-rat IgG-alkaline phosphatase (AP) conjugates. In the last case, 5-bromo-4-chloro-3-indolyl-phosphate (BCIP)/nitroblue tetrazolium (NBT) is used as the AP substrate. The enzymatic activity will release a blue to purple insoluble NBT diformazan end product (some washes with dH_2O will stop the reaction).
22. Each CBM probe has to be chosen regarding the target of interest within the plant section. We choose to work with a monoclonal secondary antibody that recognizes specifically the Histidine-Tag from the CBM. The tertiary species-specific labeled antibody will target the secondary antibody. In each experiment, to have details on the background, controls need to be performed with the same labeling except the first probe.
23. Usually, CBMs are used at a concentration comprised between 10 and 100 μg/mL [23] but dilutions should be tested to optimize the signal/background ratio.
24. Remove coverslips gently after each antibody incubation to avoid a section tearing. For this, allow the coverslip to fall by itself or dip the slide in a $TTBS_{500}$ bath to facilitate coverslip fall over. Do not remove the coverslip by hand since the applied pressure could result in section damage.
25. Many samples can be accommodated within a slide as in Fig. 1a, constituting so-called tissue microarrays [38]. This multiplication of samples secures final analysis if one sample is lost from the slide or damaged during the labeling process, and this allows comparing more samples (*e.g.*, mean ± standard deviation,

various developmental stages, organs, or species) between treatment/labeling conditions.

26. This step is very important. To compare each treatment or labeling, sections of interest need to be aligned very precisely on all directions to compare the exact same zone on serial sections differentially treated.
27. NDP.view2 has a lot of functions (zoom, scale bar addition, work with only one channel, exportation of the image for example). It allows annotations (rectangle to delimit the area where the work is done, color) and a comment (sample number for example) to avoid working two times on the same area.
28. The threshold will delimit the fluorescence quantification. The image without the first probe can be used to set the threshold.

Acknowledgments

We acknowledge Alain Jauneau and Aurelie Le Ru for technical assistance on Nanozoomer RS provided by the TRI-genotoul facility (http://trigenotoul.com/).

References

1. Himmel ME, Ding S-Y, Johnson DK, Adney WS, Nimlos MR, Brady JW et al (2007) Biomass recalcitrance: engineering plants and enzymes for biofuels production. Science 315:804–807
2. Velickovic D, Ropartz D, Guillon F, Saulnier L, Rogniaux H (2014) New insights into the structural and spatial variability of cell-wall polysaccharides during wheat grain development, as revealed through MALDI mass spectrometry imaging. J Exp Bot 65(8):2079–2091
3. Scheller HV, Ulvskov P (2010) Hemicelluloses. Annu Rev Plant Biol 61(1):263–289
4. McCartney L, Ormerod AP, Gidley MJ, Knox JP (2000) Temporal and spatial regulation of pectic (1-4)-beta-D-galactan in cell walls of developing pea cotyledons: implications for mechanical properties. Plant J 22(2):105–113
5. Freshour G, Clay RP, Fuller MS, Albersheim P, Darvill AG, Hahn MG (1996) Developmental and tissue-specific structural alterations of the cell-wall polysaccharides of Arabidopsis thaliana roots. Plant Physiol 110(4):1413–1429
6. McCartney L, Marcus SE, Knox JP (2005 Apr) Monoclonal antibodies to plant cell wall xylans and arabinoxylans. J Histochem Cytochem 53(4):543–546
7. Boraston AB, Bolam DN, Gilbert HJ, Davies GJ (2004) Carbohydrate-binding modules: fine-tuning polysaccharide recognition. Biochem J 781:769–781
8. Black GW, Hazlewood GP, Millward-Sadler SJ, Laurie JI, Gilbert HJ (1995) A modular xylanase containing a novel non-catalytic xylan-specific binding domain. Biochem J 307(1):191–195
9. Van Tilbeurgh H, Tomme P, Claeyssens M, Bhikhabhai R, Petterson G (1986) Limited proteolysis of the cellobiohydrolase I from *Trichoderma reesei*. FEBS Lett 204(2): 223–227
10. Coutinho JB, Gilkes NR, Warren RAJ, Kilburn DG, Miller RC (1992) The binding of *Cellulomonas fimi* endoglucanase C (CenC) to cellulose and Sephadex is mediated by the N-terminal repeats. Mol Microbiol 6(9): 1243–1252
11. Stoll D, Boraston A, Stålbrand H, McLean BW, Kilburn DG, Warren RAJ (2000) Mannanase Man26A from *Cellulomonas fimi* has a mannan-binding module. FEMS Microbiol Lett 183(2):265–269
12. Sorimachi K, Le Gal-Coëffet M-F, Williamson G, Archer DB, Williamson MP (1997) Solution structure of the granular starch binding domain

of *Aspergillus niger* glucoamylase bound to β-cyclodextrin. Structure 5(5):647–661

13. Ruel K, Joseleau JP (1984) Use of enzyme-gold complexes for the ultrastructural localization of hemicelluloses in the plant cell wall. Histochemistry 81(6):573–580
14. Bendayan M, Banhamou N (1987) Ultrastructural localization of glucoside residues on tissue sections by applying the enzyme-gold approach. J Histochem Cytochem 35(10): 1149–1155
15. Hilden L, Daniel G, Johansson G (2003) Use of a fluorescence labelled, carbohydrate-binding module from *Phanerochaete chrysosporium* Cel7D for studying wood cell wall ultrastructure. Biotechnol Lett 25:553–558
16. McCartney L, Gilbert HJ, Bolam DN, Boraston AB, Knox JP (2004) Glycoside hydrolase carbohydrate-binding modules as molecular probes for the analysis of plant cell wall polymers. Anal Biochem 326:49–54
17. Tomme P, Boraston A, McLean B, Kormos J, Creagh AL, Sturch K et al (1998) Characterization and affinity applications of cellulose-binding. J Chromatogr 715:283–296
18. Hervé C, Rogowski A, Gilbert HJ, Knox JP (2009) Enzymatic treatments reveal differential capacities for xylan recognition and degradation in primary and secondary plant cell walls. Plant J 58(3):413–422
19. McCartney L, Blake AW, Flint J, Bolam DN, Boraston AB, Gilbert HJ et al (2006) Differential recognition of plant cell walls by microbial xylan-specific carbohydrate-binding modules. Proc Natl Acad Sci U S A 103(12):4765–4770
20. Blake AW, McCartney L, Flint JE, Bolam DN, Boraston AB, Gilbert HJ et al (2006 Sep) Understanding the biological rationale for the diversity of cellulose-directed carbohydrate-binding modules in prokaryotic enzymes. J Biol Chem 281:29321–29329
21. Mondolot L, Roussel JL, Andary C (2001) New applications for an old lignified element staining reagent. Histochem J 33(7):379–385
22. Sauter M, Seagull RW, Kende H (1993) Internodal elongation and orientation of cellulose microfibrils and microtubules in deep-water rice. Planta 190(3):354–362
23. Knox JP (2012) *In situ* detection of cellulose with carbohydrate-binding modules. In: Methods in Enzymology, 1st edn. Elsevier Inc., Amsterdam, pp 233–245
24. Hoch HC, Galvani CD, Szarowski DH, Turner JN (2005) Two new fluorescent dyes applicable for visualization of fungal cell walls. Mycologia 97(3):580–588
25. Marcus SE, Blake AW, Benians TAS, Lee KJD, Poyser C, Donaldson L et al (2010 Oct) Restricted access of proteins to mannan polysaccharides in intact plant cell walls. Plant J 64:191–203
26. Ordaz-Ortiz JJ, Marcus SE, Knox PJ (2009) Cell wall microstructure analysis implicates hemicellulose polysaccharides in cell adhesion in tomato fruit pericarp parenchyma. Mol Plant 2(5):910–921
27. Lamed R, Setter E, Bayer EA (1983) Characterization of a cellulose-binding, cellulase-containing complex in *Clostridium thermocellum*. J Bacteriol 156(2):828–836
28. Shoham Y, Lamed R, Bayer EA (1999) The cellulosome concept as an efficient microbial strategy for the degradation of insoluble polysaccharides. Trends Microbiol 7(7):275–281
29. Leroux O, Leroux F, Bagniewska-Zadworna A, Knox JP, Claeys M, Bals S et al (2011) Ultrastructure and composition of cell wall appositions in the roots of *Asplenium* (Polypodiales). Micron 42(8):863–870
30. Hervé C, Rogowski A, Blake AW, Marcus SE, Gilbert HJ, Knox JP (2010) Carbohydrate-binding modules promote the enzymatic deconstruction of intact plant cell walls by targeting and proximity effects. Proc Natl Acad Sci U S A 107(34):15293–15298
31. Matos DA, Whitney IP, Harrington MJ, Hazen SP (2013) Cell walls and the developmental anatomy of the *Brachypodium distachyon* stem internode. PLoS One 8(11):1–9
32. Moller I, Sørensen I, Bernal AJ, Blaukopf C, Lee K, Øbro J et al (2007) High-throughput mapping of cell-wall polymers within and between plants using novel microarrays. Plant J 50:1118–1128
33. Zhang M, Wang B, Xu B (2013) Measurements of single molecular affinity interactions between carbohydrate-binding modules and crystalline cellulose fibrils. Phys Chem Chem Phys 15:6508–6515
34. Bertrand I, Chabbert B, Kurek B, Recous S (2006) Can the biochemical features and histology of wheat residues explain their decomposition in soil? Plant and Soil 281(1–2):291–307
35. Schneider CA, Rasband WS, Eliceiri KW (2012) NIH image to ImageJ: 25 years of image analysis. Nat Methods 9(7):671–675
36. Burlat V, Oudin A, Courtois M, Rideau M, St-Pierre B (2004 Apr) Co-expression of three MEP pathway genes and geraniol 10-hydroxylase in internal phloem parenchyma of *Catharanthus roseus* implicates mul-

ticellular translocation of intermediates during the biosynthesis of monoterpene indole alkaloids and isoprenoid-derive. Plant J 38:131–141

37. Ruzin BSE (1999) Plant microtechnique and microscopy. Oxford Univ Press, Oxford

38. Francoz E, Ranocha P, Pernot C, Le Ru A, Pacquit V, Dunand C et al (2016) Complementarity of medium- throughput in situ RNA hybridization and tissue-specific transcriptomics : case study of *Arabidopsis* seed development kinetics. Sci Rep 6:24644

Chapter 15

Determining the Localization of Carbohydrate Active Enzymes Within Gram-Negative Bacteria

Richard McLean, G. Douglas Inglis, Steven C. Mosimann, Richard R.E. Uwiera, and D. Wade Abbott

Abstract

Investigating the subcellular location of secreted proteins is valuable for illuminating their biological function. Although several bioinformatics programs currently exist to predict the destination of a trafficked protein using its signal peptide sequence, these programs have limited accuracy and often require experimental validation. Here, we present a systematic method to fractionate gram-negative cells and characterize the subcellular localization of secreted carbohydrate active enzymes (CAZymes). This method involves four parallel approaches that reveal the relative abundance of protein within the cytoplasm, periplasm, outer membrane, and extracellular environment. Cytoplasmic and periplasmic proteins are fractionated by lysis and osmotic shock, respectively. Outer membrane bound proteins are determined by comparing cells before and after exoproteolytic digestion. Extracellularly secreted proteins are collected from the media and concentrated. These four different fractionations can then be probed for the presence and quantity of target proteins using immunochemical methods such as Western blots and ELISAs, or enzyme activity assays.

Key words Signal peptideSecretion Subcellular localization Cell fractionation Osmotic shock Whole cell dot blot Gram-negative bacteria

1 Introduction

Signal peptides are short amino acid sequences appended to the N-terminus, C-terminus, or within the internal sequence of a protein [1]. They are employed across all domains of life and serve to signal the subcellular localization of secreted proteins [2, 3]. Accurate subcellular trafficking of proteins is essential for the function of many different classes of proteins, including virulence factors, transporters, and catabolic enzymes. Thus, determining the subcellular destination of a protein can help to elucidate its function. In gram-negative cells, signal peptides traffic proteins into the periplasm, onto the surface of or into cellular membranes, or into the local environment surrounding the cell. Conventionally, proteins that lack signal peptides are not secreted and are destined

D. Wade Abbott and Alicia Lammerts van Bueren (eds.), *Protein-Carbohydrate Interactions: Methods and Protocols*, Methods in Molecular Biology, vol. 1588, DOI 10.1007/978-1-4939-6899-2_15, © Springer Science+Business Media LLC 2017

to function within the cytoplasm [4] (Fig. 1a). Although several bioinformatics programs exist to predict the destination of a secreted protein from signal peptide sequence, they have limited accuracy and often require experimental validation [5]. Established protocols for the lysis of gram-negative bacteria are diverse [6–8]; however, protocols for accurately fractionating the periplasmic contents from the cytoplasmic contents are less common. Most periplasmic protein protocols aim to maximize yields [9, 10], which may result in contamination by cytoplasmic proteins. Here, we present an optimized osmotic shock protocol adapted from [9, 10] which improves the purity of the periplasmic fraction by minimizing contamination by cytoplasmic proteins (Fig. 1b). The remaining cytoplasmic compartment is subsequently lysed to extract the cytoplasmic contents.

Most established protocols for the purification of membrane bound CAZymes are performed to extract functional membrane proteins. This is effective; however, it is time consuming and requires expensive equipment such as ultracentrifuges and complex density gradients [4]. The protocol presented here, adapted from [11–13], has been used for the identification of surface-exposed CAZymes without purification (Fig. 1b). Cells are incubated in the presence or absence of Proteinase K and blotted onto PVDF membrane. The dots are then bound by specific antibodies to determine if the protein signal was removed in the proteolytically treated sample (Fig. 2). Below we provide a detailed description of how these methods are performed and how they can be used to enable accurate comparison between periplasmic and cytoplasmic trafficking of a target CAZyme.

2 Materials

Prepare all solutions using deionized water (dH_2O) and analytical grade reagents. The following reagents are adequate for most species of gram-negative bacteria; however, optimization may be required for the osmotic shock step (*see* Subheading 2.3).

2.1 Protein Expression

1. *E. coli* BL21 DE3 (*see* **Note 1**).
2. pET-21b Vector (Novagen 69741-3) (*see* **Note 2**).
3. 1.0 M Isopropyl β-D-1-thiogalactopyranoside (IPTG). Add 2.38 g IPTG to 9 mL of dH_2O and dissolve. Adjust total volume to 10 mL.
4. Lysogeny Broth (LB): Add 5 g yeast extract, 10 g Bacto-tryptone, and 10 g NaCl to 900 mL of dH_2O and dissolve. Adjust the total volume to 1 L and sterilize in an autoclave.

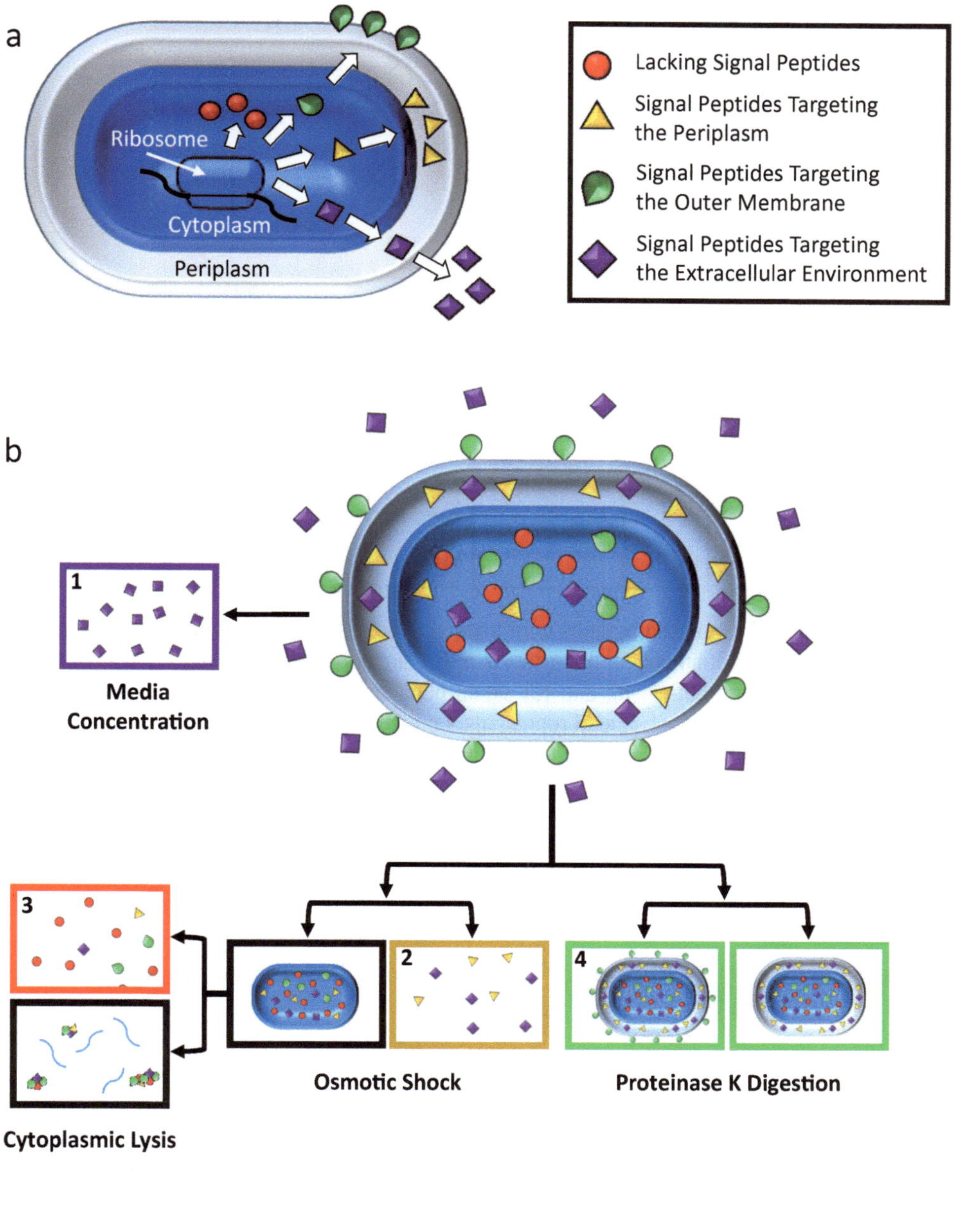

Fig. 1 Schematic work flow for fractionation of secreted proteins in a gram-negative cell. **(a)** General structure of a gram-negative cell. Proteins are synthesized in the cytoplasm, where they can remain (*red circles*) or be targeted to the periplasm (*yellow triangles*), outer membrane surface (*green teardrop*), or the environment (*purple diamonds*). **(b)** Cell fractionation protocol: (1) Spent media is removed from the culture and concentrated allowing for the identification of secreted proteins (*purple box*), (2) Osmotic shock fractionates the periplasmic contents (*yellow box*) and leaves the inner membrane intact, (3) Cytoplasmic compartment remaining after osmotic shock is lysed into soluble cytoplasmic (*red box*) and insoluble fractions, (4) Proteinase K digestion followed by full cell dot blot allows for the identification of proteins on the cell surface (*green boxes*). The purple panel represents the fraction containing secreted proteins, yellow panel represents periplasmic proteins, red panel represents cytoplasmic proteins, and green panels were used to verify surface-exposed proteins

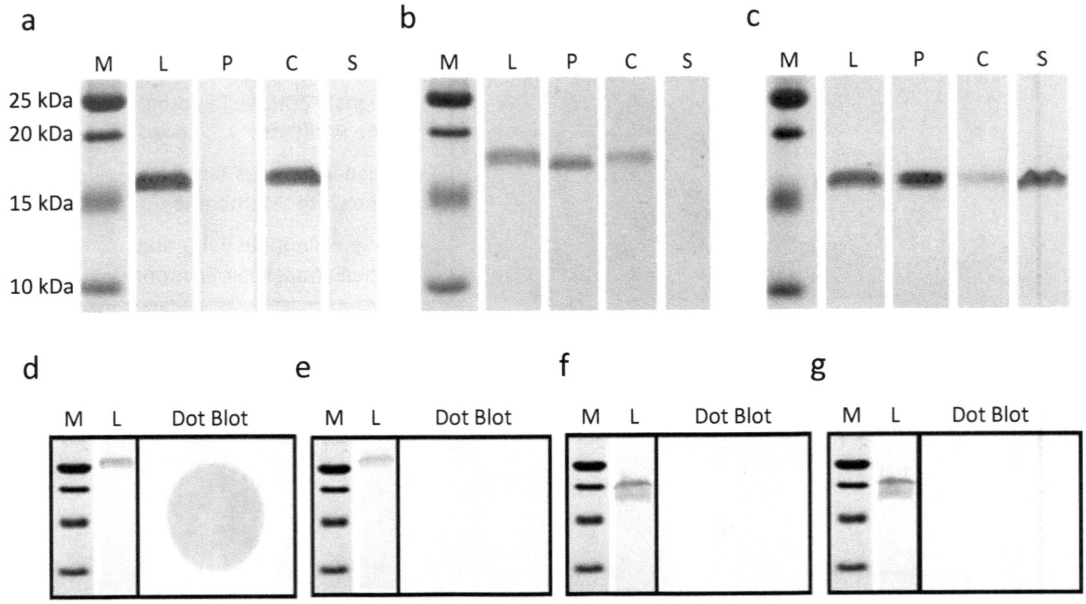

Fig. 2 Immunodetection of expressed proteins from fractionated cells. Western blots of constructs: **(a)** lacking a signal peptide, showing protein only in the cytoplasm; **(b)** signal peptide targeting the periplasm, showing protein in the cytoplasm and periplasm; and **(c)** signal peptide targeting the extracellular space, showing protein in all fractions. In each panel, lane M represents a protein size standard, L is the complete cell lysate, P is the periplasmic fraction, C is the cytoplasmic fraction, and S is the media. Dot blots of live cells expressing protein: **(d)** targeted to the outer membrane; **(e)** targeted to the outer membrane after Proteinase K digestion; **(f)** lacking a signal peptide; and **(g)** lacking a signal peptide, after Proteinase K digestion. In each panel, M represents a protein size standard, L is a complete cell lysate demonstrating that intracellular protein is intact, and the dot blot shows whether or not proteins are present of the surface of the cell

2.2 Cell Lysis

1. BugBuster® Primary Amine-Free Extraction Reagent (Novagen 70923) (*see* **Note 3**).

2.3 Osmotic Shock

1. 10% (w/v) Sucrose, 20 mM Tris pH 8.0, 1 mM EDTA (STE Solution). Add 100 g of sucrose and 292 mg of EDTA to ~800 mL of dH_2O and dissolve. Add 20 mL of 1 M Tris pH 8.0 and cool to 4 °C. Ensure the pH is 8.0 and adjust the total volume 1 L (*see* **Note 4**). Store at 4 °C.
2. 100 mM $MgCl_2$. Add 9.52 g of $MgCl_2$ to ~900 mL dH_2O and dissolve. Adjust volume to 1 L (*see* **Note 4**). Store at room temperature.

2.4 Whole Cell Dot Blot

1. Tris Buffered Saline (TBS Solution). Add 8.76 g NaCl to 800 mL dH_2O and dissolve. Add 5 mL of 1.0 M Tris pH 8.0. Ensure the pH is 8.0 and adjust the total volume 1 L. Store at room temperature.
2. Tris Buffered Saline with Tween-20 (TBST Solution). Prepare as with TBS solution except add 1 mL Tween-20 prior to bringing the solution to 1 L.

3. Proteinase K. Suspend Proteinase K powder to a final concentration of 100 μM In TBS (*see* **Note 5**).
4. Methanol 100% (w/v). Store at room temperature.
5. Polyvinylidene fluoride (PVDF membrane). Store at room temperature.
6. Blocking Buffer. Suspend either 2.5 g Casein or 1.5 g Bovine Serum Albumin in 50 mL of TBS.
7. Antibody solution. Suspend either 2.5 g Casein or 1.5 g Bovine Serum Albumin in 50 mL of TBST. Add antibody concentrate to appropriate dilution (*see* **Notes 6** and **7**).
8. Opti-4CN™ Substrate Kit (Bio-Rad 170-8235) (*see* **Note 8**).

3 Methods

The following protocols have been optimized for evaluating signal peptides appended to the N-terminus of proteins during heterologous expression in *E. coli.* Although over-expression protocols have been established for *E. coli* which can increase the total amount of protein produced and, by extension, simplify detection of proteins, nonnative signal peptides in *E. coli* may behave abnormally. In this work, we demonstrate that signal peptides from species unrelated at the level of phylum can function properly; however, when possible, signal peptides should be verified in their native species as well.

3.1 Protein Production

1. Vectors containing the gene of interest (GOI) can be generated using established molecular biology techniques or be commercially synthesized.
2. Transform vectors containing GOI into *E. coli* according to selected manufacturers' instructions (*see* **Note 9**).
3. Inoculate 5 mL of LB containing appropriate antibiotic. Incubate at 37 °C with agitation until a cell density of OD_{600nm} ~0.8 is reached.
4. Subculture 1 mL of culture into 1 L of LB containing appropriate antibiotic. Incubate at 37 °C with agitation until a cell density of OD_{600nm} ~0.8 is reached.
5. Reduce temperature of the culture to 16 °C with continued agitation (~1 h) (*see* **Note 10**).
6. Induce the expression of the GOI with the addition of 200 μL IPTG (*see* **Note 11**).
7. Incubate at 16 °C overnight with continued agitation (*see* **Note 10**).

3.2 Cell Lysis

1. Centrifuge 12.5 mL of culture at OD_{600nm} ~0.8 for 10 min at 13,000 RCF while refrigerated.
2. Aspirate the media and resuspend the pellet in 500 μL of room temperature BugBuster® Solution. Incubate at room temperature; occasionally gently invert the tubes to mix. Incubate for 10 min.
3. Centrifuge the samples for 10 min at 13,000 RCF while refrigerated. Carefully aspirate all BugBuster® solution and save (*see* **Note 12**).

3.3 Osmotic Shock

1. Centrifuge 12.5 mL of culture at OD_{600nm} ~0.8 for 10 min at 13,000 RCF while refrigerated.
2. Aspirate the media and resuspend the pellet in 1 mL of 4 °C STE Solution. Incubate on ice; occasionally gently invert the tubes to mix. Incubate for 10 min.
3. Centrifuge the samples for 10 min at 13,000 RCF while refrigerated. Carefully aspirate all STE Solution and discard.
4. Resuspend the pellet in 500 μL of room temperature 100 mM $MgCl_2$ solution. Incubate at room temperature; occasionally gently invert the tubes to mix. Incubate for 10 min.
5. Centrifuge the samples for 10 min at 13,000 RCF while refrigerated. Carefully aspirate all $MgCl_2$ solution and save (*see* **Note 13**).
6. Resuspend the pellet in 500 μL BugBuster® Solution. Incubate at room temperature occasionally; gently invert the tubes to mix. Incubate for 10 min (*see* **Note 14**).
7. Centrifuge the samples for 10 min at 13,000 RCF while refrigerated. Carefully aspirate all BugBuster® solution and save the supernatant (*see* **Note 15**).

3.4 Whole Cell Dot Blot

1. Centrifuge 12.5 mL of culture at OD_{600nm} of ~0.8 10 min at 13,000 RCF while refrigerated.
2. Aspirate the media and resuspend the pellet in 500 μL of room temperature TBS Solution. Repeat centrifugation and resuspension in TBS twice (or more) to remove any proteins in the media.
3. Prior to the final wash, the sample should be split evenly into two and centrifuged. One will be resuspended in TBS; the other should be resuspended in Proteinase K solution.
4. Incubate both samples at 37 °C for 1 h (*see* **Note 5**).
5. Aspirate and resuspend as in **step 2** to remove Proteinase K (*see* **Note 16**).
6. Create a grid on a sheet of PVDF membrane with 1 cm intervals using a pencil.

7. Add 5 μL methanol to a single 1 × 1 cm^2 and immediately add 5–20 μL of a sample to the same square.
8. Repeat with variable volumes of both digested and undigested samples in triplicate.
9. Allow the PVDF membrane to completely dry at 30 °C.
10. Briefly dip the membrane in methanol, allow excess to drip off, and submerge the membrane in Blocking Buffer. Incubate for 2 h at room temperature or at 4 °C overnight (*see* **Note 17**).
11. Remove the membrane from Blocking Solution and submerge in Antibody Solution. Incubate for 2 h at room temperature or at 4 °C overnight (*see* **Note 17**).
12. Rinse membrane thoroughly with TBST solution at least twice.
13. Prepare Opti-4CN™ according to manufacturer's guidelines. Add 1 mL of diluent to 9 mL of dH_2O. Add 200 μL Opti-4CN™ and mix well. Remove all TBST solution from the blot and add the Opti-4CN™ solution.
14. Allow the blot to develop to preferred intensity and stop the reaction by removing Opti-4CN™ solution and repeatedly washing with dH_2O (*see* **Note 18**).

4 Notes

1. Several stains may be used for expression, *E. coli* BL21 was chosen for this protocol due to the relative ease of expression of large quantities of protein. Signal peptides from unrelated species (e.g., Bacteroidetes) have been validated in *E. coli*; however, care should be taken to ensure that *E. coli* is an appropriate species for expression. Signal peptides from gram-positive bacteria or Eukaryotes will require other expression systems and may necessitate expression in the native species.
2. Several vectors are available for this purpose that may be more suitable depending on the protein and expression system chosen. pET21b is chosen here due to the presence of a C-terminal poly-histidine tag (HIS-tag), which minimizes potential interactions with N-terminal signal peptides. Many other vectors contain N-terminal tags; however, care should be taken to avoid any manipulation 5′ of N-terminal signal peptides and 3′ of C-terminal signal peptides as this may obstruct interaction between the tag and associated proteins.
3. Many techniques exist for the lysis of bacterial cells which may also be used including French-press, sonication, and lysis buffers including detergents, urea/guanidine-HCl, and/or Lysozyme. BugBuster® was chosen for this protocol because

proteins remain folded upon lysis and it does not solubilize protein aggregates (i.e., minimizes solubilization of misfolded proteins).

4. These concentrations can be modified to alter how aggressively the outer membrane is lysed. Increasing the sucrose concentration to as much as 20% (w/v) and reducing the $MgCl_2$ concentration to as little as 5 mM will result in the release of more periplasmic proteins. Care must be taken however, because these changes can also result in the release of some cytoplasmic proteins which will result in a false positive where proteins from the cytoplasm are found in the periplasmic fraction.
5. Concentration of Proteinase K and length of incubation may need to be optimized for different proteins and different species. These times and concentrations were found to adequate in our cases.
6. Use of the vectors containing HIS-tags allows for the use of anti-His6 antibodies for all proteins which minimizes variability between antibody binding. If vectors with no epitope tag or native protein expression are being evaluated, custom antibodies will need to be generated.
7. In the case of protein overexpression as presented here, a primary antibody conjugate is often sufficient. If difficult to express proteins or native expression is being evaluated, a secondary antibody conjugate should be used. In this case, prepare the secondary antibody in the same way as the primary antibody. After the application of primary antibody to the membrane, rinse thoroughly a minimum of twice before adding the secondary antibody solution. Rinse the membrane thoroughly again before adding the substrate.
8. Depending on antibody conjugate chosen, this will be variable. Horse radish peroxidase conjugates and Opti-4CN™ substrate were chosen for its ease of use though other conjugates and a variety of substrates are equally appropriate. If fluorescent conjugates are chosen, it must be established that the cells are not autofluorescent prior to use.
9. Several techniques serve this purpose including chemically competent transformation or electroporation. It is imperative to ensure any cells not containing the vector and GOI are removed.
10. Reducing the temperature of the culture may improve the expression of minimally soluble proteins [14]. Other temperatures may prove useful such as 25 °C and 37 °C. If possible, expression should be carried out as close to physiological temperatures as possible to ensure proper function of transporters and signal peptide associated proteins.

11. This volume results in a final concentration of 200 μM. Concentrations between 100 μM and 1 mM may improve expression level and should be optimized.
12. The supernatant from this step contains all soluble cellular protein which is important for demonstrating that the amount of protein found in periplasmic and cytoplasmic fractions is consistent with total protein. The pellet that is formed can be resuspended in 250 μL of dH_2O and visualized. This can be particularly useful when over-expressing protein and expressing protein in nonnative species as appreciable amounts of protein may be lost to aggregates and inclusion bodies. If other techniques are used to lyse the cells such as sonication or detergents, these aggregates may be solubilized and give an inaccurate picture as to how much functional protein is remaining in the cytoplasm.
13. This sample contains periplasmic proteins and should be retained.
14. This sample may be applied to Subheading 3.4 to determine if a protein is bound to the inner membrane and exposed within the periplasm.
15. This sample contains cytoplasmic proteins and should be retained.
16. Incomplete removal of the Proteinase K may cause the degradation of antibody in later steps and result in inaccurate data.
17. These solutions can be saved and reused as long as the membranes were thoroughly washed prior to submersion. Improper washing may result in the release of live cells into the buffers, contaminating further experiments.
18. Pictures should be taken as soon as the membranes dry. Membranes are stable in the dark for at least a few days but will eventually discolor and will quickly discolor in the light.

Acknowledgments

This work was supported by an Alberta Innovates BioSolutions grant #BIOFS026 and Agriculture and Agri-Food Canada.

References

1. Pfeffer SR, Rothman JE (1987) Biosynthetic protein transport and sorting by the endoplasmic reticulum and Golgi. Annu Rev Biochem 56(1):829–852
2. Bernstein HD, Poritz MA, Strub K, Hoben PJ, Brenner S, Walter P (1989) Model for signal sequence recognition from amino-acid sequence of 54K subunit of signal recognition particle. Nature 340(6233):482–486
3. Bernstein HD, Zopf D, Freymann DM, Walter P (1993) Functional substitution of the signal recognition particle 54-kDa subunit by its

Escherichia coli homolog. Proc Natl Acad Sci 90(11):5229–5233

4. Ivankov DN, Payne SH, Galperin MY, Bonissone S, Pevzner PA, Frishman D (2013) How many signal peptides are there in bacteria? Environ Microbiol 15(4):983–990
5. Petersen TN, Brunak S, von Heijne G, Nielsen H (2011) SignalP 4.0: discriminating signal peptides from transmembrane regions. Nat Methods 8(10):785–786
6. Thein M, Sauer G, Paramasivam N, Grin I, Linke D (2010) Efficient subfractionation of gram-negative bacteria for proteomics studies. J Proteome Res 9(12):6135–6147
7. Feliu JX, Villaverde A (1994) An optimized ultrasonication protocol for bacterial cell disruption and recovery of β-galactosidase fusion proteins. Biotechnol Tech 8(7):509–514
8. Jaschke PR, Drake I, Beatty JT (2009) Modification of a French pressure cell to improve microbial cell disruption. Photosynth Res 102(1):95–97
9. Neu HC, Heppel LA (1965) The release of enzymes from *Escherichia coli* by osmotic shock and during the formation of spheroplasts. J Biol Chem 240(9):3685–3692
10. Nossal NG, Heppel LA (1966) The release of enzymes by osmotic shock from *Escherichia coli* in exponential phase. J Biol Chem 241(13):3055–3062
11. Sharp JA, Echague CG, Hair PS, Ward MD, Nyalwidhe JO, Geoghegan JA, Foster TJ, Cunnion KM (2012) *Staphylococcus aureus* surface protein SdrE binds complement regulator factor H as an immune evasion tactic. PLoS One 7(5):e38407
12. Cuskin F, Lowe EC, Temple MJ, Zhu Y, Cameron EA, Pudlo NA et al (2015) Human gut Bacteroidetes can utilize yeast mannan through a selfish mechanism. Nature 517(7533):165–169
13. Rogowski A, Briggs JA, Mortimer JC, Tryfona T, Terrapon N, Lowe EC et al (2015) Glycan complexity dictates microbial resource allocation in the large intestine. Nat Commun 6(7481)
14. Schein CH (1989) Production of soluble recombinant proteins in bacteria. Nat Biotechnol 7(11):1141–1149

Chapter 16

Analysis of Complex Carbohydrate Composition in Plant Cell Wall Using Fourier Transformed Mid-Infrared Spectroscopy (FT-IR)

Ajay Badhan, Yuxi Wang, and Tim A. McAllister

Abstract

Fourier transformed mid-infrared spectroscopy (FTIR) is a powerful tool for compositional analysis of plant cell walls (Acebes et al., Front Plant Sci 5:303, 2014; Badhan et al., Biotechnol Biofuels 7:1–15, 2014; Badhan et al., BioMed Res Int 2015: 562952, 2015; Roach et al., Plant Physiol 156:1351–1363, 2011). The infrared spectrum generates a fingerprint of a sample with absorption peaks corresponding to the frequency of vibrations between the bonds of the atoms making up the material. Here, we describe a method focused on the use of FTIR in combination with principal component analysis (PCA) to characterize the composition of the plant cell wall. This method has been successfully used to study complex enzyme saccharification processes like rumen digestion to identify recalcitrant moieties in low-quality forage which resist rumen digestion (Badhan et al., BioMed Res Int 2015: 562952, 2015), as well as to characterize cell wall mutant lines or transgenic lines expressing exogenous hydrolases (Badhan et al., Biotechnol Biofuels 7:1–15, 2014; Roach et al., Plant Physiol 156:1351–1363, 2011). The FTIR method described here facilitates high-throughput identification of the major compositional differences across a large set of samples in a low cost and nondestructive manner.

Key words Fourier transformed infrared spectroscopy , Plant cell wall, Carbohydrate, Cellulose, Xylan, Lignin

1 Introduction

The plant cell wall is a heterogeneous, complex matrix composed of different polymers like cellulose, hemicellulose, and lignin [5]. In brief, cellulose can be described as a linear polymer of glucose connected via β-1, 4- linkages that form a flat, linear chain. Whereas the backbone of hemicellulose is composed of a variety of sugars (i.e., glucose, xylose, and mannose) with a number of different types of branches, while lignin is a complex polymer of aromatic alcohols that fills the spaces in the cell wall between cellulose, hemicellulose, and pectin [6–7]. Carbohydrate composition of the cell wall can be assayed with great precision by gas-chromatography.

D. Wade Abbott and Alicia Lammerts van Bueren (eds.), *Protein-Carbohydrate Interactions: Methods and Protocols*, Methods in Molecular Biology, vol. 1588, DOI 10.1007/978-1-4939-6899-2_16, © Springer Science+Business Media LLC 2017

However, it involves a series of labor intensive, long, and tedious solvent extraction steps to isolate different components of the cell wall and to derivatize them to alditol acetates, making this approach less favorable for studies with large sample sets. Infra-red spectroscopy relies on absorption of the infra-red radiation by interatomic bonds. Absorption of incident infrared causes bonds to bend and vibrate. The frequency of vibration that corresponds to absorption can be detected and converted into spectrum by using Fourier transformed algorithms which represents the molecular architecture of the sample. Furthermore, the combination of attenuated total reflection (ATR) with infrared spectroscopy allows biomass samples to be analyzed in its innate state without sample extensive preparation [8]. As Fourier transformed infrared spectroscopy (FTIR) analysis does not involve extensive sample preparation and derivatization procedures, it allows for a high-throughput nondestructive and low-cost analysis of complex carbohydrate constituents and their organization within biomass as well as identification of functional groups including carboxylic esters, phenolics esters, protein amides, and carboxylic acids [9–10].

The spectral information generated from FTIR spectroscopy is generally complex and extensive, requiring sophisticated data reduction techniques such as principal component analysis. Principal component analysis can identify the most distinctive features of the collected spectra and reduce the dimensionality of the data from several hundred data points in the original spectra to a fewer number of dimensions [11]. Alternatively, digital subtraction of averaged spectrum from different samples can provide valuable information regarding differences in cell wall composition. We were able to identify the abundance of undigested arabino-xylan and esterified hemicellulose components by differential spectral analysis of barley silage (feed) versus total track indigested fiber residue (isolated from fecal samples from cows fed a barley silage diet) by FTIR analysis (Fig. 1).

2 Materials

2.1 Plant Biomass Sample and Sample Preparation

1. Depending on research need any tissue from any plant part can be used.
2. Mill grinder with 0.1 mm sieve.
3. Deionized water.
4. Chloroform/methanol (1:1 v/v).
5. 70% (v/v) aqueous ethanol.
6. 100% Acetone.
7. 20% sodium azide stock solution.

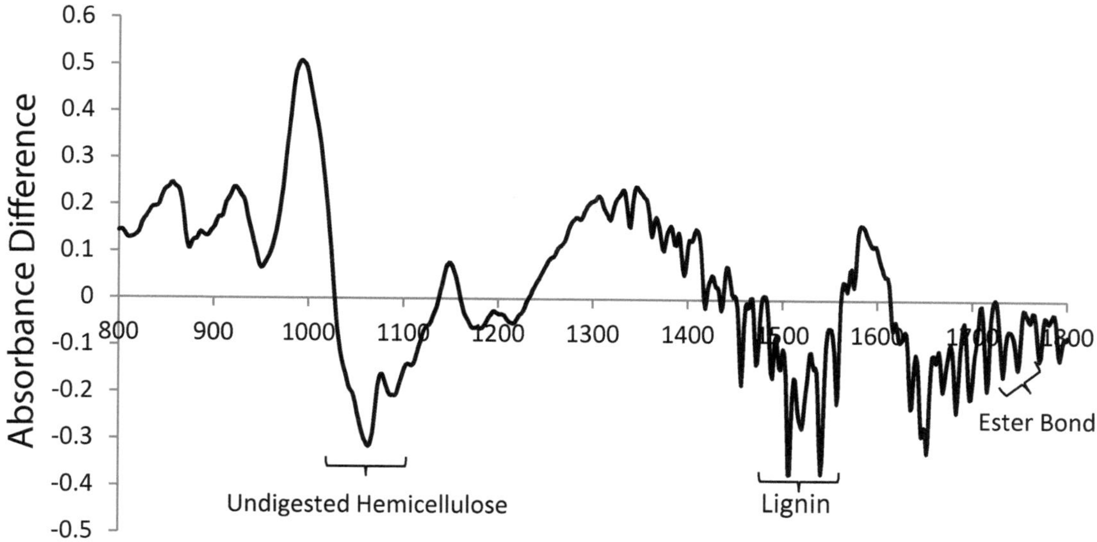

Fig. 1 FTIR spectral difference for feed versus tract indigested fiber residues (TIFR) showing major undigested plant cell wall components after rumen digestion. Barley silage was used as feed, whereas TIFR were recovered as NDF washed residual fiber from feces of cattle fed a barley silage based diet. Spectral difference peaks ranging between 1020 and 1130 cm^{-1} corresponded to undigested arabionoglucuronoxylan, xyloglucan, arabinan, and pectin and suggested an abundance of cross-linked hemicellulose within TIFR. Likewise, cross-linked esterified xylan and pectin in undigested residue were evident at peak 1714 cm^{-1} (C=0 from xylan) and 1738–1747 cm^{-1} (unconjugated C=0 stretch in xylan from acetic acid ester and pectin). Abundance of lignin in TIFR is evident by spectral differences at 1508, 1541, 1653, and 1688 cm^{-1}

8. α-Amylase from porcine pancreas DFP treated, type I-A, saline suspension, 20 mg protein/ml; 1000 units/mg protein (Sigma-Aldrich; approximately 500 units/100 mg cell wall AIR).
9. 0.1 M ammonium formate buffer (pH 6.0).
10. Water bath.
11. Fume hood.
12. pH meter.
13. Vortex mixture.
14. Shaker incubator.
15. Falcon tubes (50 ml).
16. Freezer and freeze dryer.
17. Centrifuge and centrifuge tubes (50 ml).

2.2 ATR-FTIR Spectroscopic Analysis

1. ALPHA FT-IR spectrometer (Bruker Optics, Ettlingen, Germany) equipped with a platinum diamond attenuated total reflectance (ATR) attachment. OPUS was used as spectral acquisition software provided by the instrument manufacturer.
2. XLSTAT/Minitab can be used for data analysis (i.e., PCA, subtractive analysis).

3 Methods

3.1 Biomass Preparation: Preparation of Alcohol Insoluble Residues

1. Grind plant biomass using grinding mill to pass through a 1 mm screen. Quick freeze the ground samples at −80 °C and lyophilize biomass samples for a week to remove moisture (*see* **Notes 1** and **2**).
2. Take approximately 5 g of dried and ground sample for sequential extraction over a sintered glass funnel under vacuum with two volumes of 100 ml of ice cold 80% ethanol, 100% ethanol, chloroform: methanol (1:1) mixture and 100% acetone (*see* **Notes 3** and **4**).
3. Dry the sample in fume hood at room temperature (*see* **Note 4**).
4. Prepare ammonium formate buffer (pH 6.0) by dissolving 6.306 mg of ammonium formate into 500 ml of deionized water and adjust pH to 6.0 with formic acid. Add 1 ml of 20% sodium azide stock and make final volume to 1 L with deionized water.
5. De-starch samples by suspending 1 g dried biomass in 100 ml of 0.1 M ammonium formate buffer, pH 6.0. Add 25 μl of type-1 porcine α-amylase (50 Units per 100 mg cell wall) and incubate at 25 °C for 48 h with shaking at 110 rpm.
6. Deactivate the amylase by incubating the reaction solution in water bath set at 95 °C for 15 min. Cool down the samples on ice and centrifuge at 900 × *g* for 10 min and discard supernatant.
7. Wash pellet at least thrice with distilled water and twice with acetone to get rid of residual amylase. Allow samples to dry in a fume hood overnight.

3.2 Acquisition of Spectra

1. Open the instrument-operating software (Opus software). Clean the sample platform with tissue wipe and ethanol (*see* **Notes 5** and **6**).
2. Set up measurement parameters by clicking on Measurement icon to open Measurement menu. Name your sample using sample description function. Under the advanced setting, set the file name and path for the folder files for data to be stored, set the scan mode to transmittance and the scan range between 4000 and 800 cm^{-1} with a resolution of 4 cm^{-1} and 32 repetitious scans. Once completed, save and exit the settings mode.
3. Click the measure icon once more. Acquire background spectrum by clicking start background scan and wait for background scanning to be completed as shown by the green bar at the bottom of the scan window (*see* **Note** 7).
4. Place the sample on the sample platform and close the sample holder so that the sample is in complete contact with ATR-Internal reflection element (IRE).

5. Acquire sample spectra by clicking on Start sample measurement and wait till spectral acquisition is completed. Take at least three spectrums for each sample (*see* **Notes 8** and **9**).
6. Click on Manipulate menu and select base line correction followed by the normalization icon to perform baseline correction and spectra normalization.
7. Save spectral data in excel file format (CSV) for further analysis. Select the data analysis approach that is appropriate to your goal.
8. Copy paste all spectral data from excel files to a single excel file so that the combined data can be used as the input file for PCA analysis.
9. Averaged spectra from two samples can be used for accessing digital subtraction spectra showing major compositional differences among two samples, as shown in Fig. 1. Spectral difference between feed (barley silage) and total tract undigested residue from fecal samples from cows fed a barley silage diet [3] identified esterified hemicellulose cross linked to lignin as a major recalcitrant component for rumen digestion (Fig. 1).

4 Notes

1. Use of an oven to dry the samples should be avoided as this can change the structure and composition of the plant cell wall.
2. The time for freeze drying varies depending on the sample and its source. Samples should be lyophilized for a prolonged period to ensure complete dryness.
3. Chloroform is highly toxic by all exposure routes. Do not allow pregnant staff to handle chloroform. Avoid skin contact. Avoid breathing the vapor; all work should be conducted in a fume hood. Keep away from heat or naked flames.
4. Always work with acetone in fume hood as vapors from it can cause dizziness. Never use oven to dry large sample set of acetone-soaked biomass, as it can lead to an explosion. Always dry acetone washed samples in a hood overnight at room temperature.
5. The crystal should be thoroughly cleaned and dried before acquiring background spectra.
6. Do not use gloves when you touch keyboard or mouse to minimize contamination. Solvents should not be applied directly on the instrument as they can damage it; they should be first applied to a chem wipe that can be used to clean the crystal.
7. Background spectrum should be taken before every sample. During background scanning the sample holder should be kept open.

8. During sample spectra acquisition, the ATR crystal should be completely covered by sample.
9. For ATR-FTIR, sample thickness should be at least 3 μm or higher (three to four fold times the depth of wave penetration), there is no maximum thickness as samples with millimeter thickness can still be analyzed.

Acknowledgment

The authors wish to acknowledge the support of Cellulosic Biofuel Network of Agriculture and Agri-Food Canada, Genome Canada, Genome Alberta, and Genome Quebec for their generous support of their research program.

Reference

1. Acebes JL, Largo-Gosens A, Hernández-Altamirano M, García-Calvo L, Alonso-Simón A, Álvarez JM (2014) Fourier transform mid infrared spectroscopy applications for monitoring the structural plasticity of plant cell walls. Front Plant Sci 5:303. doi:10.3389/fpls.2014.00303
2. Badhan A, Jin L, Wang Y, Han S, Kowalczys K, Brown DC, Ayala CJ, Latoszek-Green M, Miki B, Tsang A, McAllister T (2014) Expression of a fungal ferulic acid esterase in alfalfa modifies cell wall digestibility. Biotechnol Biofuels 7:1–15
3. Badhan A, Wang YX, Gruninger R, Patton D, Powlowski J, Tsang A, TA MA (2015) Improvement in saccharification yield of mixed rumen enzymes by identification of recalcitrant cell wall constituents using enzyme fingerprinting. BioMed Res Int 2015:562952. doi:10.1155/2015/562952
4. Roach MJ, Mokshina NY, Badhan A, Snegireva AV, Hobson N, Deyholos MK, Gorshkova TA (2011) Development of cellulosic secondary walls in flax fibers requires β-galactosidase. Plant Physiol 156:1351–1363
5. Ribeiro GO, Gruninger RJ, Badhan A, McAllister TA (2016) Mining the rumen for fibrolytic feed enzymes. Anim Front 6:20–26
6. Badhan A, McAllister TA (2015) Designer plants for biofuel; a review. Curr Metabolomics 4(1):49–55
7. Badhan AK, Chadha BS, Kaur J, Sonia KG, Saini HS, Bhat MK (2007) Role of transglycosylation products in the expression of multiple xylanases in myceliophthora sp. IMI 387099. Curr Microbiol 54:405–409
8. Baker MJ, Trevisan J, Bassan P, Bhargava R, Butler HJ, Dorling KM, Fielden PR, Fogarty SW, Fullwood NJ, Heys KA, Hughes C, Lasch P, Martin-Hirsch PL, Obinaju B, Sockalingum GD, Sulé-Suso J, Strong RJ, Walsh MJ, Wood BR, Gardner P, Martin FL (2014) Using Fourier transform IR spectroscopy to analyze biological materials. Nat Protoc 9:1771–1791
9. Allison GG, Thain SC, Morris P, Morris C, Hawkins S, Hauck B, Barraclough T, Yates N, Shield I, Bridgwater AV, Donnison IS (2009) Quantification of hydroxycinnamic acids and lignin in perennial forage and energy grasses by Fourier-transform infrared spectroscopy and partial least squares regression. Bioresour Technol 100:1252–1261
10. Carpita NC, McCann MC (2015) Characterizing visible and invisible cell wall mutant phenotypes. J Exp Bot 66:4145–4163
11. Chen L, Carpita NC, Reiter WD, Wilson RH, Jeffries C, McCann MC (1998) A rapid method to screen for cell-wall mutants using discriminant analysis of Fourier transform infrared spectra. Plant J 16:385–392

Chapter 17

Separation and Visualization of Glycans by Fluorophore-Assisted Carbohydrate Electrophoresis

Mélissa Robb, Joanne K. Hobbs, and Alisdair B. Boraston

Abstract

Fluorophore-assisted carbohydrate electrophoresis (FACE) is a method in which a fluorophore is covalently attached to the reducing end of carbohydrates, thereby allowing visualization following high-resolution separation by electrophoresis. This method can be used for carbohydrate profiling and sequencing, as well as for the determination of the specificity of carbohydrate-active enzymes. Here, we describe and demonstrate the use of FACE to separate and visualize the glycans released following digestion of oligosaccharides by glycoside hydrolases (GHs) using two examples: (1) the digestion of chitobiose by the streptococcal β-hexosaminidase GH20C, and (2) the digestion of glycogen by the GH13 member SpuA.

Key words Fluorophore, ANTS, Electrophoresis, Glycoside hydrolases, Glycan digestion, CAZymes

1 Introduction

Fluorophore-assisted carbohydrate electrophoresis (FACE) is a sensitive, fast, and inexpensive electrophoretic method for carbohydrate analysis [1–8]. Carbohydrates are covalently labeled at their reducing end using a fluorophore, such as 8-aminonaphthalene-1,3,6-trisulphonic acid (ANTS), and the resulting labeled sugars are separated in a high percentage polyacrylamide gel. In contrast to proteins and DNA, oligosaccharides can be relatively small; therefore, the gel percentage required for separation is usually between 20 and 40%. Gels can be then visualized and imaged using a standard UV light imager. Given that a sugar must be charged to migrate during electrophoresis, the choice of fluorophore depends on the normal charge state of the sugar to be visualized. Many sugars found in human glycans–such as glucose, galactose, fucose, and N-acetylglucosamine (GlcNAc)–are uncharged; therefore, the fluorophore ANTS, which possesses three negative charges, is particularly suited to the labeling of these sugars. Other fluorophores suitable for use in FACE include 2-aminoacridone (AMAC; uncharged) and sodium 4-amino-napthalene

D. Wade Abbott and Alicia Lammerts van Bueren (eds.), *Protein-Carbohydrate Interactions: Methods and Protocols*, Methods in Molecular Biology, vol. 1588, DOI 10.1007/978-1-4939-6899-2_17,

disulfonate (ANSA; singly charged) [7]. Separation of labeled mono- and oligosaccharides occurs on the basis of charge as well as size; therefore larger, more highly charged oligosaccharides may migrate faster than smaller, less charged sugars. In the case of fluorophore-labeled uncharged sugars, the uniform charge state means that migration is based largely on hydrodynamic size [2]. Derivatization of carbohydrates with a FACE fluorophore occurs in two steps. First, the C1 aldehyde of the saccharide at the reducing end interacts with the primary amine of the fluorophore to create a Schiff base [9]. Second, using sodium cyanoborohydride, this base is reduced to the mixed aryl/aliphatic secondary stable amine [7]. Fluorophore labeling of glycans offers a high level of sensitivity, with reports of detection down in the low picomolar range [2, 7, 10]. This allows the use of relatively small amounts of oligosaccharides compared to other common methods of carbohydrate analysis, such as TLC and HPLC.

FACE has been used to perform oligosaccharide profiling and sequencing [3, 7, 11], as well as for the analysis of various types of carbohydrate moieties such as glycoproteins [12, 13], glycolipids [5, 14], plant and bacterial polysaccharides [6, 8, 15], and glycosaminoglycans [4, 16]. It can also be used to determine the activity and specificity of carbohydrate-active enzymes (CAZymes) [17–19]. Here, we describe the labeling, separation, and visualization of uncharged mono- and oligosaccharides with ANTS. We also demonstrate the use of FACE to assess the substrate and linkage specificity of glycoside hydrolases, using the digestion of chitobiose and glycogen by the streptococcal enzymes GH20C and the GH13 member SpuA, respectively (Fig. 1).

2 Materials

Prepare all solutions using ultrapure water and analytical grade reagents. Prepare and store all reagents at room temperature (unless indicated otherwise).

2.1 Sample Labeling

1. 0.2 M ANTS solution: Weigh 85 mg of ANTS and dissolve in 1 mL 15% acetic acid. Cover the tube with aluminum foil to protect from light (*see* **Note 1**).
2. 1 M $NaBH_3CN$ solution: Weigh 63 mg of $NaBH_3CN$ and dissolve in 1 mL of DMSO. Resuspend by vortexing (*see* **Notes 1** and **2**).
3. Ice cold 95% ethanol.

2.2 Polyacrylamide Gel

1. Native running buffer: 25 mM Tris-base, 0.2 M glycine. Add about 100 mL of water to a 1 L graduated cylinder, then add 3 g of Tris-base and 15 g of glycine. Add about 800 mL of

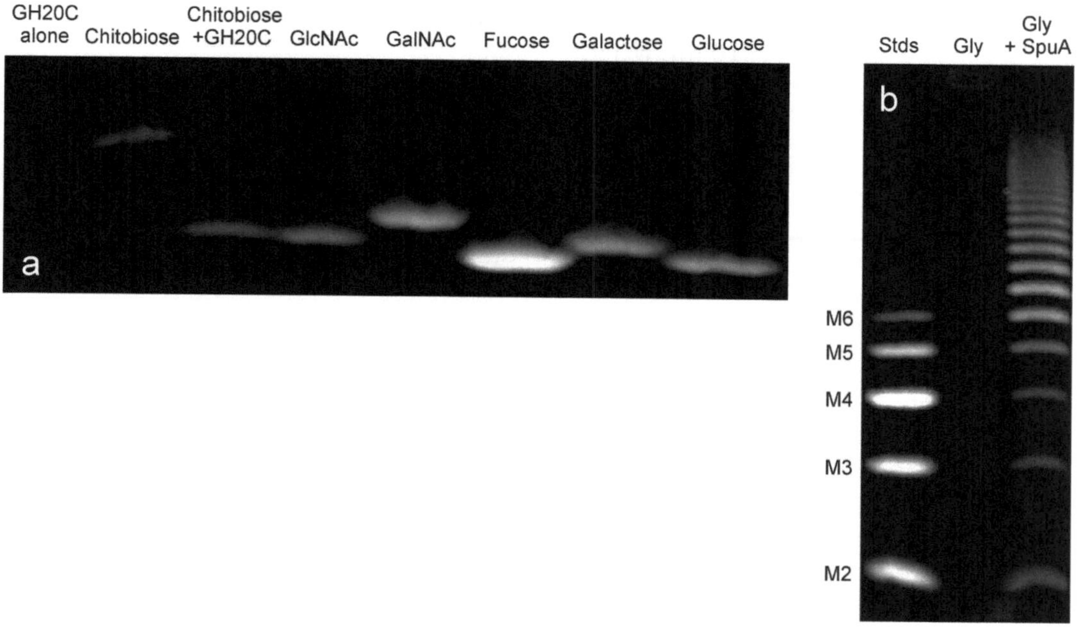

Fig. 1 FACE analysis of glycans released from oligo- and polysaccharides by glycoside hydrolases. (**a**) Digestion of chitobiose (β-D-GlcNAc-(1→4)-β-D-GlcNAc) by the streptococcal β-hexosaminidase GH20C [20]. Chitobiose was treated with 1 μM GH20C at 37 °C for 20 h in 100 mM sodium phosphate buffer pH 6.5, 45 mM β-mercaptoethanol. Labeled products and standards were separated on a 35% polyacrylamide gel. The expected product of GH20C digestion (GlcNAc) and other labeled monosaccharides are shown for comparison and to demonstrate the separation of different monosaccharides. (**b**) Degradation of glycogen (α-linked glucose polysaccharide; Gly) by the streptococcal glycoside hydrolase SpuA [17]. A 1% solution of glycogen was digested with 30 μg SpuA in PBS. Labeled products and standards were separated on a 28% polyacrylamide gel. Maltooligosaccharide standards (Stds), ranging from di- to hexasaccharide, are shown for comparison. Panel B is reproduced from Abbott et al. 2010 with permission from John Wiley and Sons

water and stir. Once everything is dissolved, adjust the volume of water to 1 L. For this buffer, the pH does not need to be adjusted. Transfer to a 1 L glass bottle and store at 4 °C.

2. Resolving gel buffer: 2 M Tris-base, pH 8.8. Add 242.3 g of Tris-base to a graduated cylinder. Add 900 mL of water. Stir until dissolved and adjust pH with concentrated HCl. Make up to 1 L with water. Transfer to a glass bottle and store at 4 °C.
3. Stacking gel buffer: 1 M Tris-base, pH 6.8. Add 121.2 g of Tris-base to a graduated cylinder. Add 900 mL of water. Stir and adjust pH with concentrated HCl. When everything is dissolved, adjust the volume to 1 L with water. Transfer to a glass bottle and store at 4 °C.
4. 40% acrylamide/Bis solution (19:1): store at 4 °C.
5. Ammonium persulfate (APS): 10% solution in water (*see* **Note 3**).
6. N, N, N, N′-Tetramethyl-ethylenediamine (TEMED): Store at room temperature.

7. Bromophenol blue (BPB) solution: Dissolve 0.1 g BPB in 10 mL water.
8. FACE loading dye: Combine 310 μL stacking gel buffer, 70 μL of BPB solution and 500 μL glycerol. Make up to 5 mL by adding 4.12 mL water.

3 Methods

3.1 Carbohydrate Labeling

1. Prepare all carbohydrate stock solutions (e.g., 10 mM) in dH_2O.
2. Set up the desired reactions/samples containing 10 μg carbohydrate (*see* **Note 4**) in an appropriate buffer, plus enzyme if required, and incubate at an appropriate temperature for 1–20 h depending on the degree of digestion desired (*see* **Note 5**).
3. Stop the reactions by adding 0.5 mL ice cold 95% ethanol.
4. Dry the samples in a speed vacuum for approximately 2 h at medium heat, until samples are completely dry.
5. To label the dried samples, resuspend each sample in 5 μL fresh ANTS solution and 5 μL fresh $NaBH_3CN$ solution by vortexing well.
6. Briefly spin the samples to bring them to the bottom of the tube and incubate them overnight at 37 °C wrapped in foil.
7. Dry the labeled samples in a speed vacuum for 2–4 h with suitable ventilation at medium heat or until completely dry (*see* **Note 6**).
8. Resuspend the dried pellet in 25 μL FACE loading dye and run on gel immediately or store at −20 °C wrapped in foil (*see* **Note 7**).

3.2 35% Polyacrylamide Gel Electrophoresis

1. For two gels, mix 1.8 mL resolving buffer, 12.8 mL 40% acrylamide, and 75 μL 10% APS in a 50 mL Erlenmeyer and mix gently by swirling. Add 7.5 μL of TEMED, mix briefly and cast gels within 8 cm × 8 cm × 1 mm gel cassettes. Allow space for stacking gel, gently overlay with isopropanol, and allow to polymerize for at least 30 min (*see* **Notes 8** and **9**).
2. Prepare the stacking gel solution by mixing 1.0 mL stacking buffer, 1.9 mL 40% acrylamide solution, 4.8 mL water, and 39 μL 10% APS in a 50 mL Erlenmeyer. Mix gently.
3. Once the resolving gel is polymerized, remove the isopropanol, flush with water, and remove excess water by tapping upside down.
4. Add 7.8 μL TEMED to the stacking solution mixture, mix gently, and quickly add on the top of the solidified resolving gel. Immediately add a 12-well gel comb without introducing air bubbles. Allow to polymerize for at least 30 min (*see* **Note 10**).

5. When ready, carefully remove the comb and clean up any loose acrylamide from around the wells.
6. Fill an ice box with ice and pack it around an empty gel tank.
7. Put the gel(s) in the tank, fill the middle reservoir with running buffer, and pour the remainder of the buffer outside and over the electrodes.
8. Flush the wells well with running buffer using a Pasteur pipette.
9. Load the required volume of each sample onto the gel(s) (*see* **Notes 11** and **12**).
10. Run the gel(s) for 30 min at 100 V followed by 1 h at 300 V (*see* **Notes 13** and **14**).
11. Wet the surface of the UV lightbox with water, carefully open the gel cassette, and push the gel out from the bottom onto the imager, adding more water on the top of the gel to stop it from drying out (*see* **Notes 15–17**).
12. Image the gel and adjust the exposure adequately.

4 Notes

1. The concentration of ANTS and $NaBH_3CN$ used can be optimized to ensure efficient labeling while keeping the background fluorescence low. We have found that GlcNAc, and oligosaccharides with GlcNAc at the reducing end, require a higher concentration of ANTS and $NaBH_3CN$ for labeling than other sugars, such as galactose and glucose. We routinely use 0.2 M ANTS and 1 M $NaBH_3CN$ for labeling GlcNAc, and 1/10 these concentrations for all other sugars.
2. Wear a face mask when weighing out $NaBH_3CN$ and wear gloves when handling the solution.
3. APS solution should be prepared fresh before each use or stored at −20 °C in aliquots. Thawed aliquots can be used for up to 1 week.
4. How much carbohydrate is used depends on the availability and expense of the sugar, as well as the number of expected products of digestion. A minimum of 2.5 μg is suggested. For most sugars that are a hexasaccharide or smaller, we have found 10 μg to be a suitable amount. Up to 100 μg may be required for large polysaccharides, especially if they are expected to be digested into many different sized products.
5. The duration of incubation depends on the amount of enzyme used and the degree of digestion desired. We have found that complete digestion is typically achieved following 20 h incubation with 1 μM enzyme.

6. When 0.2 M ANTS is used for labeling, the dried pellet obtained should be large and yellow.
7. After resuspending the pellet with the FACE loading dye, the samples will become blue-green in color (depending on the amount of ANTS used). Following storage at −20 °C, the samples will turn yellow but this does not affect their use in any way.
8. Because of the high acrylamide concentration of these gels, we find it is best to make them fresh before each experiment as they tend to dry out.
9. Overlaying the resolving gel with isopropanol prevents contact with atmospheric oxygen (which inhibits acrylamide polymerization) and helps to level the resolving gel.
10. Provided the stacker has fully polymerized, we do not suggest waiting more than 1 h to run gels as we have found that they dry out quickly resulting in a poor quality gel image.
11. The volume of each sample that should be loaded depends on the concentration of the sugar in the reaction and its molecular weight. For reactions containing 10 μg sugar and resuspended in 25 μL loading dye, we suggest loading 1 μL for mono-, di-, and trisaccharides, and 2–5 μL for larger sugars.
12. Avoid using the outer most two wells on each end as the gels tend to "smile."
13. The current at 300 V should be around 20–25 mA (for one gel).
14. Depending on the volume of sample loaded, it may or may not be visible as it migrates through the gel. For samples labeled with 0.2 M ANTS, a yellow band corresponding to excess ANTS should be visible approx. two-thirds of the way down the gel after running. If you are not sure if your samples have fully resolved, the gel can be visualized under UV while still in the cassette and run for longer if necessary.
15. If running two gels simultaneously, visualize them one at a time as they tend to dry out. While imaging one gel, leave the other in its cassette and place it in a large beaker or container filled with the used running buffer to avoid it from drying out.
16. If the corners of the gel curl up, flood the surface of the gel with water and lay a glass plate on the top with no air bubbles, then visualize.
17. Depending on the amount of ANTS used during labeling, a bright band corresponding to excess ANTS may be observed toward the bottom of the gel. This can interfere with the visualization of the gel by creating too much background noise; if so, this part of the gel can be cut off and discarded prior to imaging.

Acknowledgments

This work has been supported by the Canadian Institute of Health Research operating grant MOP 130305.

References

1. Jackson P (1993) Fluorophore-assisted carbohydrate electrophoresis: a new technology for the analysis of glycans. Biochem Soc Trans 21:121–125
2. Jackson P (1990) The use of polyacrylamide-gel electrophoresis for the high-resolution separation of reducing saccharides labelled with the fluorophore 8-aminonaphthalene-1,3,6-trisulphonic acid. Detection of picomolar quantities by an imaging system based on a cooled cha. Biochem J 270:705–713
3. Jackson P (1996) The analysis of fluorophore-labeled carbohydrates by polyacrylamide gel electrophoresis. Mol Biotechnol 5:101–123
4. Calabro A, Midura R, Wang A et al (2001) Fluorophore-assisted carbohydrate electrophoresis (FACE) of glycosaminoglycans. Osteoarthritis Cartilage 9:16–22
5. Gao N (2005) Fluorophore-assisted carbohydrate electrophoresis: a sensitive and accurate method for the direct analysis of dolichol pyrophosphate-linked oligosaccharides in cell cultures and tissues. Methods 35:323–327
6. Morell MK, Samuel MS, O'Shea MG (1998) Analysis of starch structure using fluorophore-assisted carbohydrate electrophoresis. Electrophoresis 19:2603–2611
7. Starr CM, Masada RI, Hague C et al (1996) Fluorophore-assisted carbohydrate electrophoresis in the separation, analysis, and sequencing of carbohydrates. J Chromatogr A 720:295–321
8. Young KD (1996) A simple gel electrophoretic method for analyzing the muropeptide composition of bacterial peptidoglycan. J Bacteriol 178:3962–3966
9. Cordes EH, Jencks WP (1962) On the mechanism of schiff base formation and hydrolysis. J Am Chem Soc 84:832–837
10. Kooy FK, Ma M, Beeftink HH et al (2009) Quantification and characterization of enzymatically produced hyaluronan with fluorophore-assisted carbohydrate electrophoresis. Anal Biochem 384:329–336
11. Frado LY, Strickler JE (2000) Structural characterization of oligosaccharides in recombinant soluble human interferon receptor 2 using fluorophore-assisted carbohydrate electrophoresis. Electrophoresis 21:2296–2308
12. Boraston AB, Sandercock LE, Warren RAJ, Kilburn DG (2003) O-glycosylation of a recombinant carbohydrate binding module mutant secreted by Pichia pastoris. J Mol Microbiol Biotechnol 5:29–36
13. Bardor M, Cabanes-Macheteau M, Faye L, Lerouge P (2000) Monitoring the N-glycosylation of plant glycoproteins by fluorophore-assisted carbohydrate electrophoresis. Electrophoresis 21:2550–2556
14. Basu SS, Dastgheibhosseini S, Hoover G et al (1994) Analysis of glycosphingolipids by fluorophore-assisted carbohydrate electrophoresis using ceramide glycanase from *Mercenaria mercenaria*. Anal Biochem 222:270–274
15. De Rezende CE, Anriany Y, Carr LE et al (2005) Capsular polysaccharide surrounds smooth and rugose types of salmonella enterica serovar typhimurium DT104 capsular polysaccharide surrounds smooth and rugose types of salmonella enterica serovar typhimurium DT104. Appl Environ Microbiol 71:7245
16. Oonuki Y, Yoshida Y, Uchiyama Y, Asari A (2005) Application of fluorophore-assisted carbohydrate electrophoresis to analysis of disaccharides and oligosaccharides derived from glycosaminoglycans. Anal Biochem 343:212–222
17. Abbott DW, Higgins MA, Hyrnuik S et al (2010) The molecular basis of glycogen breakdown and transport in streptococcus pneumoniae. Mol Microbiol 77:183–199
18. Gregg KJ, Zandberg WF, Hehemann J-H et al (2011) Analysis of a new family of widely distributed metal-independent alpha-mannosidases provides unique insight into the processing of N-linked glycans. J Biol Chem 286:15586–15596
19. Hehemann J-H, Kelly AG, Pudlo NA et al (2012) Bacteria of the human gut microbiome catabolize red seaweed glycans with carbohydrate-active enzyme updates from extrinsic microbes. Proc Natl Acad Sci U S A 109:19786–19791
20. Robb M, Robb CS, Higgins MA et al (2015) A second β-hexosaminidase encoded in the streptococcus pneumoniae genome provides an expanded biochemical ability to degrade host glycans. J Biol Chem 290:30888–30900

Chapter 18

A Rapid Procedure for the Purification of 8-Aminopyrene Trisulfonate (APTS)-Labeled Glycans for Capillary Electrophoresis (CE)-Based Enzyme Assays

Hayden J. Danyluk*, Leona K. Shum*, and Wesley F. Zandberg

Abstract

Purified glycan standards are required for glycan arrays, characterizing substrate specificities of glycan-active enzymes, and to serve as retention-time or mobility standards for various separation techniques. This chapter describes a method for the rapid separation, and subsequent desalting, of glycans labeled with the highly fluorescent fluorophore 8-aminopyrene 1,3,6-trisulfonate (APTS). By using fluorophore-assisted carbohydrate electrophoresis (FACE) on polyacrylamide gels, which utilizes equipment readily available in most molecular biology laboratories, many APTS-labeled glycans can be simultaneously resolved. Excising specific gel bands containing the desired APTS-labeled glycans, followed by glycan elution from the gel and subsequent solid-phase extraction (SPE), yields single glycan species free of excess labeling reagents and buffer components. This chapter describes a FACE/SPE procedure ideal for preparing glycans for capillary electrophoresis (CE)-based enzyme assays, as well as for the purification of rare, commercially unavailable glycans from tissue culture samples.

Key words Glycan-purification; Capillary electrophoresis (CE); Solid-phase extraction (SPE); Graphite; Fluorophore-assisted carbohydrate electrophoresis (FACE)

Abbreviations

APTS	8-Aminopyrene-1,3,6-trisulfonate
CE	Capillary electrophoresis
SPE	Solid-phase extraction
FACE	Fluorophore-assisted carbohydrate electrophoresis
PGC	Porous graphitized carbon
ACN	Acetonitrile
TFA	Trifluoroacetic acid
PNGaseF	Peptide *N*-glycosidase F
AcOH	Acetic acid

*These authors contributed equally to this work.

D. Wade Abbott and Alicia Lammerts van Bueren (eds.), *Protein-Carbohydrate Interactions: Methods and Protocols*, Methods in Molecular Biology, vol. 1588, DOI 10.1007/978-1-4939-6899-2_18, © Springer Science+Business Media LLC 2017

1 Introduction

The development of techniques to identify and quantitate glycan structures present in biological samples is essential if continued advancements in the field of glycoscience are to be realized [1]. Electrophoretic separation methods are inherently well suited for separating very polar analytes, such as glycans. In past decades, capillary electrophoresis (CE) has emerged as a popular complement to existing chromatographic methods [2]. When applied to the analysis of glycans, CE has consistently demonstrated very high separation efficiencies, often resolving isomeric structures. CE's outstanding resolving power is in part owed to the fact that peak broadening, which results from multiple available pathways through the packing material of liquid chromatography columns (i.e., eddy diffusion), does not occur in the open tubular capillaries utilized by CE. These capillary tubes also lack stationary phase particles, permitting the direct injection of glycans from enzyme-containing reaction mixtures that would otherwise plug common chromatography columns. This feature allows for the characterization of glycan-active enzymes, or the structural elucidation of glycans, by exo-glycosidase sequencing. Another advantage of CE-based enzyme assays is that minimal sample injection volumes are required (low nanoliter range), allowing for repeat assay analysis. Thus, CE's powerful resolution, high separation efficiency, ability to rapidly resolve glycans in enzyme- and buffer-containing mixtures, and low sample consumption make CE an attractive technique for characterizing the substrate specificities of glyco-active enzymes [3–7].

In nearly all cases, CE analysis of glycans borne by glycoproteins requires either their chemical or enzymatic release from the protein component (Fig. 1a). Glycans *N*-linked to asparagine residues of glycoproteins are readily liberated using commercially available enzymes such as peptide *N*-glycosidase F (PNGaseF). PNGaseF initially produces glycosylamine products that, in neutral solution, rapidly hydrolyze to yield free reducing sugars [8]. These sugars can then be labeled with aldehyde-reactive chromophores or fluorophores, permitting their detection by various optical techniques [9]. Unlike the case of *N*-linked glycans, analysis of glycans *O*-linked to serine or threonine residues is hampered by the lack of a general *O*-glycanase enzyme; thus, chemical deglycosylation strategies must be employed to obtain these analytes. One of the more widely used techniques to chemically cleave *O*-linked glycans from their serine or threonine glycoprotein anchors is an ammonia-based β-elimination reaction [10]. This method, although time-consuming and dependent on fastidious desalting, nevertheless yields reducing glycans amenable to fluorophore derivatization.

Currently, the fluorophore most widely used for CE glycan analysis is 8-aminopyrene-1,3,6-trisulfonate (APTS; Fig. 1a) [11–15].

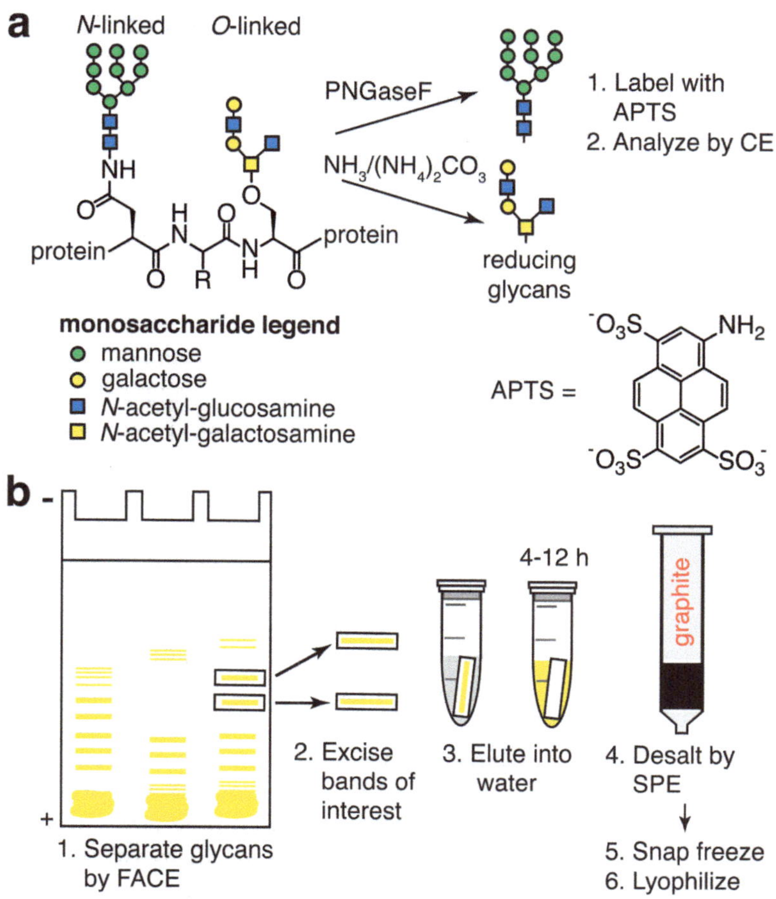

Fig. 1 Overview of the FACE/SPE glycan purification method. (**a**) Protein-linked glycans must be cleaved from proteins and purified before labeling with APTS; for *N*-linked glycans this is routinely accomplished by peptide *N*-glycanase F (PNGaseF) digestion while ammonia-catalyzed β-elimination may be used to obtain reducing *O*-glycans. CE analysis is performed to initially characterize the APTS-labeled glycans. (**b**) Several glycan samples may be separated in parallel by fluorophore-assisted carbohydrate electrophoresis (FACE) on polyacrylamide slab gels. Resolved glycans may be visualized using a UV transilluminator and specific glycans of interest are cut out of the gels using a scalpel. These gel slices are placed in centrifuge tubes and immersed in water to elute the APTS-labeled glycans, which are rapidly desalted using solid-phase extraction cartridges containing porous graphitized carbon. Glycans purified in this way are amenable to sensitive CE-based assays

APTS contains three sulfonate groups, which ensures that the labeled glycan species will uniformly possess a net negative charge regardless of whether the parent glycan is charged or neutral. APTS is among the brightest of all glycan derivatization reagents with a λ_{EX} of 455 nm—close to the output (488 nm) of an argon ion laser that is commonly used by CE. Low attomole detection limits have been achieved for APTS-labeled glycans using CE instruments equipped with laser-induced fluorescence (LIF) detectors [14]. APTS-labeled glycans have been resolved by both capillary zone electrophoresis [11, 12] and capillary gel electrophoresis [13–15]. They may also be resolved on polyacrylamide gels using electrophoresis equipment widely available in a molecular biology laboratory [16, 17]. This separation of fluorescently labeled glycans by slab gel electrophoresis is commonly known as fluorophore-assisted carbohydrate electrophoresis (FACE)—a technique that has been successfully applied to analyze diverse sets of glycan samples [18–20].

Here, we present a simple and inexpensive method in which FACE is used to separate and purify APTS-labeled glycans (Fig. 1b). In short, FACE-resolved glycans are visualized on a UV transilluminator, allowing for the excision of specific gel bands and subsequent elution of the glycans of interest. Solid-phase extraction (SPE) conducted using porous graphitized carbon (PGC) SPE cartridges subsequently yields pure, salt-free, APTS-labeled glycans. Glycans of interest are physically retained on the SPE cartridge through strong electrostatic interactions between PCG and the APTS label. In order to effectively release APTS-labeled glycans from the PCG SPE cartridges, eluents containing millimolar (mM) concentrations of a strong, volatile ion-pairing reagent, such as trifluoroacetic acid (TFA), are required. We have employed this FACE/SPE procedure to remove derivatization reagents and excess APTS from commercially available *N*-glycan standards to assess the substrate specificity of bacterial glycosidases (Fig. 2) [4]. This FACE/SPE purification method has also been used successfully to purify commercially unavailable *O*- and *N*-linked glycans produced in tissue cultured mammalian cell lines (Fig. 3) [3]. Rapid CE-based exo-glycosidase assays were used to confirm the structures of these glycans—these assays do not require any sample cleanup, and can be injected directly into the CE from reaction vials. By using FACE/SPE-purified glycans in these exo-glycosidase assays, the substrate—product relationships are made very clear, as there is no overlap between glycosidase digestion products and glycans that may have been present in the initial sample upon subsequent CE analysis. Hence, this simple and inexpensive FACE/SPE procedure allows for both the rapid purification of APTS-labeled glycans from a variety of sources in addition to enabling the establishment of glycan structures by CE-based exo-glycosidase sequencing.

2 Materials

Prepare all solutions using Milli-Q purified water (18.2 MΩ cm^{-1}). Solutions should be stored at 20 °C unless otherwise specified. When possible, analytical grade reagents and solvents should be used.

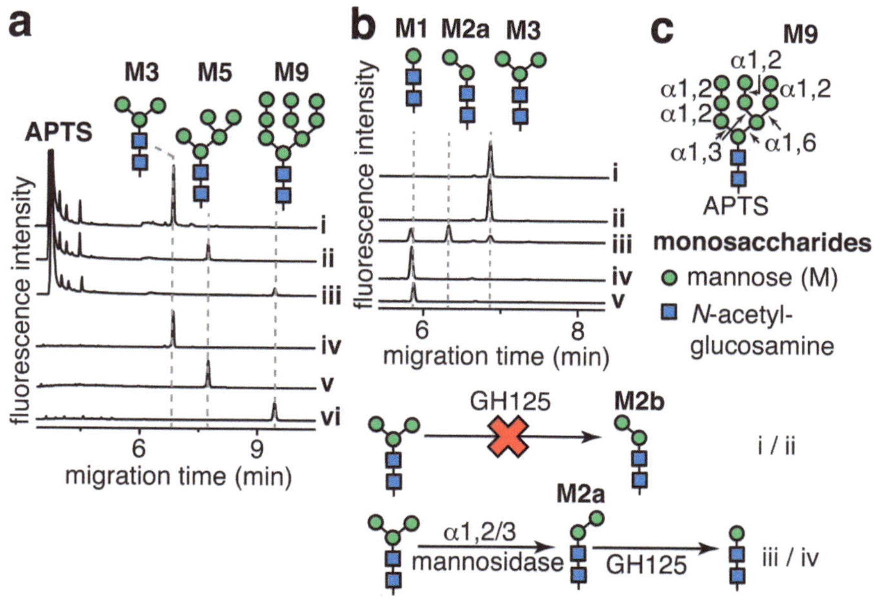

Fig. 2 FACE/SPE may be used to rapidly remove excess APTS and other labeling reagents (such as borate salts and acetic acid) from commercially available *N*-glycan standards [4]. (**a**) *N*-glycans standards M3 (*i*), M5 (*ii*) and M9 (*iii*) are labeled with APTS and desalted by FACE/SPE (*iv–vi*). CE analysis confirms the success of these procedures. (**b**) FACE/SPE purified M3 was used to demonstrate the substrate-specificity of previously uncharacterized family 125 glycoside hydrolyses (GH125) from *Streptococcus pneumoniae* (SpGH125) and *Clostridium perfringens* (CpGH125). Neither SpGH125 nor CpGH125 could hydrolyze substrates containing an α1,3-mannose residue (*i* and *ii*, respectively). However, pretreatment of glycans with an α1,2/3-mannosidase (*iii*) formed a glycan (M2a) in which the remaining α1,6-linked mannose residue could be efficiently hydrolyzed by SpGH125 (*iv*) and CpGH125 (*v*). (**c**) Structure of an intact high mannose *N*-linked glycan. The symbols are recommended by the Consortium for Functional Glycomics (http://www.functionalglycomics.org).

2.1 Preparation of Glycan Samples

1. Glycan standards are available from Glyko/Prozyme (Hayward, CA, USA) or V-Labs (Covington, LA, USA).
2. Suitable cell culture cell lines (American Type Culture Collection, Manassas, VA, USA). Chinese hamster ovary (CHO) K1 cells are a well-characterized source of both high mannose (Fig. 3b) and complex-type *N*-glycans; these cells also produce rare *O*-glucose-linked glycans (Fig. 3a).
3. CHO K1 cells were typically maintained in a 1:1 mixture of Dulbeco's modified Eagle's medium (DMEM) and Ham's F12 containing 5% fetal bovine serum.
4. 10-cm cell culture dishes.
5. Phosphate-buffered saline (PBS), pH 7.4.
6. Cell scrapers.
7. Peptide *N*-glycanase F (New England Biolabs, Ipswich, MA, USA).
8. Ethanol (−20 °C).

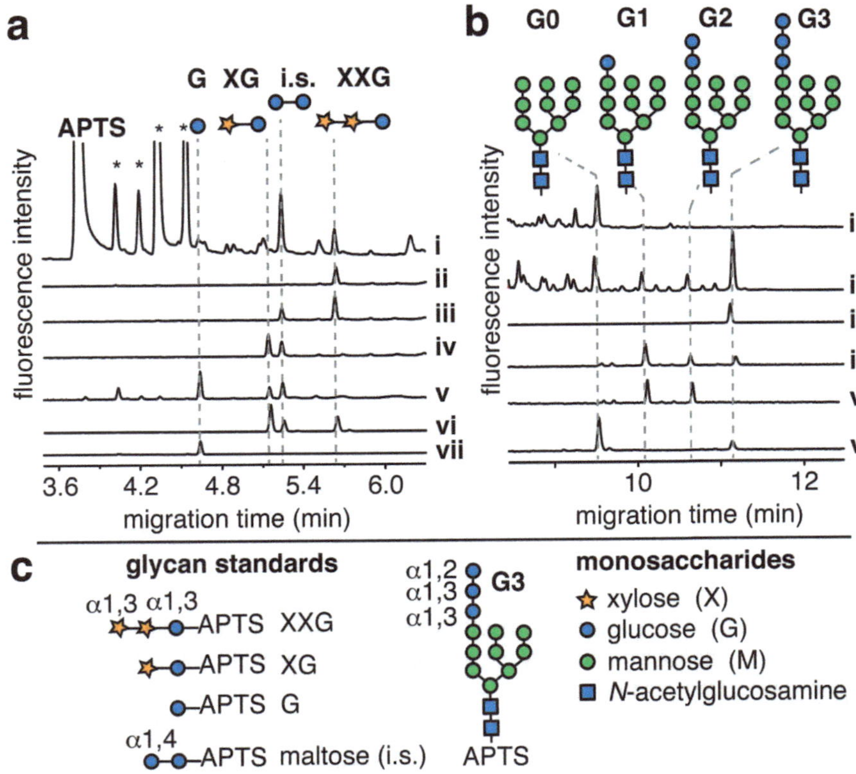

Fig. 3 FACE/SPE permits the purification of rare, commercially unavailable glycans from proteins secreted by cell cultured cells or from cell lysates and allows them to be used in CE-based glycosidase assays. (**a**) Developmentally important *O*-linked glycans borne on a recombinantly produced portion of the cell surface receptor Notch [3] were released by β-elimination and derivatized with APTS (*i*). A trisaccharide (*ii*) of the same mobility as a chemically synthesized standard, D-Xyl-α1-3-D-Xyl-α-1-3-D-Glc (XXG; *vi*), was purified by FACE/SPE and spiked with maltose (*iii*) as an internal standard (i.s.). α-Xylosidase (XylS) digestion and chemically defined standards were used to confirm the structure of XXG by CE. After a 3 h reaction with XylS, XXG was completely converted into XG (*iv*). Tris, a buffer salt used in the FACE purification procedure, completely blocks the ability of XylS to cleave the second xylose residue from XG; however, purification by SPE permits complete hydrolysis of XXG into APTS-labeled glucose (G; *v*) which was confirmed based on its co-migration with a FACE-purified standard (*vii*). (**b**) CE analysis of the *N*-linked glycans from Chinese hamster ovary (CHO) K1 cells grown in the absence (*i*) or presence (*ii*) of castanospermine demonstrated that this inhibitor blocked the processing of these glycans by endoplasmic reticulum (ER) glucosidase I and II. These glucosylated glycans ($GlcNAc_2Man_9Glc_{1\text{-}3}$; G1-G3) cannot currently be purchased. G1–G3 were individually purified by FACE/SPE (*iii*, only G3 is shown) and mixed together (*iv*) to have their structures confirmed by exo-glycosidase sequencing. ER glucosidase I removes a single α1,2-linked Glc residue from G3, leaving G2 and G1 intact (*v*). In contrast, G3 is protected from hydrolysis by an α1,3-glucosidase (Megazyme, Chicago, IL, USA) while G1 and G2 are converted into G0 (*vi*). The identity of G0 was confirmed based on its susceptibility to jack bean mannosidase and also its co-migration with a commercially available standard (not shown). (**c**) Structures of glycan standards.

9. W-375 Ultrasonic processor.
10. Savant SpeedVac concentrator.
11. Modulyo freeze dryer or equivalent.
12. 28% Ammonium hydroxide.

13. Ammonium carbonate.
14. 1.5 mL screw-cap centrifuge tubes or equivalent.
15. Water bath sonicator.
16. Vortex mixer.
17. Heating block.
18. 50 mg/mL HyperCarb HyperSep solid phase extraction (SPE) cartridges (Thermo-Fisher) [21].
19. 1 M Sodium hydroxide.
20. 30% Acetic acid (AcOH).
21. Wash solvent: 5% Acetonitrile (ACN), 0.1% trifluoroacetic acid (TFA).
22. Eluting solvent: 50% ACN, 0.1% TFA.
23. 0.2-μm Syringe filters.

2.2 APTS Derivatization

1. 200 μL Microtubes.
2. APTS (Sigma-Aldrich).
3. 15% AcOH.
4. 1 M Sodium cyanoborohydride ($NaBH_3CN$) in tetrahydrofuran (THF) (Sigma-Aldrich).
5. Heating block.

2.3 Analysis by CE-LIF

1. ProteomeLab PA800 capillary electrophoresis instrument (CE) equipped with an argon ion laser-induced fluorescence (LIF) detector (Beckman-Coulter, Brea, CA, USA) containing a 488-nm notch filter and a long pass emission filter at 520 nm ± 10 nm (*see* **Note 1**).
2. Capillaries thermostated at 22 °C and sample compartment set to 10 °C.
3. *N*-linked carbohydrate separation gel buffer (Beckman-Coulter): 25 mM sodium acetate, pH 4.75, 2% polyethelyene oxide (*see* **Note 2**) [13, 14].
4. Zero Flow neutral coated capillary (MicroSolv, Leland, NC, USA) or equivalent. 50 μm × 50 cm; 40 cm to detector.

2.4 FACE Separation of Glycans [16]

1. UV transilluminator.
2. Mini Protean 3 gel casting and electrophoresis equipment (Bio-Rad, Hercules, CA, USA).
3. Scalpels.
4. Cold room or 4 °C refrigerated cabinet.
5. 40% acrylamide solution (acrylamide/bis-acrylamide: 37.1 to 1). Store at 4 °C.
6. 8× Resolving gel buffer: 1.5 M Tris hydrochloride, pH 8.8.

7. 8× Stacking gel buffer: 1.0 M Tris hydrochloride, pH 6.8.
8. 10× Running buffer: 1.92 M glycine, 0.25 M Tris (base), pH 8.3.
9. N,N,N′,N′-tetramethylethylenediamine (TEMED). Store at 4 °C.
10. 10% Ammonium persulfate (APS). Store 1 mL aliquots at −20 °C.
11. 2× Sample loading buffer: 0.01% thorin I (Sigma-Aldrich) in 20% glycerol. Store aliquots at −20 °C.

2.5 SPE of APTS-Labeled Glycans

1. 50 mg/mL HyperCarb HyperSep SPE cartridges.
2. 1 M Sodium hydroxide.
3. 30% AcOH.
4. Elution solvent: 50% ACN, 50 mM TFA (*see* **Note 3**).
5. 0.2-μm syringe filters (Merck-Millipore).

3 Methods

3.1 Preparation of Glycan Samples

Pure glycans, such as those commercially available (Fig. 2), may be directly labeled with APTS and purified by FACE. If glycoproteins or cell lysates are to be used as a source of glycan standards, the glycans must first be removed from proteins and desalted before they are fluorescently labeled, as excess salt and protein will impair the derivatization procedure. Methods are listed for preparing both *N*- and *O*-linked glycans from cell culture lines.

3.1.1 N-Linked Glycans

1. Grow the cells until they are about 80% confluent.
2. Aspirate the media and wash the cells with 2 × 5 mL 4 °C PBS.
3. Add 1 mL 4 °C PBS to the plates and use a cell scraper to remove the adherent cells from the plate. Pellet the cells by centrifugation (800 × *g*, 5 min, 4 °C), then remove the PBS.
4. Suspend cells in 180 μL PBS and add 20 μL 10× glycoprotein denaturing buffer (supplied with the PNGaseF deglycosylation kit; *see* **Note 4**).
5. Remove an aliquot and heat at 100 °C for 10 min to denature the proteins.
6. Allow samples to cool and add 1/10th volume 10% NP40.
7. Add 1–5 μL PNGaseF and digest at 37 °C overnight. Digestion may be extended after the addition of fresh PNGaseF if necessary.
8. Precipitate proteins by adding 3 volumes of ice cold ethanol. Store samples for 30 min at −20 °C.
9. Pellet proteins by centrifugation (14,000 × *g*, 10 min, 4 °C); *N*-glycans are contained in the supernatant, which should be

transferred into clean centrifuge tubes, concentrated on a SpeedVac concentrator for several hours, snap frozen in liquid nitrogen, and subsequently lyophilized (*see* **Note 5**).

10. Resuspend the dried glycans in 100 μL H_2O, and then desalt on PGC SPE cartridges. Cartridges must first be conditioned by sequentially washing with 1 mL 1 M NaOH, 2 mL H_2O, 1 mL 30% AcOH, 2 mL H_2O, 1 mL elution solvent, and 3 mL H_2O before applying glycan samples. After applying samples the SPE cartridges are sequentially washed with 1 mL H_2O, 1 mL wash solvent, and eluted with 4 × 0.5 mL elution solvent. Filter all eluates through a syringe filter to remove small pieces of the graphite SPE material.
11. Concentrate samples on a SpeedVac to approximately 100–150 μL, transfer to 200 μL tubes, snap freeze and lyophilize the desalted glycan samples. Once dry, samples are ready for APTS-labeling.

3.1.2 O-Linked Glycans

1. Prepare cells as described above (Subheading 3.1.1, **steps 1–3**); pellet the washed cells in a screw-cap micro centrifuge tube (*see* **Note 6**).
2. In a fume hood, add 500 μL ammonium hydroxide saturated with ammonium carbonate. Add 50 mg of additional ammonium carbonate and tightly cap tubes. Briefly (5 min) heat (60 °C) to dissolve ammonium carbonate. Vortex vigorously, and sonicate until cell pellets are completely suspended (*see* **Note 7**).
3. Incubate at 60 °C for 40 h to completely β-eliminate glycans (*see* **Note 8**).
4. Cool tubes (−20 °C) for 30 min to reduce the pressure. Add 0.5 mL H_2O and dry using a SpeedVac. Repeat several times to remove the majority of ammonium salts (*see* **Note 9**).
5. Resuspend pellet in 100–200 μL H_2O, pellet insoluble material by centrifugation (14,000 × *g*, 5 min, room temperature), and desalt the glycan-containing supernatant by SPE (Subheading 3.1.1, **steps 10** and **11**).

3.2 Labeling Glycans with APTS

1. To each dried sample, add 2 μL 50 mM APTS in 15% AcOH, then 2 μL 1 M $NaBH_3CN$ in THF (*see* **Note 10**).
2. Incubate at 60 °C for 4 h in the dark (*see* **Note 11**).

3.3 Glycan Purification by FACE

1. Assemble gel-casting apparatus.
2. Prepare 20% acrylamide resolving gel solution by mixing the following stock solutions: 1 mL 8× resolving gel buffer, 3 mL water, and 4 mL 40% acrylamide solution. Initiate polymerization by adding 20 μL APS and 20 μL TEMED; these volumes are sufficient for preparing two mini gels. Mix well, and immediately pipette into the casting apparatus. Overlay the gel solution with 1 mL ethanol. Allow 30–60 min for polymerization.

3. Prepare stacking gel solution by mixing the following stock solutions: 0.25 mL 8× stacking gel buffer, 1.5 mL water, and 0.25 mL 40% acrylamide solution. Initiate polymerization by adding 6 μL APS and 5 μL TEMED; these volumes are sufficient for preparing two stacking gels.
4. Pour off the ethanol overlay from the resolving gel. Rinse well with 18 MΩ H_2O. Add the stacking gel solution and insert comb immediately. Polymerization of the stacking gel is usually complete within 15–30 min.
5. Assemble the electrophoresis apparatus and add 1× FACE running buffer. This buffer should be prepared in advance and chilled to 4 °C.
6. Mix the APTS-labeled glycan samples with an equal volume of 2× FACE loading buffer. With care up to 15 μL of each sample can be loaded per well (*see* **Notes 12** and **13**).
7. Connect to an appropriate power supply and run at a constant voltage of 200 V at 4 °C in the dark (*see* **Note 14**).
8. Disconnect the power supply, remove the gel, and take off the top glass plate. Place FACE gel, glass side down, on a UV trans-illuminator and visualize using the long wave (360 nm) setting. Using a clean scalpel, mark the position of APTS-labeled glycan bands to be excised (*see* **Notes 14–16**).
9. Switch off the UV and quickly cut out the marked glycans, transferring each into a 1.5 mL centrifuge tube (*see* **Notes 17** and **18**).
10. Completely immerse the gel slices in H_2O. Allow APTS-labeled glycans to elute at 4 °C overnight in the dark.
11. Centrifuge samples to pellet pieces of polyacrylamide gel. The supernatant, containing APTS-labeled glycans may be desalted without further sample concentration.

3.4 Desalting Using Graphite SPE

1. Desalt the APTS-labeled glycans on conditioned PGC SPE cartridges. These should be conditioned by sequentially washing with 1 mL 1 M NaOH, 2 mL H_2O, 1 mL 30% AcOH, 2 mL H_2O, 1 mL elution solvent, and 3 mL H_2O before applying APTS-labeled glycan samples. SPE cartridges are washed with 1 mL H_2O, and glycan samples are eluted with 4 × 0.5 mL elution solvent (use less elution solvent if possible). Filter all eluates through a syringe filter to remove small pieces of the graphite SPE material (*see* **Notes 19** and **20**).
2. Concentrate eluted samples on a SpeedVac to approximately 25–50 μL, transfer to 200 μL tubes, then snap freeze and lyophilize the desalted APTS-labeled glycan samples. Exposure of the samples to light should be minimized.
3. Resuspend the sample in 100 μL H_2O. The samples are ready for capillary electrophoresis analysis (*see* **Note 21**).

4 Notes

1. Methods have been described elsewhere for using DNA sequencers for the analysis of APTS-labeled glycans [22].
2. Other CE buffers may be used. In our experience, the *N*-linked carbohydrate separation gel buffer (Beckman-Coulter) produces the best peak resolution.
3. Do not use a micropipette to dispense concentrated TFA.
4. The PNGaseF deglycosylation kit (New England Biolabs) is supplied with a 10× protein denaturing buffer and 10% NP-40 (a nonionic detergent). Protocols for the use of this kit have been optimized for purified glycoproteins. For the PNGaseF digestion of cell lysates, it may be preferable to sonicate cell pellets in PBS containing 0.5% sodium dodecylsulfate (SDS), and then denature at 100 °C after adding 1/10th volume of 400 mM dithiothreitol (DTT). NP-40 should be added to the denatured proteins to a final concentration of 1% before adding the PNGaseF. More thorough deglycosylation protocols amenable to tissues have been published elsewhere [23].
5. The concentration of ethanol must be reduced before samples are lyophilized to prevent thawing.
6. The ammonia-catalyzed β-elimination procedure produces significant pressure. The use of screw-cap tubes prevents the lids from popping open during this procedure.
7. Larger cell samples tend to form an insoluble residue if care is not taken to ensure that they are initially suspended in the ammonium hydroxide/ammonium carbonate mixture.
8. These conditions will also eliminate *N*-linked glycans, although at lower yields than those obtained using PNGaseF [24]. Nevertheless, β-elimination is easily scaled up to accommodate larger samples than the enzymatic procedure.
9. When removing ammonium salts from the samples, a cold trap should be used between the pump and SpeedVac. Alternatively, the pump exhaust should be vented into a fume hood.
10. If the 1 M $NaBH_3CN$ solution is to be prepared from the pure reagents, care should be taken as this process may produce toxic HCN gas. $NaBH_3CN$ should be weighed in a fume hood. A 1 M solution in dimethylsulfoxide (DMSO) will also yield reproducibly labeled glycans. This reagent may be stored at −20 °C; however, a fresh solution is typically prepared after one month.
11. APTS-labeled glycans are stable for years if they are properly stored at −20 °C with minimal light exposure.
12. It is advisable to load alternating wells in the gel to eliminate possible contamination due to the horizontal diffusion of APTS-labeled glycans when excising bands.

13. Detection limits using UV transilluminators for visualization are substantially higher than those obtained by CE-LIF. Thus, it is advisable to run mobility markers, such as dextran oligomers, on the outer lanes of the FACE gels. A mobility marker will permit the approximate position of a glycan to be determined even if it is not readily visualized using the transilluminator. We use a dextran ladder ranging from 1 to approximately 30 glucose units (Beckman-Coulter, part number: 477613); other glucose polymers such as a 0.1 M HCl hydrolysate of starch may also be used.
14. Excess APTS will be both the furthest migrating and most intense band on the gel, unless it has been run off completely. If present, this band must be removed immediately to prevent it from diffusing into the thin layer of running buffer between the gel and the glass plate.
15. Protective equipment, such as a UV face shield, should be worn to prevent skin exposure to the UV light from the transilluminator. A lab coat or long-sleeved shirt is recommended to reduce UV forearm exposure as well.
16. While cutting out the bands, the FACE gel should remain on whichever glass plate it resides post gel-apparatus disassembly; gel transfer will introduce water between the plate and gel surface, which will begin to elute and mix the APTS-labeled glycans.
17. Minor changes in the relative order of glycan migration times have been observed when comparing the FACE and CE separations. For complex cell-derived samples, it is advisable to remove several extra glycan bands on either side of the glycan of interest.
18. Glycans should be excised from FACE gels as quickly as possible to minimize their diffusion after electrophoresis. As the gels continue to sit, they will dry out and curl along their edges, making it very difficult to accurately excise the desired glycan bands. In addition, dry gels begin to crumble, and become very difficult to pick up with forceps.
19. Glycine present in the FACE running buffer neutralizes the TFA required to elute APTS-labeled glycans from the graphite SPE cartridges. Thus, instances where large pieces of gel were used to obtain glycans may require eluent containing up to 250 mM TFA to counteract this glycine neutralizing effect.
20. Samples eluted from SPE may be stored in liquid nitrogen, while the other samples are still being prepared, in order to avoid the hydrolysis of acid labile glycans. Sialic acids residues are preserved if care is taken to ensure immediate freezing of the samples after elution.
21. A freeze-dryer with a −84 °C cold trap is recommended for lyophilizing solutions containing ACN and TFA.

Acknowledgments

We thank Dr. David Rose for providing the ER glucosidase I and the National Sciences and Engineering Research Council of Canada (NSERC) for funding the CE-LIF.

References

1. National Research Council of the National Academies (2012) Transforming glycoscience: a roadmap for the future. The National Academies Press, Washington, DC
2. Mechref Y (2011) Analysis of glycans derived from glycoconjugates by capillary electrophoresis-mass spectrometry. Electrophoresis 32:3467–3481
3. Whitworth GE, Zandberg WF, Clark T, Vocadlo DJ (2010) Mammalian notch is modified by D-Xyl-alpha1-3-D-Xyl-alpha1-3-D-Glc-beta1-O-Ser: implementation of a method to study O-glucosylation. Glycobiology 20:287–299
4. Gregg KJ, Zandberg WF, Hehemann J, Whiworth GE, Deng L, Vocadlo DJ, Boraston AB (2011) Analysis of a new family of widely distributed metal-independent alpha-mannosidases provides unique insight into the processing of N-linked glycans. J Biol Chem 286:15586–15596
5. Zhao JY, Dovichi NJ, Hindsgaul O, Gosselin S, Palcic MM (1994) Reaction, Detection of 100 molecules of product formed in a fucosyltransferase. Glycobiology 4:239–242
6. Zhang Y, Le X, Dovichi NJ, Compston CA, Palcic MM, Diedrich P, Hindsgaul O (1995) Monitoring biosynthetic transformations of N-acetyllactosamine using fluorescently labeled oligosaccharides and capillary electrophoretic separation. Anal Biochem 227:368–376
7. Craig D, Arriaga EA, Banks P, Zhang Y, Renborg A, Palcic MM, Dovichi NJ (1995) Fluorescence-based enzymatic assay by capillary electrophoresis laser-induced fluorescence detection for the determination of a few beta-galactosidase molecules. Anal Biochem 226:147–153
8. Plummer TH, Elder JH, Alexander S, Phelan AW, Tarentino AL (1984) Demonstration of peptide:N-glycosidase F activity in endo-beta-N-acetylglucosaminidase F preparations. J Biol Chem 259:10700–10704
9. Harvey DJ (2011) Derivatization of carbohydrates for analysis by chromatography; electrophoresis and mass spectrometry. J Chromatogr B Analyt Technol Biomed Life Sci 879:1196–1225
10. Huang Y, Mechref Y, Novotny MV (2001) Microscale nonreductive release of O-linked glycans for subsequent analysis through MALDI mass spectrometry and capillary electrophoresis. Anal Chem 73:6063–6069
11. Chen FT, Evangelista RA (1995) Analysis of mono- and oligosaccharide isomers derivatized with 9-aminopyrene-1,4,6-trisulfonate by capillary electrophoresis with laser-induced fluorescence. Anal Biochem 230:273–280
12. Evangelista RA, Liu MS, Chen FA (1995) Characterization of 9-aminopyrene-1, 4, 6-trisulfonate derivatized sugars by capillary electrophoresis with laser-induced fluorescence detection. Anal Chem 67:2239–2245
13. Guttman A, Chen FA, Evangelista RA, Cooke N (1996) High-resolution capillary gel electrophoresis of reducing oligosaccharides labeled with 1-aminopyrene-3,6,8-trisulfonate. Anal Biochem 233:234–242
14. Guttman A (1996) High-resolution carbohydrate profiling by capillary gel electrophoresis. Nature 380:461–462
15. Guttman A, Chen FA, Evangelista RA (1996) Separation of 1-aminopyrene-3,6,8-trisulfonate-labeled asparagine-linked fetuin glycans by capillary gel electrophoresis. Electrophoresis 17:412–417
16. Gao N, Lehrman MA (2006) Non-radioactive analysis of lipid-linked oligosaccharide compositions by fluorophore-assisted carbohydrate electrophoresis. Methods Enzymol 415:3–20
17. Starr CM, Masada RI, Hague C, Skop E, Klock JC (1996) Fluorophore-assisted carbohydrate electrophoresis in the separation, analysis, and sequencing of carbohydrates. J Chromatogr A 720:295–321
18. Gao N, Lehrman MA (2002) Analyses of dolichol pyrophosphate-linked oligosaccharides in cell cultures and tissues by fluorophore-assisted carbohydrate electrophoresis. Glycobiology 12:353–360
19. O'Shea MG, Samuel MS, Konik CM, Morell MK (1998) Fluorophore-assisted carbohydrate electrophoresis (FACE) of oligosaccharides: efficiency of labelling and high-resolution separation. Carbohydr Res 307:1–12

20. Stack RJ, Sullivan MT (1992) Electrophoretic resolution and fluorescence detection of N-linked glycoprotein oligosaccharides after reductive amination with 8-aminonaphthalene-1,3,6-trisulphonic acid. Glycobiology 2:85–92
21. Packer NH, Lawson MA, Jardine DR, Redmond JW (1998) A general approach to desalting oligosaccharides released from glycoproteins. Glycoconj J 15:737–747
22. Laroy W, Contreras R, Callewaert N (2006) Glycome mapping on DNA sequencing equipment. Nat Protoc 1:397–405
23. Jang-Lee J, North SJ, Sutton-Smith M, Goldberg D, Panico M, Morris H, Haslam S, Dell A (2006) Glycomic profiling of cells and tissues by mass spectrometry: fingerprinting and sequencing methodologies. Methods Enzymol 415:59–86
24. Schulz BL, Packer NH, Karlsson NG (2002) Small-scale analysis of *O*-linked oligosaccharides from glycoproteins and mucins separated by gel electrophoresis. Anal Chem 74:6088–6097

Part IV

CAZyme Structure, Discovery, and Prediction Methods

Chapter 19

Probing the Complex Architecture of Multimodular Carbohydrate-Active Enzymes Using a Combination of Small Angle X-Ray Scattering and X-Ray Crystallography

Mirjam Czjzek and Elizabeth Ficko-Blean

Abstract

The various modules in multimodular carbohydrate-active enzymes (CAZymes) may function in catalysis, carbohydrate binding, protein-protein interactions or as linkers. Here, we describe how combining the biophysical techniques of Small Angle X-ray Scattering (SAXS) and macromolecular X-ray crystallography (XRC) provides a powerful tool for examination into questions related to overall structural organization of ultra multimodular CAZymes.

Key words Small-angle X-ray scattering, SAXS, X-ray Crystallography, Carbohydrate-active enzymes, CAZymes, Carbohydrate binding module, CBM, Glycoside hydrolase, Cohesin, Dockerin, Multimodular, Structure, Dissect and build

1 Introduction

Carbohydrate-active enzymes (CAZymes) often have the interesting feature of being multimodular, multifunctional assemblages. That is, they are composed of an independently folded catalytic module appended to one or more discretely folded modules within a contiguous amino acid sequence. This assembly is believed to be required to tackle recalcitrant, complex, and often solid substrates within their natural context. These additional modules, also referred to as auxiliary modules, retain their function when produced recombinantly as independent modular entities. CAZyme catalytic modules may function as glycoside hydrolases, glycosyl transferases, polysaccharide lyases, or lytic oxygenases; auxiliary modules may function as linkers, carbohydrate binding modules (CBMs) or for recruitment of other enzymes through protein-protein interactions (non-exhaustive list). Ultra multimodular

D. Wade Abbott and Alicia Lammerts van Bueren (eds.), *Protein-Carbohydrate Interactions: Methods and Protocols*, Methods in Molecular Biology, vol. 1588, DOI 10.1007/978-1-4939-6899-2_19,

CAZymes are considered here as enzymes having three or more independently folded modules.

Structural characterization using X-ray crystallography (XRC) has been indispensable for understanding the fine molecular determinants of interaction, such as between catalytic modules and substrate or carbohydrate binding modules and ligand. The determination of modular structure and function provides insight into the role of the individual modules but it does not provide an understanding of how these modules are coordinated spatially to one another and the implications on full-length enzyme functionality with substrate. Further complexities arise in determining the overall X-ray crystal structures of enzymes that are ultra multimodular. Producing the full-length proteins recombinantly is challenging; they are large (often over 100 kDa) and may be prone to proteolysis. Crystal formation, required for structural characterization by XRC, is quite challenging for full-length ultra multimodular CAZymes presumably in part due to flexibility between the modules. Although several full-length (or near full-length) ultra multimodular CAZyme structures have been determined using XRC [1–7], these remain the exception, not the rule.

XRC and low-resolution structural methods (such as cryo-electron microscopy, transmission electron microscopy, atomic force microscopy, electron tomography) can be complementary in determining the overall structure of proteins and their complexes. A particularly powerful combination is between XRC and the low-resolution solution structure method Small Angle X-ray Scattering (SAXS) (Fig. 1). In general, high-resolution X-ray crystal structures of the individual modules are fit into the low-resolution SAXS molecular envelopes (~12–20 Å) generating pseudo-atomic resolution models of the ultra multimodular enzymes and their complexes [8–11]. Determining ultra multimodular protein structures using the combination of XRC and SAXS is an emerging frontier in structural biology research, providing important fundamental insights into full-length CAZymes with industrial, biomedical, and environmental significance.

Performing the pseudo-atomic resolution structural analysis of ultra multimodular CAZymes can be done using a "dissect and build" approach. This approach involves producing different single and multimodular protein constructs from the CAZyme of interest (dissect) and then piecing them together to obtain the full-length structure (build). The development of this methodology was first performed on the complex binding interactions in the cellulosome, a multi-protein, multimodular CAZyme complex produced by environmental bacteria for the efficient degradation of cellulose [12]. Here, a minimalist cellulosome-like structure was constructed to simplify SAXS data interpretation [13]. Further expansion and development of the "dissect and build" methodology has been done on the ultra multimodular CAZymes produced by *Clostridium*

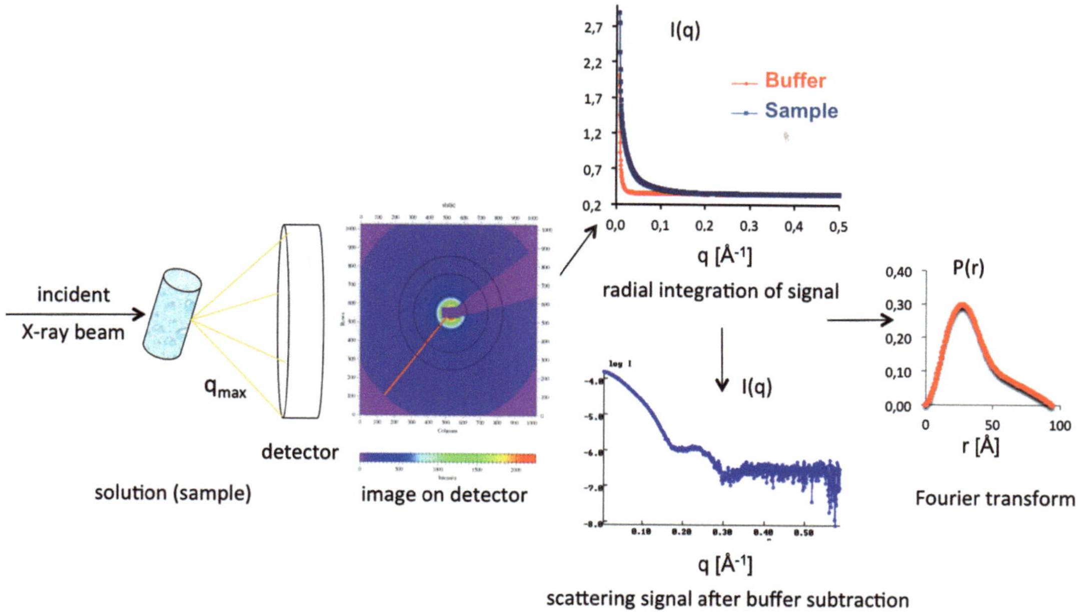

Fig. 1 Schematic representation of an experimental setup for SAXS experiments. From left to right: The incident X-ray beam is guided onto the capillary containing the sample solution, the rays are diffused and measured by transmission on an X-ray sensitive bidimensional detector. Then the 2D image of diffuse scattering is integrated in a radial manner leading to the experimental scattering curve. The scattering is measured as well for a buffer only solution (*red*) before and/or after the sample solution (*blue*) and the buffer signal is subsequently subtracted. The buffer-subtracted scattering curve is the starting point of all subsequent data analyses that may consist in fitting of the experimental curve with structural models. Since the scattering curve is registered in reciprocal space, to interpret a scattering profile in terms of a structure, it is useful to Fourier transform the scattering profile to obtain the interatomic distance distribution function, P(r) (also referred to as the pair distance or vector length distribution function)

perfringens, a bacterial pathogen, with a seminal paper on the methodology published in 2009 [14]. *Clostridium perfringens*, a notorious bacterial pathogen, produces an array of secreted ultra multimodular CAZymes with exquisite specificities for animal extracellular glycans; among these are the μ-toxin (GH84A, NagH, 1627 amino acids) [15] and GH84C (NagJ, 1001 amino acids) (Fig. 2a). GH84C, having only four modules, is one of the "simpler" ultra multimodular CAZymes produced by the organism, consisting of an N-terminal glycoside hydrolase catalytic module (GH84C_cat) [16], a family 32 carbohydrate binding module (CBM32) [17], a cohesin module implicated in ultratight protein-protein interactions with its μ-toxin dockerin module binding partner [(18),] and a C-terminal FN3 module of unknown function [14]. Recombinant protein-production trials were unsuccessful for full-length GH84C and thus the "dissect and build" methodology combining XRC and SAXS was essential for the determination of the full-length pseudo-atomic resolution structure of GH84C and the interaction with the μ-toxin via the cohesin:dockerin interaction.

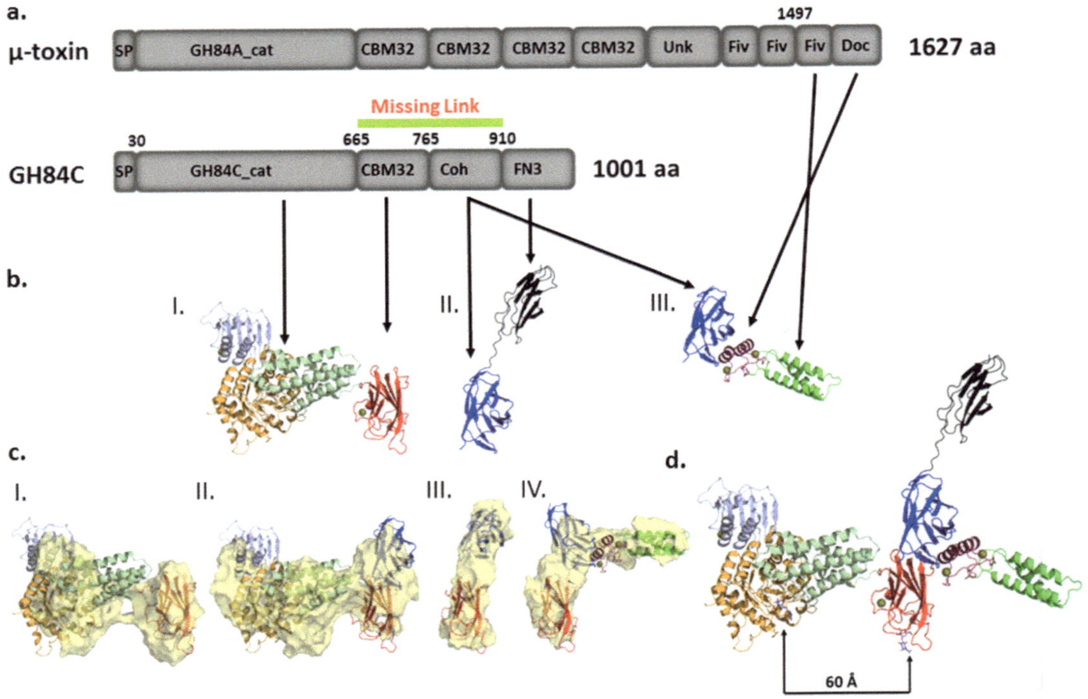

Fig. 2 Steps involved in determining the structure of ultra multimodular proteins. (**a**) The bioinformatically determined modular structure of the μ-toxin and GH84A. (**b**) Multimodular crystal structures from GH84C and the μ-toxin. The GH84C N-terminal domain is shown in gray, the catalytic module in orange, the 5-helical bundle linker region in green, the CBM32 in *red*, the cohesin in blue, and the C-terminal FN3 module in black. I. GH84C_cat-CBM32, II. Cohesin-FN3. III. The crystal structure of the FIVAR (lime green) and dockerin (purple) modules from the μ-toxin is shown in a high affinity protein-protein interaction with the GH84C cohesin module. (**c**) Fits of the modules into the SAXS generated molecular envelopes, for I. GH84C_cat-CBM32, II. GH84C_cat-CBM32-Cohesin, III. CBM32-Cohesin and IV. CBM32-Cohesin:FIVAR-dockerin. (**d**) Pseudo-atomic resolution construction of a multimodular carbohydrate-active enzyme. The GH84C catalytic site and CBM32 ligand binding site are indicated with arrows

In this chapter, the "dissect and build" methodology will be described for generating a full-length pseudo-atomic structure for the CAZyme GH84C and how to demonstrate the protein-protein interaction between the GH84C cohesin module and the dockerin module from GH84A. This protocol assumes the reader has purified protein samples of the different modular constructs and PDB coordinates for these modules obtained via XRC. The purpose of this protocol is to serve as a basis for performing similar structural characterizations of multimodular enzymes using the complimentary methods of XRC and SAXS.

2 Materials

2.1 Chemicals

Prepare all solutions using ultrapure water (prepared by purifying deionized water, to attain a sensitivity of 18 MΩ-cm at 25 °C) and analytical grade reagents. Prepare all reagents at room temperature and store at 4 °C.

1. 50 μg/ml Kanamycin. Weigh 50 μg Kanamycin per ml of solution and add to each preparation where needed.
2. 1 M isopropyl-β-D-1-thiogalactopyranoside (100 ml stock-solution). Add 90 ml water to a 100 ml graduated cylinder or a glass beaker. Weigh 23.83 g isopropyl-β-D-1-thiogalactopyranoside and transfer to the cylinder. Add water to a volume of 100 ml.
3. 50 mM Tris pH 7.5 and 100 mM NaCl (1 L). Add about 100 ml water to a 1 L graduated cylinder or a glass beaker. Weigh 6.06 g Tris–HCl and transfer to the cylinder. Add water to a volume of 900 ml. Mix and adjust pH with HCl. Make up to 1 L with water.

2.2 Constructs for Recombinant Protein Production

In order to build the pseudo-atomic resolution structure of GH84C, genetic constructs of the following multimodular constructions were made in pET28-a(+) for production and purification of 6His tagged protein in *E. coli* (*see* Fig. 2a for modular boundaries) (*see* **Notes 1** and **2**):

1. GH84C_cat_CBM32.
2. CBM32-Cohesin.
3. GH84C_cat-CBM32-Cohesin.
4. Cohesin-FN3.
5. In order to structurally characterize the Cohesin:Dockerin interaction between GH84C and μ-toxin, a construction of the FIVAR-Dockerin modules from the μ-toxin is necessary (Fig. 2a).

2.3 PDB Coordinates

PDB Codes for GH84C_cat_CBM32 [14], Cohesin-FN3 [(14),] and the Cohesin:FIVAR-Dockerin complex [18] (Fig. 2b-I–III) are 2v5d, 2w1n, and 2ozn, respectively.

The "missing piece" of the full-length GH84C structure, to be characterized by SAXS, is the spatial relationship between CBM32 and cohesin (Fig. 2a).

2.4 SAXS Requirements

1. Dynamic light scattering instrument (for example the DynaPro Plate Reader II from Wyatt Technology).
2. Beamline Instrumentation requirements. The SSRL BL4–2 has a high photon flux multilayer monochromator, a fast read-out photon counting detector (Pilatus 300 K), a stopped-flow device, and a pin-hole geometry X-ray scattering camera. Blu-Ice/DCS software (customized) is used to control the entire instrument on BL4–2. Alternatively, the Soleil Beamline SWING is positioned with an In-vacuum U20 undulator, equipped with Diaphragm and Fixed exit DCM Si111 optics, leading to a beamsize of 450 × 20 μm² FWHM in the experimental hutch. On this beamline both a stopped flow device

combined to a high-throughput sample changer, and an online HPLC (Agilent) for proteins in solution are available. This setup is controlled by the customized software MXcube.

3. Computer Software requirements. Download the ATSAS suite of programs [19, 20] from http://www.embl-hamburg.de/biosaxs/download.html. All programs mentioned come from the ATSAS suite of program, unless otherwise stated.
4. Computer Hardware requirements. Windows, Linux, and Mac computers may be used for SAXS. The following operating systems are supported by the ATSAS suite of programs:
   - Windows 32bit and 64bit.
   - Mac OS 10.9.
   - Ubuntu 12.04 LTS 32bit and 64bit.
   - Ubuntu 14.04 LTS 32bit and 64bit.
   - Debian 6x 64bit.
   - Debian 7x 32bit and 64bit.
   - Red Hat/CentOS 6x 32bit and 64bit.
   - Red Hat/CentOS 7x 64bit.
   - openSUSE 13x 64bit.

3 Methods

3.1 Sample Preparation

1. Produce and purify the different modular constructs as fresh as possible using immobilized metal affinity chromatography and size exclusion chromatography into 50 mM Tris pH 7.5 and 100 mM NaCl (*see* **Note 3**).
2. Concentrate the purified multimodular protein constructs to ~ 10 mg/ml and buffer exchange the modular constructs into 1 L of 50 mM Tris pH 7.5 and 100 mM NaCl for 24 h, set aside 100 ml of the dialysate buffer (*see* **Notes 4–7**). Analyze purity and protein quality using SDS-PAGE and dynamic light scattering (DLS) to assess monodispersity of the sample in solution. The DLS analysis should display a single peak corresponding to the monodisperse non-aggregated form of the molecules in the solution. Determine exact protein concentrations using UV absorbance and calculated extinction coefficients [21].
3. For preparing the protein-protein interacting complex, mix the two purified polypeptide constructs CBM32-Cohesin and FIVAR-Dockerin with FIVAR-Dockerin in ~1.5 M excess. Incubate overnight at 4 °C and then purify the complex in 50 mM Tris pH 7.5 and 100 mM NaCl using SEC to a final concentration of 10 mg/ml and a final volume of 200 μL.

3.2 SAXS Data Collection

1. Beamline setup (this should be discussed with the beamline operator before the start of the experiment). The following beamline setup may be used: wavelength of 1.127 Å, a sample to detector distance of 1.7 m, and a data collection series of at least 10 frames with 3 s exposures.
2. First collect SAXS data on standards. Collect SAXS data on different concentrations of standard solutions from 1, 2, 4, 6, 8, and 10 mg/ml samples in 50 mM Tris pH 7.5 and 100 mM NaCl for both BSA and hen egg white lysozyme (Fig. 1). Collect data on buffer only sample as well, which will be subtracted from the sample curves to account for background scattering. In addition, collecting a scattering curve on water is important for calibration and allows the calculation of absolute values of intensity at zero scattering. Volumes will vary depending on the beamline.
3. Collect SAXS data on samples: GH84C_cat-CBM32 (Fig. 2c-I), CBM32-Cohesin (Fig. 2c–II), GH84C_cat-CBM32-Cohesin (Fig. 2c–III), and the CBM32-Cohesin:FIVAR-Dockerin complex (Fig. 2c–IV). Collect SAXS data on 1, 2, 4, 6, 8, and 10 mg/ml concentrations for all samples in 50 mM Tris pH 7.5 and 100 mM NaCl (*see* **Notes 8–10**). In between each sample collect buffer measurements for subtraction (just 50 mM Tris pH 7.5 and 100 mM NaCl), for example: buffer measurement, 1 mg/ml sample collection, buffer measurement, 2 mg/ml sample collection, buffer measurement, 4 mg/ml sample collection, buffer measurement, 6 mg/ml sample collection, buffer measurement, 8 mg/ml sample collection, buffer measurement, 10 mg/ml sample collection, buffer measurement.

3.3 SAXS Analysis

To determine the spatial relationship between CBM32 and Cohesin, perform SAXS analysis of CBM32-Cohesin, GH84C_cat-CBM32-Cohesin and of the CBM32-Cohesin:FIVAR-Dockerin complex.

1. Initial examination and buffer subtraction of the curves should be done using the program PRIMUS [22] or equivalent (*see* **Note 11**). After the buffer subtracted scattering curves have been determined, the plots of $I(q)$ vs q at each concentration should be identical, where $I(q)$ is the scattered intensity and q is the scattering vector or momentum transfer with the units Å^{-1} (defined q=4πsinθ, where 2θ is the scattering angle). Use the buffer subtracted curves for all further analyses. These files are in ASCII format and generally have the extension name.dat.
2. Determine Rg and I(0) using the Guinier approximation in PRIMUS. There are two size and shape parameters for the scattering particle that can be easily calculated: the first is I(0),

the intensity at which q is equal to 0, the second is the radius of gyration, Rg, the square root of the average squared distance of each scatterer from the center of the particle [23]. At small values of q (that is at low scattering angles) the Guinier approximation [$\ln I(q)=\ln I(0) - q^2Rg^2/3$] can be used to describe the scattering and a linear fit of the plot of $\ln I(q)$ vs q^2 gives $I(0)$ from extrapolating to the y-intercept and Rg from the square root of the slope. Globular proteins should have a qRg less than 1.3 and more elongated shapes should have an even lower qRg [24], *see* Jacques and Trewhella [25] for the effects of interparticle interference and aggregation on SAXS data.

3. Determine Rg and $I(0)$ using GNOM [26]. $I(0)$ and Rg should also be determined using the program GNOM from the pair distance distribution function $P(r)$. Dmax should be chosen in GNOM with careful inspection of $P(r)$, which provides real space information on the probable frequency of interatomic vector lengths (r) within a macromolecule.
4. Calculate molecular weight of the samples. Molecular weight should be calculated by multiple methods to demonstrate reliability and to provide confidence in the scattering data. One such method depends on using the experimental data from a single scattering curve measured on a relative scale and within a defined high q (high angle) range, *see* Fischer et al. [27]. Another method relies on the relative $I(0)$ after scaling for concentration and using the BSA or lysozyme standards for calibration [28]. Here, the molecular weight of the experimental protein sample (MWp) is calculated from the known molecular weight of the standard (MWst) using the following formula: $MWp = \frac{I(0)p * \mathrm{MWst}}{Cp^{*}I(0)st / Cst}$, where I(0)p and I(0)st are the scattering intensities at zero angle of the experimental protein and the standard, respectively, and Cp and Cst are the concentrations. Another way of calculating the molecular weight is based on a calibration of the scattering data with water to allow normalizing the forward scattering to an absolute scale. In that case the following formula may be used: $M_w = \frac{N_A I(0)/c}{(\Delta\rho_M)^2}$, where NA is the Avogadro constant $6.02*10^{23}$, $I(0)$ is the forward scattering of the protein at an absolute scale, c is the concentration of the molecule in solution, and $\Delta\rho M = \rho M - \rho_S$, the scattering length density difference; for a protein $\Delta\rho M$ on average equals $2.086*10^{10}$ cm^{-2}. Thus, from a water calibrated curve the molecular weight can be estimated by Mw= $1385*I(0)/c$, [29].
5. Run 20 independent DAMMIF runs and align and average with DAMAVER to generate initial low-resolution *ab initio*

envelopes [30, 31]. The DAMAVER averaging provides NSD (normalized spatial discrepancy) values that tend toward zero if molecules closely resemble one another, or over one if they systematically differ.

6. Model the 3D structures into the averaged DAMMIF molecular envelope (*see* **Notes 12** and **13**). The pdb models of the modules should then be placed manually into the averaged DAMMIF molecular envelope using the program PyMOL [32]. SUBCOMB [33] can also be used to superimpose the 3D models into the averaged DAMMIF molecular envelope. If flexibility of the molecules is given, SASFIT is the most adapted software to use.
7. Determine goodness of fit using CRYSOL from the ATSAS suite of programs. CRYSOL [34] should be used to predict the solution scattering for the pseudo–atomic resolution models and to calculate the quality of fit to the actual SAXS experimental curve, goodness of fit is represented by the χ_{crysol} statistic, a lower χ_{crysol} indicating a better fit.
8. Determine the structure of the multimodular constructs using BUNCH [35] (*see* **Notes 14–17**). BUNCH uses a combined rigid body and *ab initio* approach to modeling. The 3D structures of the modules are treated as rigid bodies and any missing amino acids are modeled in as dummy atoms. In BUNCH the models are generated by fitting to one or more of the scattering curves and the goodness of fit is represented by χ_{BUNCH}. The χ_{crysol} should also be calculated for the BUNCH models using CRYSOL. The BUNCH generated models should be then superimposed into the averaged DAMAVER generated molecular envelope manually in PyMOL or using the program SUBCOMB.

3.4 Piecing Together the Multimodular Pieces

1. The orientation of the CBM32 relative to the catalytic module is known based on the X-ray crystal structure of GH84C_cat_CBM32 (Fig. 2b-I). In PyMOL, overlay the CBM32-Cohesin component from the averaged DAMMIF manually fitted models (with the best χ_{crysol} values) for both CBM32-Cohesin and the CBM32-Cohesin:FIVAR-Dockerin complex onto the GH84C_cat_CBM32 crystal structure into the averaged DAMMIF molecular envelope for GH84C_cat-CBM32-Cohesin, this will generate two models for GH84C_cat-CBM32-Cohesin. A third model should be generated by manually fitting the X-ray crystal structure of GH84C_cat_CBM32 and the cohesin module into GH84C_cat-CBM32-Cohesin. Further models should be generated for GH84C_cat-CBM32-Cohesin using 10–20× BUNCH. Evaluate the goodness of fit for all the models to the GH84C_cat-CBM32-Cohesin scattering curve using CRYSOL and determine the best model (usually that with the lowest χ_{crysol} value).

2. One important consideration when fitting the individual CBM32 and cohesin modules into the SAXS envelopes containing the missing link from GH84C (CBM32-Cohesin) is that both the CBM and the cohesin module appear radially symmetric in the low-resolution envelopes, thus, while the relative angles between the long axes of the modules can be established, the rotational orientations of the modules relative to one another are impossible to determine. This has important implications in the final model as the orientation of the dockerin interaction site is not resolvable, though sterically the enzyme is most likely to have the dockerin interaction site facing away from the CBM or the catalytic module to accommodate the protein-protein interaction with dockerin containing enzymes.
3. To build the full-length model of GH84C, the cohesin module from the crystal structure of Cohesin-FN3 should be overlaid onto the cohesin module of the best GH84C_cat-CBM32-Cohesin model. This will generate a reasonable model for the GH84C full-length structure, GH84C_cat-CBM32-Cohesin-FN3 (Fig. 2d).
4. Interpret the results. The pseudo-atomic resolution full-length structure of GH84C (Fig. 2d) should reveal that catalysis and carbohydrate binding, while distal (60 Å), are coordinated in the same line of direction in the enzyme. Thus, adherence and catalysis may be optimized for terminal sugars, such as those found on the host cell or mucosal surface. The model should also place the Cohesin:Dockerin interaction site distal to the CBM32 ligand binding site, permitting concurrent adherence to carbohydrate ligand and enzyme recruitment. The positioning of the cohesin module relative to the CBM and catalytic module facilitates the placement of recruited enzymes in line themselves for attack on the carbohydrate surface. The recruitment of other ultra multimodular CAZymes, themselves having CBMs, raises the interesting possibility of avid binding to a carbohydrate bearing surface mediated by complex protein-carbohydrate interactions shared between two interacting CAZymes.

4 Notes

1. Characterization of the modular structure requires careful bioinformatic analysis of the boundaries between the individual modules. The potential signal peptides and transmembrane domains need to be predicted using SignalP and TMHMM, for example [36, 37]. The modularity of the mature enzyme or protein can then be examined using BLAST queries against UniProt database, as well as domain searches with the InterPro server [38, 39]. Coincidence of modular boundaries of at least three similar sequences gives a good estimate of where an indi-

vidual module may start and end. A more precise delineation of each module can then be refined using Hydrophobic Cluster Analysis (HCA) [40]. It may also be useful to do a secondary structure prediction to avoid splitting within an alpha helix or beta strand that could destabilize the protein [41].

2. When determining multimodular structures using the "dissect and build" method it is best to have as many different modular constructs as reasonably possible. Therefore, clone as many combinations of the modular constructs as is possible, as determined through careful bioinformatics analysis, into an appropriate expression vector, e.g., pET-28a(+) for production in *E. coli*. This redundancy will optimize the confidence in the constructed model and provide a backup in case of construct instability. The importance here is to have a surplus of soluble recombinant protein, so choose the expression strain and the expression vector carefully.

3. Lyse the cells using an appropriate method such as the French pressure cell or chemical lysis. It is important to have a good lysis protocol for maximum yields.

4. Accurate structural parameters of a macromolecule in solution are deduced from the scattering signal only if measured from a sample of monodisperse, identical particles. SAXS thus requires pure protein and any degradation products will influence the SAXS curves and the molecular envelopes generated will be biased by these impurities. Since the scattering signal increases proportional to the square of the molecular weight (MW), small amounts of contamination by large species (such as aggregates) will contribute disproportionately to the scattering signal and bias the data to larger structural parameters than those expected for the molecule of interest. At minimum, a two-step purification, such as purification on an IMAC (immobilized metal affinity chromatography) column through a 6his tag followed by a SEC (size exclusion chromatography) column to remove aggregates and other impurities, is recommended. Protein purity should not only be estimated by SDS-PAGE, a Dynamic Light Scattering (DLS) analysis should be performed prior to sample collection to confirm the absence of aggregates.

5. The scattering signal of a macromolecule in the solution is measured as the difference to the scattering from that of its surrounding solvent. It is thus essential to match the buffers (e.g., 50 mM Tris–HCl pH 7.5 and 100 mM NaCl) exactly for accurate buffer subtraction; a single dialysis between 24 and 48 h is usually sufficient and the dialysate should be used for all dilutions and measurements. Filter a stock of buffer immediately (approx. 50–100 ml) into a clean container for future use, dilutions, etc. It is simplest if all the constructs are buffer exchanged into the same buffer, though this may not always be possible.

6. If freezing samples for transport to the synchrotron, flash freezing in liquid nitrogen is recommended.
7. Check with the beamline scientist, before preparing the samples, on how much sample volume is required per measurement. Have more than enough protein prepared, just in case.
8. Data should be collected on several (at least five) different concentrations of protein; typically a good data collection is obtained on samples that are between 1 mg ml^{-1} and 10 mg ml^{-1} with volumes between 10 and 50 μL. SAXS data should also be collected on one or more characterized standards, such as BSA or hen egg white lysozyme. In between each protein sample collection a buffer sample should be collected for buffer subtraction.
9. Major problems in SAXS experiments are denaturation, degradation, and aggregation of the protein sample. With problem proteins we have had some success with collecting SAXS data in several different buffers with the hopes of landing on the "right one" for protein stability. Another solution has been to take both fresh (4 °C) and frozen samples to the synchrotron and thawing the frozen sample just before collection. If aggregation cannot be avoided it is probably best to collect data at a beamline that couples size exclusion chromatography with SAXS. In all cases centrifuge all samples before use to minimize aggregate (e.g., 20,000 × *g* for 10 min).
10. The development of small-angle scattering instrumentation is not devoid of technical challenges; however, schematically the experimental setup is relatively simple. A highly collimated X-ray beam, traditionally of a single wavelength, is used to illuminate the sample contained within a capillary (Fig. 1). The scattered radiation is recorded on a detector, while the direct beam is usually absorbed by a beam stop, the size and position of which are key factors for determining the minimum angle measured in an experiment. Online size exclusion chromatography has become available at recent beamlines (e.g., ID-18 BioSAXS/APS, SWING/SOLEIL, BL4–2/SSRL, BM-29/ESRF, I22/Diamond, P12/Petra, SR13 ID01/Australian Synchrotron), where SAXS data is collected on the sample as it emerges from a size exclusion column, minimizing any aggregative contributions to the SAXS curve.
11. A detailed explanation on the physics behind SAXS and crystallography has been elaborated extensively by Putnam et al. [23] and the more practical aspects for good SAXS data collection and treatment by Jacques and Trewhella [25]. The ATSAS software suite provides a powerful group of programs [19, 20] for SAXS data treatment. It is important to note here that SAXS data analysis gives rise to measures of Rg, I(0) and Dmax

values, and all further modeling is associated with ambiguity. There can be several models that equally well describe a given scattering curve.

12. Pdb files should be cleaned up by deleting any waters or ligands from the text file, any alternative conformations of amino acids should also be deleted. Oftentimes if one of the ATSAS modeling programs is unable to start or finish, it is due to a problem with mismatching amino acid sequences between the pdb file(s) and the input sequence.
13. In many cases, it may be useful to construct a 3D model of a module if no pdb is publicly or privately available, in this case the Protein Homology/analogY Recognition Engine V 2.0 (Phyre2) [42] is practical.
14. Flexible systems. If flexibility in a system is suspected the SAXS data should be analyzed by the ensemble optimization method (EOM) [43], which also fits the modules as rigid bodies but outputs an ensemble of models to fit the SAXS scattering curve. Briefly, the program RanCH generates a pool of 10,000 models and selection of the optimized ensemble is realized by GAJOE. An overall χ_{EOM} is determined; however, we also find it informative to determine the individual χ_{crysol} values for each of the models in the ensemble.
15. Produce several models using different methodologies to ensure reproducibility which will provide confidence in the final model. For example, one model may be a crystal structure, another model may be generated by manually placing modules into the 20× DAMMIF DAMAVER averaged molecular envelope and other models generated by 10–20× BUNCH. EOM has on some occasions provided the best χ_{crysol} value from one model in its ensemble, even relative to BUNCH. Thus, it may be useful to generate "static" models using EOM even if flexibility is not suspected in the system.
16. Good notetaking skills are essential. Note down what you do and how you obtained your models while it is still fresh. SAXS experiments involve a massive amount of data analysis and without good notetaking skills it is very easy to lose the thread.
17. Finally, there is a very useful forum for any inevitable SAXS-related questions you have found here: http://www.saxier.org/forum.

References

1. Lammerts van Bueren AL et al (2011) The conformation and function of a multimodular glycogen-degrading pneumococcal virulence factor. Structure 19(5):640–651
2. Moustafa I et al (2004) Sialic acid recognition by *Vibrio cholerae* neuraminidase. J Biol Chem 279(39):40819–40826
3. Malecki PH, Vorgias CE, Petoukhov MV, Svergun DI, Rypniewski W (2014) Crystal structures of substrate-bound chitinase from the psychrophilic bacterium *Moritella marina* and its structure in solution. Acta Crystallogr D Biol Crystallogr 70(Pt 3): 676–684

4. Trastoy B et al (2014) Crystal structure of *Streptococcus pyogenes* EndoS, an immunomodulatory endoglycosidase specific for human IgG antibodies. Proc Natl Acad Sci U S A 111(18):6714–6719
5. Turkenburg JP et al (2009) Structure of a pullulanase from *Bacillus acidopullulyticus*. Proteins 76(2):516–519
6. Koropatkin NM, Smith TJ (2010) SusG: a unique cell-membrane-associated alpha-amylase from a prominent human gut symbiont targets complex starch molecules. Structure 18(2):200–215
7. Mikami B et al (2006) Crystal structure of pullulanase: evidence for parallel binding of oligosaccharides in the active site. J Mol Biol 359(3):690–707
8. da Silva VM et al (2014) Modular hyperthermostable bacterial endo-beta-1,4-mannanase: molecular shape, flexibility and temperature-dependent conformational changes. PLoS One 9(3):e92996
9. Suits MD et al (2014) Conformational analysis of the *Streptococcus pneumoniae* hyaluronate lyase and characterization of its hyaluronan-specific carbohydrate-binding module. J Biol Chem 289(39):27264–27277
10. Higgins MA, Ficko-Blean E, Meloncelli PJ, Lowary TL, Boraston AB (2011) The overall architecture and receptor binding of pneumococcal carbohydrate-antigen-hydrolyzing enzymes. J Mol Biol 411(5):1017–1036
11. Albesa-Jove D et al (2010) Four distinct structural domains in *Clostridium difficile* toxin B visualized using SAXS. J Mol Biol 396(5):1260–1270
12. Smith SP, Bayer EA (2013) Insights into cellulosome assembly and dynamics: from dissection to reconstruction of the supramolecular enzyme complex. Curr Opin Struct Biol 23(5):686–694
13. Czjzek M, Fierobe HP, Receveur-Brechot V (2012) Small-angle X-ray scattering and crystallography: a winning combination for exploring the multimodular organization of cellulolytic macromolecular complexes. Methods Enzymol 510:183–210
14. Ficko-Blean E et al (2009) Portrait of an enzyme, a complete structural analysis of a multimodular beta-N-acetylglucosaminidase from *Clostridium perfringens*. J Biol Chem 284(15):9876–9884
15. Canard B, Garnier T, Saint-Joanis B, Cole ST (1994) Molecular genetic analysis of the nagH gene encoding a hyaluronidase of *Clostridium perfringens*. Mol Gen Genet 243(2):215–224
16. Ficko-Blean E, Boraston AB (2005) Cloning, recombinant production, crystallization and preliminary X-ray diffraction studies of a family 84 glycoside hydrolase from *Clostridium perfringens*. Acta Crystallogr Sect F 61:834–836
17. Ficko-Blean E, Boraston AB (2006) The interaction of a carbohydrate-binding module from a *Clostridium perfringens* *N*-acetyl-beta-hexosaminidase with its carbohydrate receptor. J Biol Chem 281(49):37748–37757
18. Adams JJ, Gregg K, Bayer EA, Boraston AB, Smith SP (2008) Structural basis of *Clostridium perfringens* toxin complex formation. Proc Natl Acad Sci U S A 105(34):12194–12199
19. Petoukhov MV et al (2012) New developments in the ATSAS program package for small-angle scattering data analysis. J Appl Cryst 45:342–350
20. Konarev PV, Petoukhov MV, Volkov VV, Svergun DI (2006) ATSAS 2.1, a program package for small-angle scattering data analysis. J Appl Cryst 39:277–286
21. Mach H, Middaugh CR, Lewis RV (1992) Statistical determination of the average values of the extinction coefficients of tryptophan and tyrosine in native proteins. Anal Biochem 200(1):74–80
22. Konarev PV, Volkov VV, Sokolova AV, Koch MHJ, Svergun DI (2003) PRIMUS: a Windows PC-based system for small-angle scattering data analysis. J Appl Cryst 36(5):1277–1282
23. Putnam CD, Hammel M, Hura GL, Tainer JA (2007) X-ray solution scattering (SAXS) combined with crystallography and computation: defining accurate macromolecular structures, conformations and assemblies in solution. Q Rev Biophys 40(3):191–285
24. Hjelm RP (1985) The small-angle approximation of X-Ray and neutron scatter from rigid rods of non-uniform cross-section and finite length. J Appl Cryst 18(Dec):452–460
25. Jacques DA, Trewhella J (2010) Small-angle scattering for structural biology--expanding the frontier while avoiding the pitfalls. Protein Sci 19(4):642–657
26. Svergun DI (1992) Determination of the regularization parameter in indirect-transform methods using perceptual criteria. J Appl Cryst 25:495–503
27. Fischer H, Neto MD, Napolitano HB, Polikarpov I, Craievich AF (2010) Determination of the molecular weight of proteins in solution from a single small-angle X-ray scattering measurement on a relative scale. J Appl Cryst 43:101–109
28. Mylonas E, Svergun DI (2007) Accuracy of molecular mass determination of proteins in solution by small-angle X-ray scattering. J Appl Cryst 40:S245–S249

29. Orthaber D, Glatter O (2000) Synthetic phospholipid analogs: a structural investigation with scattering methods. Chem Phys Lipids 107(2):179–189
30. Volkov VVS, Svergun DI (2003) Uniqueness of *ab initio* shape determination in small-angle scattering. J Appl Cryst 36:860–864
31. DFaDI S (2009) DAMMIF, a program for rapid *ab initio* shape determination in small-angle scattering. J Appl Cryst 42(2):342–346
32. DeLano WL (2002) The PyMOL molecular graphics System. DeLano Scientific, San Carlos, CA, USA.
33. Kozin MB, Svergun DI (2001) Automated matching of high- and low-resolution structural models. J Appl Cryst 34:33–41
34. Svergun D, Barberato C, Koch MHJ (1995) CRYSOL - a program to evaluate x-ray solution scattering of biological macromolecules from atomic coordinates. J Appl Cryst 28:768–773
35. Petoukhov MV, Svergun DI (2005) Global rigid body modeling of macromolecular complexes against small-angle scattering data. Biophys J 89(2):1237–1250
36. Bendtsen JD, Nielsen H, von Heijne G, Brunak S (2004) Improved prediction of signal peptides: SignalP 3.0. J Mol Biol 340(4): 783–795
37. Krogh A, Larsson B, von Heijne G, Sonnhammer EL (2001) Predicting transmembrane protein topology with a hidden Markov model: application to complete genomes. J Mol Biol 305(3):567–580
38. Altschul SF, Gish W, Miller W, Myers EW, Lipman DJ (1990) Basic local alignment search tool. J Mol Biol 215(3):403–410
39. Quevillon E et al (2005) InterProScan: protein domains identifier. Nucleic Acids Res 33(Web Server issue):116–120
40. Lemesle-Varloot L et al (1990) Hydrophobic cluster analysis: procedures to derive structural and functional information from 2-D-representation of protein sequences. Biochimie 72(8):555–574
41. Cole C, Barber JD, Barton GJ (2008) The Jpred 3 secondary structure prediction server. Nucleic Acids Res 36:W197–W201
42. Kelley LA, Mezulis S, Yates CM, Wass MN, Sternberg MJ (2015) The Phyre2 web portal for protein modeling, prediction and analysis. Nat Protoc 10(6):845–858
43. Bernado P, Mylonas E, Petoukhov MV, Blackledge M, Svergun DI (2007) Structural characterization of flexible proteins using small-angle X-ray scattering. J Am Chem Soc 129(17):5656–5664

Chapter 20

Metagenomics and CAZyme Discovery

Benoit J. Kunath, Andreas Bremges, Aaron Weimann, Alice C. McHardy, and Phillip B. Pope

Abstract

Microorganisms play a primary role in regulating biogeochemical cycles and are a valuable source of enzymes that have biotechnological applications, such as carbohydrate-active enzymes (CAZymes). However, the inability to culture the majority of microorganisms that exist in natural ecosystems using common culture-dependent techniques restricts access to potentially novel cellulolytic bacteria and beneficial enzymes. The development of molecular-based culture-independent methods such as metagenomics enables researchers to study microbial communities directly from environmental samples, and presents a platform from which enzymes of interest can be sourced. We outline key methodological stages that are required as well as describe specific protocols that are currently used for metagenomic projects dedicated to CAZyme discovery.

Key words Metagenomics, Carbohydrate active enzymes, Microbial communities, Assembly, Binning

1 Introduction

The continuing initiative to find novel carbohydrate-active enzymes (CAZymes) is derived from societal and industrial interest in utilizing plant biomass as a substrate for biofuel production. Cellulose, the most abundant form of carbon on earth, is notoriously difficult to deconstruct using currently available enzyme technology, whereas a variety of ecosystems such as the digestive-tract of herbivores [1] or termites' [2] are able to efficiently utilize lignocellulosic biomass. This functional capacity is controlled by microorganisms that are largely irrecoverable in isolate form, which restricts direct access to their genetic and enzymatic machinery. Cultivability "bottlenecks" can be addressed by applying culture-independent methods such as metagenomics, whereby total DNA is directly extracted and analyzed from the microbial sample without any need for prior isolation. Numerous metagenomic studies have been used to find CAZymes [2–4] and can be designed using two differing approaches that are defined as either function- or

D. Wade Abbott and Alicia Lammerts van Bueren (eds.), *Protein-Carbohydrate Interactions: Methods and Protocols*, Methods in Molecular Biology, vol. 1588, DOI 10.1007/978-1-4939-6899-2_20,

sequence-based. Function-based approaches involve the ligation of environmental DNA fragments into vectors that are introduced into "pet" strains such as *Escherichia coli*. The DNA can then be expressed and the library screened for specific functions [5–7]; however, screening limitations have restricted the discovery of CAZymes that can deconstruct recalcitrant forms of cellulose. Sequence-based approaches can be divided into "targeted," i.e., PCR amplification of particular genes of interest [8], or "shotgun" approaches that randomly sequence DNA extracted from an environmental sample without any prior cloning or amplification step.

Shotgun metagenomics can theoretically generate sequences (reads) from all of the genomes present in the sample, collectively referred to as the metagenome. In particular, recent bioinformatics progressions permit the complete reconstruction of a significant fraction of constituent genomes within a sample. These methods have already shown potential to find new non-cultivable polysaccharide-degrading bacteria [6, 9], and interpret synergistic relationships between uncultured phylotypes in a cellulolytic community [10]. Despite the inherent benefits, the use of shotgun metagenomics also comes with several defined challenges. The diversity of microbial species in digestive ecosystems combined with the immense data output of next-generation sequencing (NGS) platforms produces gigabytes of data that needs to be heavily annotated before it can be meaningfully interpreted. Today, a wide variety of different NGS platforms and bioinformatic packages tailor-made for metagenome projects are available to scientists. However, careful considerations of available resources and technical challenges that are relevant for a particular sample are still required. It is the aim of this chapter to describe the application of the biological and bioinformatic methods that are currently available in this rapidly developing field and to provide exemplar step-by-step protocols to follow for samples originating from plant biomass degrading ecosystems (Fig. 1).

2 Materials

2.1 Buffers

1. Dissociation buffer: Mix 0.1% tween 80, 1% methanol, and 1% tertiary butanol (v/v) in 1 L milliQ water. Adjust to pH 2 by adding HCl.
2. Cell wash buffer: Make a solution consisting of 10 mM Tris–HCl (pH 8.0) and 1 M NaCI, sterilized by autoclaving.
3. RBB+C buffer: Make a solution consisting of 500 mM NaCl, 50 mM Tris–HCl; 50 mM EDTA, 4% SDS.
4. NaCl/CTAB buffer: Dissolve 4.1 g NaCl and 10 g Cetyltrimethylammonium Bromide (CTAB) in 80 ml of milliQ water. Heat to 68 °C to enhance the dissolving process. Add milliQ water to a final volume of 100 ml.

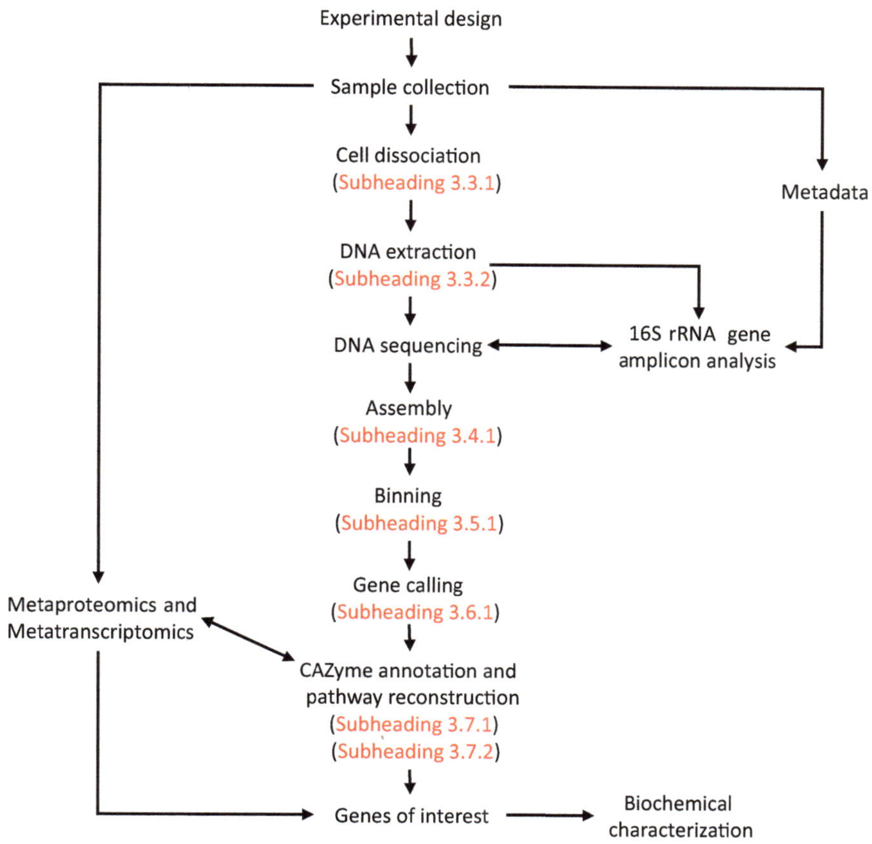

Fig. 1 Flow diagram of a typical metagenomic project dedicated to CAZyme discovery. Work flow represented by *solid lines* is discussed with specific protocols included (*red text*). *Dotted lines* indicate additional complementary techniques that are not presented in detail within this chapter

5. TE buffer: 10 mM Tris–HCl (pH 7.6) and 1 mM EDTA (pH 8.0).
6. 5 M NaCl: Dissolve 29.2 g NaCl in 100 ml milliQ water. Sterilize by autoclaving.
7. Tris–HCl: Dissolve 121.1 g Tris in 800 ml milliQ water. Adjust to pH 7.6 with HCl. Add milliQ water until a volume of 1.0 L is reached.
8. 4% SDS: Dissolve 4 g SDS in 96 ml milliQ water. Heat to 68 °C to enhance the dissolving process of SDS. Adjust to pH 7.4. Add milliQ water to a final volume of 100 ml.
9. EDTA solution: Dissolve 186.12 g EDTA•Na_2•$2H_2O$ (Molecular Weight 372.24) in 800 ml milliQ water. While stirring vigorously on a magnetic stirrer, add NaOH pellet or 10 N NaOH to adjust the solution pH 8.0. Adjust the volume to 1000 ml with deionized/Milli-Q water. Sterilize by autoclaving.

2.2 Computational Software

1. SMRT Analysis v2.3.0 (http://www.pacb.com/support/software-downloads/).
2. Sickle v1.33 (https://github.com/najoshi/sickle).
3. IDBA_UD v1.1.1 (https://github.com/loneknightpy/idba).
4. MIRA v4.0.2 (http://sourceforge.net/projects/mira-assembler/).
5. Taxator-tk v1.3.3 (https://github.com/fungs/taxator-tk/releases).
6. MetaBAT v0.26.3 (https://bitbucket.org/berkeleylab/metabat).
7. CheckM v1.0.6 (https://github.com/Ecogenomics/CheckM/releases).
8. MetaGeneMark v3.26 (http://exon.gatech.edu/genemark/license_download.cgi). MetaGeneMark can also be used via a web service. (http://exon.gatech.edu/meta_gmhmmp.cgi).
9. dbCAN (http://csbl.bmb.uga.edu/dbCAN/download.php).
10. Traitar v1.04 (https://github.com/hzi-bifo/traitar).

3 Methods

3.1 Strategy Development

For metagenomic projects dedicated to CAZyme discovery, the goal is relatively straightforward: obtain large contiguous DNA fragments that are coupled with the fewest possible misassemblies so that many complete genes and operons are available for screening. Therefore, understanding which sequencing platform and the amount of sequencing required to achieve a superior quality dataset are crucial when designing the project [11]. Different sequencing technologies have vastly different read lengths and base pair (bp) yield and require different amounts of sample DNA and downstream bioinformatic analyses (Table 1). The suitability of a specific type of NGS data to a given sample largely depends on the community structural complexity. Typically, shotgun sequencing results in deep coverage of dominant species and less reads from lower abundant species. To accommodate this discrepancy and gain greater access to the rarer community members, the sequencing effort must be increased; however, computational limitations need to be considered. Therefore, an initial evaluation of the community structure using 16S rRNA gene amplicon analysis [12, 13] is recommended to help determine the complexity, as well as the choice of NGS platform and the depth of sequencing to obtain the dataset necessary for CAZyme gene searches.

3.2 Sample Collection and Metadata

The collection and processing of environmental samples is the first stage when planning a metagenomic project. Key considerations include the number of samples needed to adequately view the

Table 1
DNA requirements and data output for Illumina HiSeq and PacBio CCS

	Illumina HiSeq	PacBio RS II P6-P4 chemistry CCS
Sequencing yield		
Read length	150 bp paired-end	500–2500 bp CCS
Number of reads	up to 4 billion × 2 per run	>200,000 CCS per 8× SMRT cells
Yield	up to 1 Tb per run	0.5–1 Gb CCS per 8× SMRT cells
DNA quality requirements		
260/280 ratio	1,8–2,1	1.8–2.0
260/230 ratio	1,8–2,4	2.0–2.2*
DNA quantity requirements		
PCR-free	2–10 μg in 130 μl 10 mM Tris–HCl or TE buffer	At least 10 μg at a concentration of at least 50–75 ng/μl in 10 mM Tris–HCl
PCR	2 μg in 130 μl 10 mM Tris–HCl or TE buffer	n/a
Comments	DNA concentration should not exceed 1 μg/μl due to viscosity issue	*PacBio sequencing is extremely sensitive to denaturants and environmental contaminants such as guanidine salts, phenol, humic acid, polyphenols

temporal dynamics or variability within the ecosystem as well as fulfill operating requirements of downstream bioinformatic software that require multiple samples (*see* Subheading 3.5). Moreover, other "omic" techniques such as metaproteomics and metatranscriptomics can be used to efficiently complement the metagenomic analysis and its outcomes (Fig. 1 and Subheading 3.8). Extra subsamples can be easily stored in standardized ways to be analyzed later if required. In addition to the number of samples, the accompaniment of "metadata" can greatly enhance the ability to interpret the sequence data and particularly for comparative, spatial, or temporal series analysis. Metadata should appropriately describe the samples and the methods used. A suite of standard languages, called the Minimum Information about any (x) Sequence checklists (MIxS) [14], provides format for recording environmental and experimental data. These standards include MIGS (Minimum Information about a Genome Sequence) and MIMS (Minimum Information about a Metagenome Sequence) checklists [15, 16]. What is recorded depends on where the samples come from but usually includes, among others, temperature, pH, substrate, sample handling, DNA extraction method, sequencing technology, and the bioinformatic methods used.

3.3 Cell Dissociation and DNA Extraction Methods

When looking for CAZymes, samples vary greatly from different environments, such as guts, soils, excrements, sediments, bioreactors, and other plant-associated biomass. These habitats present different characteristics such as the presence of host cells [2, 17], enzymatic inhibitors (such as humic acids) [18], or biofilms [19], and therefore require specific protocols. Samples of plant-biomass degrading communities often require an additional processing step to remove microbial cells that are attached to plant fibers [19]. In Subheading 3.3.1, we describe a high-yield method that enables the dissociation of the microbial cells from the substrate and the recovery of the full sample's diversity. The resulting cell biomass generated from this dissociation protocol is suitable for both kit-based DNA extractions [20, 21] and Mamur's derived protocols [19].

DNA extraction is an important part of the experimental design since it can have a major impact on the subsequent NGS platform and downstream bioinformatic analyses that are used. DNA extraction and purification is still considered a bottleneck for metagenomic analyses, compounded by the fact that there is not one common method that fits every environmental sample. Indeed, the quality and the quantity of DNA required vary from a sequencing technology to another and may influence the choice of the DNA extraction method (Table 1). For NGS platforms that require high concentrations of high molecular weight (HMW) DNA (e.g., PacBio), commercial kits may not provide adequate amounts of DNA. In these instances, gentle noninvasive methods are required (*see* Subheading 3.3.2), which produce far greater yields of HMW DNA without comprising the quality of community structural representation. In instances where a small cell biomass prohibits high nanograms or micrograms quantities of DNA, whole genome amplification of starting material can also be necessary. As with any amplification method, sequence biases can occur [22, 23] and their impacts depend on the amount of starting material and the required number of amplification rounds to produce sufficient amount of DNA.

3.3.1 Cell Dissociation for Plant-Associated Biomass Samples

The sample preparation is initially dependent on whether the sample has been stored as biomass sample at −80 °C or in 1/5 volume of phenol/ethanol (5%/95%) pH 8.0 at 4 °C (*see* **Note 1**).

1. Transfer 1.0 g of biomass or 1.5 ml sample and phenol/ethanol mix with a wide bore pipette to a 2 ml tube.
2. Centrifuge at 14,000 rcf at room temperature for 2 min.
3. Discard supernatant and resuspend biomass in 500 μl of dissociation buffer by vortexing for 30 s.
4. Centrifuge at 100 rcf for 20 s at room temperature and transfer cell-containing supernatant to a new cell-collection tube and centrifuge at 14,000 rcf for 5 min at room temperature. Discard cell-free supernatant.
5. Repeat dissociation buffer **steps 3** and **4** two to three more times transferring each cell-containing supernatant in the same cell-collection tube (*see* **Note 2**).

6. Resuspend concentrated cell pellet in 1 ml of cell wash buffer.
7. Centrifuge at 100 rcf for 20 s at room temperature and transfer cell-containing supernatant to a new tube and centrifuge at 14,000 rcf for 5 min at room temperature. Discard cell-free supernatant.
8. Resuspend cell pellet in 1 ml of cell wash buffer.
9. Count cells and dispense ~10^{10} cells per 2 ml tube.
10. Centrifuge at 14,000 rcf for 2 min at room temperature and discard supernatant. Wet cell pellet should weigh ~ 200 mg (*see* **Note 3**).
11. Proceed to DNA extraction.

3.3.2 HMW DNA Extraction Suitable for PacBio

1. Resuspend cell pellet into 1 ml RBB+C buffer.
2. Incubate for 20 min at 70 °C, mix tube by inversion every 5 min.
3. Split into 2 × 1.5 ml tubes and add NaCl to 0.7 M and 1:10 volume of CTAB Buffer.
4. Heat 70 °C for 10 min.
5. Add an equal volume of Chloroform. Mix well and centrifuge at 14,000 rcf for 15 min at room temperature. Transfer aqueous phase to new tube (*see* **Note 4**).
6. Add equal volume of Phenol:Chloroform:Isoamylalcohol (25:24:1). Mix well, centrifuge at 14,000 rcf for 15 min at room temperature, and transfer aqueous phase to new tube. !CAUTION Phenol, chloroform, and isoamylalcohol are harmful. Handle using appropriate safety equipment and measures.
7. Add 2× vol of 95% ethanol and mix gently until DNA spools. Use a sterile loop to transfer the DNA to a tube containing 200 μl 70% ethanol (*see* **Notes 5** and **6**).
8. Centrifuge at 14,000 rcf for 2 min at room temperature and carefully discard supernatant. Briefly air-dry the pellet.
9. Resuspend in 20–30 μl TE buffer (pH 8.0) and stand at room temperature for 30–60 min to allow DNA to dissolve. Using a spectrophotometer, measure the OD_{260} to estimate DNA concentration. DNA concentration should be adjusted to ~ 0.5 μg/μl (*see* **Notes** 7 and **8**).
10. Removal of sequencing inhibitors. As mentioned in Table 1, PacBio sequencing is extremely sensitive to environmental contaminants. If the DNA quality does not fit the requirements, PacBio recommends using MoBio PowerClean Pro DNA Clean-Up Kit following manufacturer's instructions to remove the contaminants.

3.4 Sequencing and Assembly

Numerous next-generation sequencing (NGS) technologies are now available, providing cheaper, faster, and higher-throughput sequencing (*see* reviews [24] and [25]). Methods that produce short reads, such as Illumina, can generate high sequencing depth at comparatively low costs (Table 1). Illumina's HiSeq can generate over 1 terabyte of bases per run (2 × 150 bp, High Output Run Mode, Dual Flow Cell), whereas the MiSeq can produce up to 15 Gbp (with 2 × 300 bp), with both platforms exhibiting mean error rates < 1% [26]. However, the high quantity of data for samples with high species complexity leads to increased difficulties for assembly, due to computational requirements. In theory, longer read sequencing technologies can overcome many of the known assembly problems associated with short reads because they have the potential to resolve complex repeats and span entire open reading frames (ORFs). However, these technologies are traditionally accompanied with other inherent issues, such as lower sequencing depth and higher error rates. Examples of third generation sequencing technology include Oxford Nanopore's MinION and single-molecule real-time sequencing (SMRT) developed by Pacific Bioscience (PacBio). In particular, PacBio can provide reads of which 50% are over 20 kb and 5% exceeding 30 kb (PacBio RS II). High sequencing costs and error rates reported as high as 15% in individual reads have previously prevented the use of raw PacBio reads in metagenomics. However, it is possible to generate accurate sequences by reading a circularized molecule multiple times (circular consensus or CCS) [27]. This approach reduces read length by a factor equal to the number of times the molecule is traversed but can ultimately result in high-quality sequences with greater than 99% accuracy and about 500–2500 nt in length (Table 1).

Assembly is a key stage required to generate large contiguous sequence fragments (contigs), which are required to maximize the number of open reading frames and operons available for downstream CAZyme screening. Assembly algorithms that process metagenomic data are highly sensitive to the read coverage for community members, which is correlated to the species complexity in a sample and the metagenomic sequencing depth. A plethora of assembly algorithms are currently available (reviewed by [24, 28]), including several that are designed to handle large metagenomic datasets such as IDBA_UD [29], MEGAHIT [30], and metaSPAdes [31]. Algorithms designed for longer reads typically use an overlap-consensus approach, where sufficiently similar reads (based on an overlapping nucleotide region) would be merged into a contig. Alternatively, many short-read assemblers use a de Bruijn graph approach and initially deconstruct each read into a series of oligomers of a set "word" length (commonly referred to as "*k*-mers"). The *k*-mer length is often a user-specified parameter, with longer *k*-mers overcoming repetitive/nonunique regions in the metagenome at a cost of reduced coverage and accuracy. In contrast, short *k*-mers generate contigs with higher coverage, but often shorter in length. Illumina data can provide high-quality metagenomic assemblies from complex microbial communities, which are conducive to CAZyme searches. Specific examples include large datasets that have been assembled from the rumen microbiome [3]. Subheading 3.4.1, details the assembly of HiSeq data using IDBA_UD [29], an iterative (iterates from a small *k* to a large *k*) de Bruijn graph de novo assembler for short read sequencing data with highly uneven sequencing depth, which is typically characteristic of many metagenomic datasets.

Alternative approaches to reduce the computational strain of metagenomic assembly include the use of taxonomic binning or normalization methods to select subsets of reads that are then assembled separately as well as hybrid assemblies that use data from multiple sequencing platforms [32, 33]. Hybrid assemblies are still infrequent, despite indications that combined approaches yield improvements in assembly contiguity and per-base accuracy, which are important in CAZyme discovery projects that seek to interrogate larger saccharolytic gene clusters (*see* Subheading 3.7.2). In particular, previous studies have shown that high confidence reads from Illumina can be used to correct the errors inherent in PacBio sequences [34]. Combination of PacBio CCS and HiSeq reads provides improvement in assembly statistics such as total assembly size and large contig size [35]. It also improves the assembly of the universal marker genes, which assists taxonomic binning and enables enhancements in genome reconstruction of uncultured microorganisms that inhabit complex communities [27]. Despite a paucity of algorithms customized for a hybrid input (to our knowledge), co-assembly can be attained using raw reads from PacBio CCS and Illumina datasets or by using a combination of several assemblers that are optimized for the different sequencing platforms [27]. Subheading 3.4.1 outlines the various stages required to assemble PacBio and Illumina raw reads using MIRA 4.0 [36].

When assembling metagenomic datasets from complex community, chimeric assemblies (misassemblies) can occur. Misassemblies can be prevented by producing paired-end reads, whereby one read of the pair may map to a common sequence or a repetitive element (ambiguous region with risk for chimeric assembly), whereas the other can potentially map to a non-ambiguous region and prevent misassembly. Additionally, assembly with more stringent parameters, such as higher identity threshold or longer minimum overlap (Subheading 3.4.1), can also help avoid chimeric assemblies. Even after applying these precautions, contigs should always be inspected. Abrupt change in GC% content and read coverage within a same contig can indicate chimeric assembly. Changes in the contigs' characteristics can be visualized using different tools such as Anvi'o [37] or MGAviewer [38].

3.4.1 Assembly

1. Filter PacBio raw reads using the SMRT Portal and retain CCS reads that produced a minimum accuracy of 0.99.
2. If necessary, trim Illumina raw reads using sickle and the following command line:

```
sickle pe -f forward_reads.fastq -r reverse_reads.fastq -t sanger -o trimmed_forward_reads.fastq -p trimmed_reverse_reads.fastq -s single_reads.fastq
```

   "pe" indicates paired-end reads. The flag "-t" represents the type of quality value. Illumina quality refers to qualities encoded with the CASAVA pipeline between versions 1.3 and 1.7. Illumina quality using CASAVA >= 1.8 is Sanger encoded. Sickle is available at https://github.com/najoshi/sickle.

Assembly of Illumina Using IDBA_UD

IDBA_UD can be run on paired-end files using the program fq2fa (bundled with IDBA_UD) with the parameters --merge --filter to generate an interleaved FASTA file, and then this file can be assembled with IDBA_UD using default parameters

1. Convert Illumina paired-end fastq files in one fasta file using the following command line:

```
fq2fa --merge --filter forward_reads.fastq reverse_reads.fastq combined_reads.fasta
```

2. Using IDBA_UD, assemble Illumina reads using the following command line:

```
idba-ud -r combined_reads.fasta -o idba_assembly --pre_correction --num_threads 15
```

Assembly of PacBio CCS Data Using MIRA

1. Use modified MIRA configuration file that includes the following:

```
project = #project name#
job = genome,denovo,accurate

parameters = COMMON_SETTINGS -NW:cmrnl=warn \
             PCBIOHQ_SETTINGS -CL:pec=yes -
             ALIGN:min_relative_score=90 -
             ALIGN:min_overlap=40 \
readgroup = PacBioData
data = #/filepath#
technology = pcbiohq
```

   Pcbiohq indicates MIRA that the sequencing technology is high-quality PacBio (CCS). CL:pec=yes is used to trim back ends of reads to an area without sequencing errors, which can happen in PacBio CCS. ALIGN parameters are more stringent than default parameters, which prevent misassemblies that can occur in some high complexity datasets.

2. Assemble PacBio reads using the following command line:

```
Mira configuration_file.conf > assembly.log
```

Co-assembly of PacBio CCS and Illumina Datasets

1. Direct co-assembly (MiSeq and PacBio CCS data) using MIRA

   Use modified Mira configuration file that includes the following:

```
project = #project name#
job = genome,denovo,accurate

parameters = COMMON_SETTINGS   -SK:mmhr=1
             -NW:cac=warn   -NW:cdrn=no
             -DI:trt=/work/users/
             -NW:cmrnl=warn \
             PCBIOHQ_SETTINGS -CL:pec=yes
             SOLEXA_SETTINGS -CL:pec=yes

readgroup = HiSeqData
data = #/filepath#
technology = solexa
template_size = 100 400
segment_placement = ---> <---
segment_naming = solexa

readgroup = PacBioData
data = #/filepath#
technology = pcbiohq
```

   The first set of parameters is technology-dependent. Solexa settings are used for Illumina datasets. Pcbiohq directs MIRA that the sequencing technology is high-quality PacBio (CCS). CL:pec=yes is used to trim back ends of reads to an area without sequencing errors, which can happen in PacBio CCS and Illumina raw reads. The second set describes the data provided. If the size of the DNA template and the orientation of the reads are known, they can be given using the parameters template_size and segment_placement.

   More parameters and information can be found on the online manual

   (http://mira-assembler.sourceforge.net/docs/DefinitiveGuideToMIRA.html#chap_reference).

2. Assemble all files using the following command line:

   Mira configuration_file.conf > assembly.log

3.5 Binning

Binning is the post-assembly taxonomic assignment of contigs into genome bins that enables the study of individual organisms (and their interactions), directly from deeply sequenced metagenomes. Therefore, the task of a binning tool is to assign an identifier to every assembled contig, with each identifier ideally representing a single genome [39]. Taxonomic binning tools, such as Megan [40], PhyloPythiaS+ [41], and taxator-tk [42], act as classifiers and label contigs with taxa from an existing taxonomy, such as the NCBI Taxonomy database [43] (Subheading 3.5.1 "Taxonomic Binning with Taxator-tk"). Unsupervised and reference-free binning tools traditionally use oligonucleotide composition to group contigs with similar usage, thus effectively differentiating between contigs of different species, in particular focusing on their tetranucleotide frequencies [44]. Today, binning tools increasingly leverage additional information to improve genome recovery even in the presence of multiple genomes from individual species in a sample, such as paired-end read linkage [45], mean contig coverage [46], per-sample (differential) coverage [47], or combinations thereof [48, 49]. High-quality genomes can be recovered (Subheading 3.5.1 "Unsupervised Binning with MetaBAT"), in particular if multiple metagenomes of the same community were generated, which subsequently can be mined for new CAZymes. In addition, binning results should be inspected carefully by, e.g., looking at taxonomic assignments of individual contigs, visualizing the underlying differential coverage information (as done in Albertsen et al. [50]), or using an automated method for assessing the quality of metagenome-derived microbial genomes [51] (Subheading 3.5.1 "Quality Assessment for Near-Complete Genomes with CheckM"). Overall, computational tool development for taxonomic binning is a very active research area. The "Critical Assessment of Metagenomic Information" (CAMI) initiative [52] continuously benchmarks tools for binning, metagenome assembly, and profiling on various benchmark datasets reflecting common experimental setups and properties of underlying microbial communities. Up-to-date evaluation results for several use cases and commonly utilized software are available at: https://data.cami-challenge.org/

3.5.1 Metagenome Binning

Taxonomic Binning with Taxator-tk

Input:

Metagenome assembly (*see* Subheading 3.4)

Taxator-tk refpack

Usage:

```
binning-last.bash <path to refpack> assembly.fa
```

Output:

Lineage and NCBI Taxonomy ID for each contig.

Taxator-tk is a very precise tool across all ranks and taxa. Therefore, it is exceptionally useful to complement and verify unsupervised binning approaches, as outlined below.

Unsupervised Binning with MetaBAT

Input:

Metagenome assembly (*see* Subheading 3.4).

Read mapping file(s) in BAM format, one file per sample.

Usage:

```
runMetaBat.sh assembly.fa sample1.bam [sample2.
   bam ...]
```

Output:

One Fasta file per genome bin

Usually, there is a tradeoff between a tool's sensitivity and specificity. MetaBAT's default settings work reasonably well for most use cases. However, for very simple or very complex communities, nondefault options might improve binning results. Please run: metabat -h

Quality Assessment for Near-Complete Genomes with CheckM

Input:

Genome bins in Fasta format

Usage:

```
checkm lineage_wf -x fa <bin folder> <output
   folder>
```

Output:

Contamination and completeness estimates for each genome bin

"Contamination" and "completeness" are estimated using a set of clade-specific marker genes that should occur only once per genome. Sometimes, contigs from multiple strains of the same organism can be found in the same genome bin and potentially contribute toward "contamination." Therefore, CheckM also lists the percentage of contamination that can be explained by "strain heterogeneity."

3.6 Gene Calling

Once a metagenomic dataset has been adequately assembled and taxonomically assigned, gene calling or ORF prediction is required to identify protein or RNA coding regions within the (meta) genome. Depending on the assembly, its feasibility, and its success, gene calling can be performed on assembled contigs or raw reads (for long read NGS data). There are two different ways for ORF prediction: the "sequence similarity-based" method and the "ab initio" gene calling method [53]. The "sequence similarity-based" method uses homology searches to identify genes similar to those already present in databases. This method possesses high specificity and ability to characterize functions of predicted genes. The "ab initio" gene calling approach relies on dependencies between codon frequencies and genome nucleotide composition to discriminate coding from noncoding regions. Frequently, metagenomic assemblies result in many genes that are partially sequenced or fragmented. In addition, metagenomic data from diverse communities can have too low similarities with sequences from databases due to evolutionary distance or short contig/read lengths, which can prevent the identification of homologs and poor detection of novel genes. Therefore, ab initio tools such as

MetaGeneMark [54] are essential for metagenomic analysis, especially when looking for novel enzymes [53]. Utilization of MetaGeneMark has been successfully used to predict ORFs from various metagenomes [55–57], and requires minimal effort to operate.

3.6.1 Gene Prediction with MetaGeneMark

MetaGeneMark (v3.26) can be used via a web service *(http://exon.gatech.edu/meta_gmhmmp.cgi)* or downloaded on the GeneMark website and run with the following command line:

```
gmhmmp -m MetaGeneMark_v1.mod -f G -a -d
```

where –f provides a gff output file and –a and –d incorporate the protein and the nucleotide sequences of the predicted genes in the output.

More detail can be found on the MetaGeneMark manual (*https://github.com/ablab/quast/blob/master/libs/genemark/macosx/README.MetaGeneMark*).

3.7 Enzymes/Pathway Annotation

Gene calling is typically followed by functional annotation, which details comparisons of predicted ORFs to previously annotated sequences present in functional databases. The objective is to generate accurate annotations to correctly identify orthologues. There are multiple approaches to annotate ORFs and numerous tools and databases are publicly available. These include, among others, COG (Clusters of Orthologous Groups) for functional grouping [58], Pfam (the protein families database) for the identification of protein families and domains [59], TIGRfam for full-length protein families [60], and Enzyme Commission (E.C.) numbers for numerical classification scheme for enzymes, based on the chemical reactions they catalyze [61]. In addition, particular databases enable reconstruction of pathway maps for cellular and organismal functions. Key examples include KEGG (Kyoto Encyclopedia for Genes and Genomes) [62] and MetaCyc (metabolic Pathway Database) [63].

Many functional annotation resources are collectively available via web-based platforms that provide support for visualization and comparative analysis of metagenomic datasets. For example, the U.S. Department of Energy (DOE)-Joint Genome Institute hosts the Integrated Microbial Genomes with Microbiome Samples—Expert Review (IMG/MER) system, which provides support for functional annotation and curation of metagenomic datasets of interest. A typical pipeline analysis in IMG/MER starts with the user uploading metagenomic contigs and/or unassembled reads. Protein-coding genes are identified using four ab initio gene-calling tools: GeneMark, Metagene, Prodigal, and FragGeneScan. Finally, predicted proteins are compared with protein families and proteomes of selected "core" genomes. Protein sequences are compared with COG using RPS-BLAST [64] and Pfam and TIGRfam using HMMER 3 [65]. Finally, protein-coding genes are associated with KEGG Orthology terms, EC numbers, and phylogeny using USEARCH [66] against a nonredundant reference database composed of the public genomes available on IMG and KEGG databases. Further information and a procedure to submit data on IMG can be found on [67] and on IMG website (*https://img.jgi.doe.gov/*).

3.7.1 CAZyDB, dbCAN and Multimodular CAZymes

For metagenomic projects dedicated to CAZyme discovery it is recommended that ORFs are annotated using a specialized database. The most comprehensive database is the CAZy database [68] (hereafter called CAZyDB), which specializes in the display and analysis of genomic, structural, and biochemical information on carbohydrate-active enzymes. The CAZyDB contains more than 300 families of catalytic and ancillary modules and is presented as glycoside hydrolases (GHs), glycosyltranferases (GTs), polysaccharide lyases (PLs), carbohydrate esterases (CEs), auxiliary activities (AAs), and the carbohydrate binding modules (CBMs). The CAZyDB identifies evolutionary-related families using the classification introduced by Bernard Henrissat [69], which are based on significant amino acid sequence similarity with at least one biochemically characterized founding member. The CAZy group actively develops tools for unambiguous high-throughput modular and functional annotation of CAZymes in sequences issued from genomic and metagenomic efforts. Annotation of unpublished datasets therefore requires collaboration with CAZy researchers; otherwise, query proteins are required to be deposited as finished entries in GenBank (or EMBL and DDBJ) and will be analyzed via their operational routines.

Several alternatives to CAZyDB currently exist for automated CAZyme annotation, including dbCAN [70] and CAT [71]; however, neither has the same levels of manual inspection by expert curators that is offered by CAZy. Importantly, dbCAN enables automated and comprehensive annotation that is based on the classification scheme of CAZy but relies on a defined signature domain model for each CAZy family [70]. In addition, signature domains of each CAZy family are represented by a hidden Markov model that is available to the public and easily amendable to local searches within unpublished metagenomic datasets. Searches against the dbCAN database can be performed either through its web platform or locally and are described below.

Search Against dbCAN Database

Web Server

1. Protein sequences can be loaded on the web server: *http://csbl.bmb.uga.edu/dbCAN/annotate.php*.

Local Search

1. Download the dbCAN Hidden Markov Models (HMMs) database (dbCAN-fam-HMMS.txt) and hmmscan-parser.sh on *http://csbl.bmb.uga.edu/dbCAN/download.php*.

2. Download and install HMMER 3.0 package (hmmer.org).
3. Format HMM db: `hmmpress dbCAN-fam-HMMs.txt`
4. Run: `hmmscan --domtblout yourfile.out.dm dbCAN-fam-HMMs.txt yourfile > yourfile.out`
5. Run: `sh hmmscan-parser.sh yourfile.out.dm > yourfile.out.dm.ps`

Complementary information and extra parameters can be found on the readme file provided by dbCAN website.

3.7.2 Identifying Plant Biomass Degrading Operons

In addition to identifying individual CAZymes, it is valuable to observe a more global picture by identifying the gene localization and organization of CAZymes encoded in the microbial community. This enables visualization of potential plant biomass degrading operons that are encoded within a metagenome. Several saccharolytic mechanisms such as polysaccharide utilization loci (PULs) and cellulosomes are encoded in large gene clusters and have been successfully recovered in metagenomes [10, 72]. Tools such as the IMG/MER are conducive to identifying gene clusters, whereby once a metagenome has been uploaded, signature protein domains for specific gene clusters can be searched and the surrounding ORFs visualized and functionally interrogated. For PULs, several pfam domains represent the archetypical outer-membrane proteins SusC and SusD [73], including the IDs; TonB_dep_Rec (PF00593), TonB_C (PF03544) SusD (PF07980), SusD-like (PF12741), SusD-like_2 (PF12771), and SusD-like_3 (PF14322). For cellulosomes, key signatures include cohesin (PF00963) and dockerin (PF00404) domains. In addition to known operons, other uncharacterized operons can be investigated by using gene identifiers from previously identified CAZymes as a search query and by interrogating surrounding genomic regions. However, such an approach requires manual intervention and is not amendable to large collections of CAZymes. Alternative approaches that use bioinformatics methods to identify uncharacterized operons are discussed below (Subheading 3.8).

3.8 Identifying New Gene Targets

Many CAZymes are multimodular with catalytic modules and one or more additional domains that are often substrate-targeting carbohydrate binding modules (CBMs). "Module walking" is a method that probes the potential CAZyme activity of unknown regions or domains of ORFs that flank annotated CAZyme domains. This method has been used by [74] to find a new LPMO family (AA11). Based on the observation that several sequences from AA9 LPMO family carry a conserved domain of unknown function (X278), Hemsworth et al. searched for other multimodular proteins containing that domain. Subsequently, they looked at the adjacent regions within an X278-encoding ORF and identified hypothetical domains with LPMO-like characteristics. Interestingly, those domains did not exhibit significant similarity to other LPMOs families (AA9 and AA10) and were thus considered a new LPMO family (AA11).

Methods that investigate bacterial genotype-phenotype inferences attempt to link genetic features across bacterial genomes from strains with a common phenotype (e.g., plant biomass degradation) and have the potential to identify novel domains and CAZyme families [75, 76]. In particular, machine learning methods can learn phenotype models by exploiting co-occurrence patterns of phenotype and protein families in genomes or across the course of evolution (shared gain and loss events) to determine sets of protein families that are predictive of a phenotype [77, 78]. These phenotype models can also be used to identify phylotypes that encode the phenotype of interest from metagenomic draft genomes. We previously developed a method that used support vector machines (SVMs) [79] to determine the protein families distinctive for plant biomass degradation from the genomes of a large curated set of plant biomass degrading bacteria and non-degraders [80]. We identified Pfam and CAZy families known to be implicated with plant biomass degradation but also uncharacterized protein families that represent potential new CAZymes. The model was also applied to draft genomes of uncultured phylotypes from a cow rumen metagenome [3] and identified Bacteroidetes-affiliated phylotypes predicted to be involved in plant biomass degradation, which was supported by biochemical analysis of enzymatically active CAZymes [9]. In a complementary approach, we identified functional modules that are distinctive for plant biomass degradation [81]. Each of the five plant biomass degradation modules incorporated protein families involved in degradation of different components of plant material such as cellulose, hemicellulose, and pectin but also the structural components of the cellulosome. Using these modules, we identified many characterized plant biomass degradation-associated operons in genomes of known lignocellulose degraders. Importantly, we also revealed many uncharacterized gene clusters that contain both known CAZy families and hypothetical proteins, which have become targets for on-going biochemical characterization studies in our lab.

Traitar is a newly developed software for the accurate prediction of 67 diverse microbial phenotypes, among them 24 traits dedicated to degradation of various mono- and oligosaccharides, from isolate genomes and genomes from metagenomes or single cells [82]. Genome characterization by Traitar accurately complemented a manual metabolic reconstruction of two uncultured phylotypes that were recovered from a digestive ecosystem. The software can be used as a first pass method to characterize novel phylotypes. For instance, it is particularly suitable for finding saccharolytic genomes and identifying the protein families within these genomes that play a role in carbohydrate degradation (*see* Subheading 3.7).

Other methodologies to consider that can assist in CAZyme identification is the incorporation of additional "meta-omic" data such as metaproteomics or metatranscriptomics, which are powerful tools when used in combination with metagenomics. In particular, mapping meta-omic data against reconstructed genomes enables the visualization of the species identity, together with the relative quantity of key carbohydrate-active enzymes and proteins that are expressed. Such techniques have the potential to detect known CAZymes that are metabolically active, as well as to identify hypothetical genes that are upregulated and/or expressed in response to growth on a particular plant biomass substrate. In such instances, these emphasized ORFs are presented as key targets for downstream biochemical characterization.

3.8.1 Traitar Phenotyping Software

1. Highlights of this software include:
   - Two prediction modes *phypat* and *phypat+PGL*.
   - Phenotyping results visualized via heatmaps.
   - Allows identification of gene targets in the input genomes associated with the predicted phenotypes.
2. Web server. Traitar can be used via a web server for small data sets: https://research.bifo.helmholtz-hzi.de/webapps/wa-webservice/pipe.php?pr=traitar

   Short instructions for the Traitar stand-alone program available for Linux/Unix

   Detailed information available at: https://github.com/hzi-bifo/traitar.
3. User Input:

   Supply amino acid FASTA file (.faa) or nucleotide FASTA file (.fna) for each genome or genome bin via a sample table (tab separated) (the user may link some meta data, e.g., the environment to the input samples).

Sample_file_name	Sample_name	Category
1457190.3.RefSeq.faa	Listeria_ivanovii_WSLC3009	environment_1
525367.9.RefSeq.faa	Listeria_grayi_DSM_20601	environment_2

4. Execution

   Requires *prodigal* for gene calling and *hmmsearch* for Pfam annotation to be available on the command line

```
>traitar phenotype <FASTA dir> <sample_table>
<from_genes/from_nucleotides> <out_dir>
```

4 Notes

1. Biomass samples can be stored at 4 °C in phenol/ethanol (5%/95%) pH 8.0 for several weeks or as a biomass sample at −80 °C for longer periods. We recommend processing the samples as quickly as possible. Longer storage may result in differential lysis of microbial cells.
2. After two to three repetitions the cell-containing supernatant in **step 4** should become much clearer as cells are removed and collected. Long exposure of samples to dissociation buffer should be avoided due to its low pH. It is recommended that not more than two to three repetitions be performed.
3. Dissociated cells may be stored at 4 °C for 1 day or at −20 °C for several weeks.
4. CTAB specifically binds to proteins at high salt concentrations. **Steps 3–5** should remove cell wall debris, denatured proteins, and polysaccharides complexed to CTAB, while retaining the nucleic acids in solution. Aqueous phase should be clear before proceeding to **step 6**. If aqueous phase still retains an opaque-yellow color, repeat **steps 3–5.**
5. Caution! Ethanol is flammable. Handle using appropriate safety equipment and measures.
6. If no DNA spools in **step 7** it implies either the DNA is of low concentration and/or DNA has sheared into relatively low-molecular weight fragments. In this instance, DNA can still be collected by centrifugation at 14,000 rcf for 30 min at 4 °C before proceeding to **step 8**.
7. DNA can take up to an hour at room temperature to resuspend since DNA is HMW. Take care not to over dry pellet as this will make resuspending much more difficult.
8. DNA can be stored at 4 °C for several weeks or at −20 °C for longer terms. It is recommended to restrict the number of freeze thaw cycles as this can degrade HWM DNA.

4.1 Current Challenges

Due to the inherent species complexity found in many microbial communities, many bottlenecks still confront metagenomic projects. A consideration when facing complex communities is to decrease species complexity while keeping its potential for new discovery. This can be achieved by enriching for a consortium that maintains a desired function, such as plant biomass deconstruction. Such an approach can allow the cultivation of microbes that cannot be isolated in pure cultures but provides information on microbial interactions in the biological system under chosen conditions. There are several ways to enrich a community for a specific functional trait. Physico-chemical parameters such as temperature, acidity, or substrate can be fine-tuned to enrich specific types of microorganisms, whereas serial dilutions (dilution-to-extinction approach) can be used to decrease the diversity. The serial dilution method has been applied on communities from soil [10] and the rumen [83], where samples have been enriched on filter paper or avicel to obtain a cellulolytic consortium. In [10], the authors managed to reconstruct the genomes of seven uncultured phylotypes that suggested cooperation for cellulose degradation in soil, an environment with high species complexity that is notoriously difficult to study with metagenomics.

Every metagenome is unique and requires specific consideration and analyses to adapt to its particular confines. Using today's molecular tool-kit together with the constant improvements in sequencing technologies and bioinformatics tools, the mining for CAZymes and novel enzymes is becoming more accessible and amendable. Data can be generated from a wider range of environments, providing a direct way to interrogate uncultivable phylotypes that constitute a microbial community and enable access to untapped sources of new and interesting CAZymes.

References

1. Morrison M et al (2009) Plant biomass degradation by gut microbiomes: more of the same or something new? Curr Opin Biotechnol 20:358–363
2. Warnecke F et al (2007) Metagenomic and functional analysis of hindgut microbiota of a wood-feeding higher termite. Nature 450:560–565
3. Hess M et al (2011) Metagenomic discovery of biomass-degrading genes and genomes from cow rumen. Science 331(6016):463–467
4. Pope PB et al (2010) Adaptation to herbivory by the Tammar wallaby includes bacterial and glycoside hydrolase profiles different to other herbivores. Proc Natl Acad Sci U S A 107:14793–14798
5. Liu J et al (2011) Cloning and functional characterization of a novel endo-beta-1,4-glucanase gene from a soil-derived metagenomic library. Appl Microbiol Biotechnol 89:1083–1092
6. Pope PB et al (2012) Metagenomics of the Svalbard reindeer rumen microbiome reveals abundance of polysaccharide utilization loci. PLoS One 7:e38571
7. Zhou Y et al (2016) A novel efficient β-glucanase from a paddy soil microbial metagenome with versatile activities. Biotechnol Biofuels 9:36
8. Ouwerkerk D et al (2005) Characterization of culturable anaerobic bacteria from the forestomach of an eastern grey kangaroo, *Macropus giganteus*. Lett Appl Microbiol 41:327–333
9. Naas AE et al (2014) Do rumen Bacteroidetes utilize an alternative mechanism for cellulose degradation? MBio 5:e01401–e01414
10. Zhou Y et al (2014) Omics-based interpretation of synergism in a soil-derived cellulose-degrading microbial community. Sci Rep 4:5288
11. Sims D et al (2014) Sequencing depth and coverage: key considerations in genomic analyses. Nat Rev Genet 15:121–132
12. Kuczynski J et al (2011) Using QIIME to analyze 16S rRNA gene sequences from microbial communities (Chapter:Unit). Curr Protoc Bioinformatics 10:7
13. Gilbert JA, Jansson JK, Knight R (2014) The Earth Microbiome project: successes and aspirations. BMC Biol 12:69
14. Yilmaz P et al (2011) Minimum information about a marker gene sequence (MIMARKS) and minimum information about any (x) sequence (MIxS) specifications. Nat Biotechnol 29:415–420
15. Yilmaz P et al (2011) The genomic standards consortium: bringing standards to life for microbial ecology. ISME J 5:1565–1567
16. Field D et al (2008) The minimum information about a genome sequence (MIGS) specification. Nat Biotechnol 26:541–547
17. Burke C, Kjelleberg S, Thomas T (2009) Selective extraction of bacterial DNA from

the surfaces of macroalgae. Appl Environ Microbiol 75:252–256

18. Delmont TO et al (2011) Metagenomic comparison of direct and indirect soil DNA extraction approaches. J Microbiol Methods 86:397–400
19. Rosewarne CP et al (2011) High-yield and phylogenetically robust methods of DNA recovery for analysis of microbial biofilms adherent to plant biomass in the herbivore gut. Microb Ecol 61:448–454
20. Denman SE et al (2015) Metagenomic analysis of the rumen microbial community following inhibition of methane formation by a halogenated methane analog. Front Microbiol 6:1087
21. Cardenas E et al (2015) Forest harvesting reduces the soil metagenomic potential for biomass decomposition. ISME J 9:2465–2476
22. Marine R et al (2014) Caught in the middle with multiple displacement amplification: the myth of pooling for avoiding multiple displacement amplification bias in a metagenome. Microbiome 2:3
23. Binga EK, Lasken RS, Neufeld JD (2008) Something from (almost) nothing: the impact of multiple displacement amplification on microbial ecology. ISME J 2:233–241
24. Bragg L, Tyson GW (2014) Metagenomics using next-generation sequencing. Methods Mol Biol 1096:183–201
25. Di Bella JM et al (2013) High throughput sequencing methods and analysis for microbiome research. J Microbiol Methods 95:401–414
26. Laehnemann D, Borkhardt A, McHardy AC (2016) Denoising DNA deep sequencing data-high-throughput sequencing errors and their correction. Brief Bioinform 17:154–179
27. Frank JA et al (2016) Improved metagenome assemblies and taxonomic binning using long-read circular consensus sequence data. Sci Rep 6:25373
28. Nagarajan N, Pop M (2013) Sequence assembly demystified. Nat Rev Genet 14:157–167
29. Peng Y et al (2012) IDBA-UD: a denovo assembler for single-cell and metagenomic sequencing data with highly uneven depth. Bioinformatics 28:1420–1428
30. Li D et al (2015) MEGAHIT: an ultra-fast single-node solution for large and complex metagenomics assembly via succinct de Bruijn graph. Bioinformatics 31:1674–1676
31. Nurk S et al (2016) MetaSPAdes: a new versatile de novo metagenomics assembler. arXiv:1604.03071
32. Scholz M, Lo CC, Chain PS (2014) Improved assemblies using a source-agnostic pipeline for MetaGenomic Assembly by Merging (MeGAMerge) of contigs. Sci Rep 4:e6480
33. Tsai YC et al (2016) Resolving the Complexity of Human Skin Metagenomes Using Single-Molecule Sequencing. MBio 7:e01948
34. Koren S et al (2012) Hybrid error correction and de novo assembly of single-molecule sequencing reads. Nat Biotechnol 30:693–700
35. Utturkar SM et al (2014) Evaluation and validation of de novo and hybrid assembly techniques to derive high-quality genome sequences. Bioinformatics 30:2709–2716
36. Chevreux B, Wetter T, Suhai S (1999) Genome Sequence Assembly Using Trace Signals and Additional Sequence Information. Comput Sci Biol 99:45–46
37. Eren AM et al (2015) Anvi'o: an advanced analysis and visualization platform for 'omics data. Peer J 3:e1319
38. Zhu Z et al (2013) MGAviewer: a desktop visualization tool for analysis of metagenomics alignment data. Bioinformatics 29:122–123
39. McHardy AC, Rigoutsos I (2007) What's in the mix: phylogenetic classification of metagenome sequence samples. Curr Opin Microbiol 10:499–503
40. Huson DH et al (2007) MEGAN analysis of metagenomic data. Genome Res 17:377–386
41. Gregor I et al (2016) PhyloPythiaS+: a self-training method for the rapid reconstruction of low-ranking taxonomic bins from metagenomes. Peer J 4:e1603
42. Dröge J, Gregor I, McHardy AC (2015) Taxator-tk: precise taxonomic assignment of metagenomes by fast approximation of evolutionary neighborhoods. Bioinformatics 31:817–824
43. Federhen S (2012) The NCBI Taxonomy database. Nucleic Acids Res 40:D136–D143
44. Teeling H et al (2004) TETRA: a web-service and a stand-alone program for the analysis and comparison of tetranucleotide usage patterns in DNA sequences. BMC Bioinformatics 5:163
45. Iverson V et al (2012) Untangling genomes from metagenomes: revealing an uncultured class of marine Euryarchaeota. Science 335:587–590
46. Wu YW et al (2014) MaxBin: an automated binning method to recover individual genomes from metagenomes using an expectation-maximization algorithm. Microbiome 2:26
47. Imelfort M et al (2014) GroopM: an automated tool for the recovery of population genomes from related metagenomes. Peer J 2:e409v1

48. Alneberg J et al (2014) Binning metagenomic contigs by coverage and composition. Nat Methods 11:1144–1146
49. Kang DD et al (2015) MetaBAT, an efficient tool for accurately reconstructing single genomes from complex microbial communities. Peer J 3:e1165
50. Albertsen M et al (2013) Genome sequences of rare, uncultured bacteria obtained by differential coverage binning of multiple metagenomes. Nat Biotechnol 31:533–538
51. Parks DH et al (2015) CheckM: assessing the quality of microbial genomes recovered from isolates, single cells, and metagenomes. Genome Res 25:1043–1055
52. Sczyrba A et al (2017) Critical assessment of metagenome interpretation—a benchmark of computational metagenomics software. bioRxiv 099127
53. Kunin V et al (2008) A bioinformatician's guide to metagenomics. Microbiol Mol Biol Rev 72:557–578
54. Zhu W, Lomsadze A, Borodovsky M (2010) Ab initio gene identification in metagenomic sequences. Nucleic Acids Res 38:e132
55. Lopez-Lopez O et al (2015) Metagenomics of an alkaline hot spring in Galicia (Spain): microbial diversity analysis and screening for novel lipolytic enzymes. Front Microbiol 6:1291
56. Mhuantong W et al (2015) Comparative analysis of sugarcane bagasse metagenome reveals unique and conserved biomass-degrading enzymes among lignocellulolytic microbial communities. Biotechnol Biofuels 8:16
57. Jimenez DJ, Chaves-Moreno D, van Elsas JD (2015) Unveiling the metabolic potential of two soil-derived microbial consortia selected on wheat straw. Sci Rep 5:13845
58. Tatusov RL, Koonin EV, Lipman DJ (1997) A genomic perspective on protein families. Science 278:631–637
59. Finn RD et al (2016) The Pfam protein families database: towards a more sustainable future. Nucleic Acids Res 44:D279–D285
60. Haft DH (2003) The TIGRFAMs database of protein families. Nucleic Acids Res 31:371–373
61. Bairoch A (2000) The ENZYME database in 2000. Nucleic Acids Res 28:304–305
62. Kanehisa M et al (2015) KEGG as a reference resource for gene and protein annotation. Nucleic Acids Res 43:1–6
63. Caspi R et al (2014) The MetaCyc database of metabolic pathways and enzymes and the BioCyc collection of Pathway/Genome Databases. Nucleic Acids Res 42:D459–D471
64. Marchler-Bauer A et al (2015) CDD: NCBI's conserved domain database. Nucleic Acids Res 43:D222–D226
65. Finn RD, Clements J, Eddy SR (2011) HMMER web server: interactive sequence similarity searching. Nucleic Acids Res 39:W29–W37
66. Edgar RC (2010) Search and clustering orders of magnitude faster than BLAST. Bioinformatics 26:2460–2461
67. Markowitz VM et al (2014) IMG/M 4 version of the integrated metagenome comparative analysis system. Nucleic Acids Res 42:D568–D573
68. Lombard V et al (2014) The carbohydrate-active enzymes database (CAZy) in 2013. Nucleic Acids Res 42:D490–D495
69. Cantarel BL et al (2009) The Carbohydrate-Active EnZymes database (CAZy): an expert resource for glycogenomics. Nucleic Acids Res 37:233–238
70. Yin Y et al (2012) dbCAN: a web resource for automated carbohydrate-active enzyme annotation. Nucleic Acids Res 40:W445–W451
71. Park BH et al (2010) CAZymes Analysis Toolkit (CAT): web service for searching and analyzing carbohydrate-active enzymes in a newly sequenced organism using CAZy database. Glycobiology 20:1574–1584
72. Rosewarne CP et al (2014) Analysis of the bovine rumen microbiome reveals a diversity of Sus-like polysaccharide utilization loci from the bacterial phylum Bacteroidetes. J Ind Microbiol Biotechnol 41:601–606
73. Martens EC et al (2009) Complex glycan catabolism by the human gut microbiota: The bacteroidetes Sus-like paradigm. J Biol Chem 284:24673–24677
74. Hemsworth GR et al (2014) Discovery and characterization of a new family of lytic polysaccharide monooxygenases. Nat Chem Biol 10:122–126
75. Liu Y et al (2006) An integrative genomic approach to uncover molecular mechanisms of prokaryotic traits. PLoS Comput Biol 2:e159
76. Korbel JO et al (2005) Systematic association of genes to phenotypes by genome and literature mining. PLoS Biol 3:e134
77. Lingner T et al (2010) Predicting phenotypic traits of prokaryotes from protein domain frequencies. BMC Bioinformatics 11:481
78. Feldbauer R et al (2015) Prediction of microbial phenotypes based on comparative genomics. BMC Bioinformatics 16(Suppl. 14):S1

79. Boser B, Guyon I, and Vapnik V (1992) A training algorithm for optimal margin classifiers. In: Fifth proceedings of the fifth annual workshop on computational learning theory, Pittsburgh, ACM, pp 144–152
80. Weimann A et al (2013) De novo prediction of the genomic components and capabilities for microbial plant biomass degradation from (meta-)genomes. Biotechnol Biofuels 6:24
81. Konietzny SG et al (2014) Inference of phenotype-defining functional modules of protein families for microbial plant biomass degraders. Biotechnol Biofuels 7:124
82. Weimann A et al (2016) From genomes to phenotypes: traitar, the microbial trait analyzer. mSystems 1:e00101–e00116
83. Wang A et al (2010) Enrichment strategy to select functional consortium from mixed cultures: consortium from rumen liquor for simultaneous cellulose degradation and hydrogen production. Int J Hydrogen Energy 35:13413–13418

Chapter 21

Identification of Genes Involved in the Degradation of Lignocellulose Using Comparative Transcriptomics

Robert J. Gruninger, Ian Reid, Robert J. Forster, Adrian Tsang, and Tim A. McAllister

Abstract

Lignocellulosic biomass represents an abundant, renewable resource that can be used to produce biofuels, low-cost livestock feed, and high-value chemicals. The potential of this resource has led to intensive research efforts to develop cost effective methods to breakdown lignocellulose. The efficiency with which the anaerobic fungi (phylum Neocallimastigomycota) degrade plant biomass is well recognized and in recent years has received renewed interest. Transcriptomics has been used to identify enzymes that are expressed by these fungi and are involved in the degradation of a range of lignocellulose feedstocks. The transcriptome is the entire complement of coding and noncoding RNA transcripts that are expressed by a cell under a particular set of conditions. Monitoring changes in gene expression can provide fundamental information about the biology of an organism. Here we outline a general methodology that will enable researchers to conduct comparative transcriptomic studies with the goal of identifying enzymes involved in the degradation of the plant cell wall. The method described here includes growth of fungal cultures, isolation and sequencing of RNA, and a basic description of data analysis for bioinformatic identification of differentially expressed transcripts.

Key words Transcriptomics, RNA-Seq, Carbohydrate active enzyme, Fungi, Neocallimastigomycota

1 Introduction

High-throughput genome sequencing provides detailed information about the complete composition of genes and genetic regulatory elements that are present in an organism or an environment. A limitation of these studies is that they do not provide information on what genes are being actively expressed under a particular set of conditions. Sequencing the transcriptome fills this knowledge gap and reveals all of the genes that are being expressed under a particular set of conditions. Importantly, altering growth conditions and monitoring the corresponding changes in gene expression enables researchers to understand how an organism or microbial population responds to environmental changes. Comparative transcriptomics has been a powerful tool for uncovering the mechanisms

D. Wade Abbott and Alicia Lammerts van Bueren (eds.), *Protein-Carbohydrate Interactions: Methods and Protocols*, Methods in Molecular Biology, vol. 1588, DOI 10.1007/978-1-4939-6899-2_21, © Springer Science+Business Media LLC 2017

that are employed by lignocellulose degrading microbes to break down the plant cell wall, and has been used to identify essential carbohydrate active enzymes (CAZymes) involved in this process [1, 2].

Anaerobic fungi within the phylum Neocallimastigomycota are known to be efficient degraders of lignocellulose and express a large contingent of CAZymes [3–5]. Many of these enzymes show low levels of sequence identity to proteins that have been characterized to date [4, 5]. Unlike with aerobic fungi, the oxidative enzymes involved in breaking down lignin and exposing cellulose and hemicellulose do not function in anaerobic environments [6]. The CAZyme cocktails that are produced by Neocallimastigomycota have been found to show similar levels of efficiency as the currently produced commercial enzyme preparations from *Trichoderma* and *Aspergillus* [5]. Due to the fastidious nature of the anaerobic gut fungi, commercial scale production of these enzymes has not yet become feasible. Transcriptomic studies of Neocallimastigomycota have led to the identification and characterization of a range of CAZymes that may allow for production of these enzymes using a system more amenable to industrial scale fermentation [4, 5].

In this chapter we outline a general methodology that will enable researchers to conduct comparative transcriptomic studies with the goal of identifying enzymes that are involved in the degradation of the plant cell wall. The method described here will include the growth of fungal cultures, isolation and sequencing of RNA, and a basic description of data analysis for bioinformatic identification of differentially expressed transcripts. Although this method will focus on anaerobic fungal cultures, this approach can also be employed to study other eukaryotes.

2 Materials

2.1 Cell Culture

1. Lowe's semi-defined anaerobic media with 1% w/v glucose, barley straw, or alfalfa hay.
2. Vacuum filter fitted with Büchner funnel.
3. Whatmann quantitative 50 fast flow filter paper.
4. Microfuge.
5. 2.0 mL Eppendorf tubes.

2.2 RNA Isolation and Purification

1. Mortar (400 mL capacity) and pestle.
2. Agilent 2100 Bioanalyzer.
3. MEGAclear Kit (Ambion Cat#: AM1908).
4. RNA 6000 Nano kit (Agilent Cat#: 5067-1511).
5. TRIzol reagent (Life Technologies Cat#: 15596026).
6. Liquid itrogen.

Table 1
Programs required for assembly and analysis of sequencing data

Package name	Minimum version	Source
Software required for de novo assembly		
Rcorrector	1.0.1	https://github.com/mourisl/Rcorrector/releases/latest
Skewer	0.2.1	https://sourceforge.net/projects/skewer/
SortMeRNA	2.1	https://github.com/biocore/sortmerna/releases
khmer	2.0	https://pypi.python.org/pypi/khmer
Megahit	1.0	https://github.com/voutcn/megahit/releases
Software required for transcriptome evaluation		
TransRate	1.0.2	https://github.com/blahah/transrate/releases
BUSCO	1.1b	http://busco.ezlab.org/
Software required for differential expression analysis		
Salmon	0.6.0	https://github.com/COMBINE-lab/salmon/releases
R	3.2.2	https://cran.r-project.org/

2.3 Strand-Specific Library Preparation

1. Spectrophotometer with fluorescence detection.
2. NanoDrop spectrophotometer.
3. High sensitivity DNA kit (Agilent Cat#: 5067-1504).
4. Quant-it PicoGreen dsDNA assay kit (Thermo Fisher Scientific Cat#: P7589).
5. TruSeq Stranded mRNA sample preparation kit (Illumina Cat#: RS-122-2201).
6. AMPure XP beads (Beckman Coulter Cat#: A63880).
7. Ambion Magnetic stand-96 (Thermo Fisher Scientific Cat#: AM10027).
8. SuperScript II Reverse Transcriptase (Invitrogen Cat#: 18064022).

2.4 Computational Analysis of Sequence Data

1. Computer with linux (PC) or unix (Mac OS) operating system with the programs listed in Table 1 installed. Minimum hardware requirements are: 32 Gb RAM, and eight processing cores.

3 Methods

3.1 Preparation of Anaerobic Media

1. Prepare enough liquid modified Lowe's media supplemented with 1% w/v carbon source to make 3 × 100 mL media bottles per carbon source (Table 2).
2. Combine media components as indicated in Table 2 and add water to a final volume of 950 mL.
3. Add 50 ml 8% w/v Na_2CO_3 to make final volume of media 1 L.

Table 2
Composition of Lowe's semi-defined media and solutions required to prepare this media. The final volume of all solutions is 1 L

Lowe's semi-defined media (per 1 L)	PO_4 solution	Macronutrient solution	Trace minerals solution[a]	Volatile fatty acid solution[b]	Vitamin mix
75 mL PO_4 solution	4.5 g KH_2PO_4	9 g KCl	0.25 g $MnCl_2$-$4H_2O$	6.85 mL Acetic acid	0.25 g 1,4-Naphthoquinone
55 mL Macronutrient solution		9 g NaCl	0.25 g $NiCl_2$-$6H_2O$	3 mL Propionic acid	0.2 g Calcium-d-panthothenate
10 mL Trace minerals solution		7.5 g $MgSO_4$-$7H_2O$	0.25 g $NaMoO_4$-$2H_2O$	1.85 mL Butyric acid	0.2 g Nicotinamide
10 mL Volitile fatty acid solution		3 g $CaCl_2$	0.25 g H_3BO_3	0.55 mL 2-Methylbutyric acid	0.2 g Riboflavin
10 mL 0.1% w/v Hemin[c]		8.1 g NH_4Cl	0.2 g $FeSO_4$-$7H_2O$	0.47 mL Isovaleric acid	0.2 g Thiamin
1 mL 0.1% w/v Resazurin[d]			0.05 g $CoCl_2$-$6H_2O$	0.55 mL n-Valeric acid	0.2 g Pyridoxine-HCl
10 mL Vitamin mix			0.07 g $NaSeO_3$		0.025 g Biotin
1 g Tryptone peptone			0.05 g NH_4VO_3		0.025 g Folic acid
0.5 g Yeast extract			0.025 g $ZnCl_2$		0.025 g Cyanocobalamin
1.5 g PIPES buffer			0.025 g $CuCl_2$-$2H_2O$		0.025 g *para*-Aminobenzoic acid
50 mL Clarified rumen fluid[e]					

[a]Trace mineral solution must be made using 0.2 M HCl to dissolve components
[b]Volatile fatty acids are added to 700 mL of 0.2 M NaOH and the pH is adjusted to 7.5 with 1 M NaOH. The solution is then diluted with water to a final volume of 1 L
[c]Hemin is dissolved in 5 mM NaOH at a concentration of 0.1% w/v
[d]Resazurin is dissolved in water at a concentration of 0.1% w/v
[e]Rumen fluid can be left out of media however anaerobic fungi tend to grow significantly better when this is included

4. Using a Hungate system on the benchtop (*see* **Note 1**), gently bubble reduced, anaerobic CO_2 into the media and bring media to a boil being careful that the media does not boil over. Add 0.1 g of cysteine to boiling media to fully reduce it.
5. Boil the media until it changes colour from pink to a pale yellow/clear.
6. Dispense 100 mL of reduced media (*see* **Note 2**) into culture bottle containing the appropriate amount of carbon source to make the final concentration 1% (w/v for solid substrates or v/v for liquid substrates).
7. Cap bottles containing reduced media and carbon source and autoclave (*see* **Note 3**).

3.2 Growth of Fungal Cultures

Grow anaerobic fungi using aseptic technique under strict anaerobic conditions. When conducting comparative transcriptomic experiments it is important to have at least two biological replicates so that statistical analysis can be carried out. If possible, three biological replicates are advised. This experimental design can be used for comparative analysis of the transcriptomes of anaerobic fungi grown with three different carbon sources: (1) glucose, (2) barley straw, (3) alfalfa hay (*see* **Note 4**). To determine basal gene expression, grow three cultures using glucose as the sole carbon source.

1. When media has cooled, inoculate cultures with fungal mycelia and incubate at 39 °C under anaerobic conditions for 72 h. Do not shake flasks (*see* **Note 5**).
2. Using a vacuum filter fitted with a Büchner funnel and Whatmann paper to carefully separate the liquid culture from the fungal mycelia and any insoluble material (*see* **Note 6**).
3. Carefully separate the fungal mycelia and associated growth substrate from culture supernatant and freeze in liquid nitrogen (*see* **Note 7**).

3.3 Extraction of RNA and Quality Control

The fungal mycelia penetrate insoluble cellulosic substrates such as straw or filter paper during fungal growth. This prevents separation of the mycelia from the growth substrate and necessitates the grinding of the entire mass.

1. Grind frozen mycelia to a fine powder with a mortar and pestle under liquid nitrogen. Ensure that mycelia remain frozen throughout the grinding procedure to prevent degradation of RNA.
2. Add TRIzol to ground mycelia using approximately 1 mL of TRIzol per 100 mg of ground mycelia. An insufficient volume of TRIzol will result in DNA contamination of isolated RNA.
3. Completely suspend ground mycelia in TRIzol. The TRIzol/fungal suspension can be stored in 1.0–1.5 mL aliquots and stored at −80 °C for later use, or RNA can be isolated from samples right away (*see* **Note 8**).

3.4 Isolation and Purification of Total RNA

The following two-step procedure for isolation and purification of total RNA routinely generates RNA suitable for RNA-seq analysis.

3.4.1 RNA Isolation

1. Purify total RNA from TRIzol/fungal suspensions according to the manufacturer's instructions (*see* **Note 9**).
2. Resuspend RNA pellet in 100 μL of RNase-free water and proceed to column purification of total RNA using MEGAclear kit (see **Note 10**).

3.4.2 Column Purification of RNA Using MEGAclear Kit (Ambion)

1. Add 350 μL of binding buffer (provided in kit) to 100 μL RNA from previous step and purify RNA according to manufacturer's instructions.
2. Elute RNA with 50 μL of elution buffer. Be sure to repeat the elution a second time so that a total of 100 μL of column cleaned RNA has been collected in the same collection tube. The final volume of purified RNA will be 100 μL (*see* **Note 11**).
3. Quality check and determination RNA concentration using an Agilent Bioanalyzer 2100 (Agilent) and 6000 RNA nano kit (Agilent) according to the manufacturer's instructions (*see* **Note 12**).

3.5 Generating a Strand-Specific Library

There are a large number of commercial and homemade methods available for users wishing to generate a library for RNA-seq experiments. Our lab uses the protocol for strand-specific library preparation optimized by Illumina, and the kits that are associated with this method. Due to our familiarity with this method, the protocol presented here will specifically refer to library preparation using the TruSeq Stranded mRNA sample preparation kit (Illumina). For additional information regarding this protocol users are encouraged to consult the Illumina documentation. Reagents that are not included will be identified. A flowchart showing the steps involved in library generation is shown in Fig. 1. Prior to starting any of the following steps, remove the appropriate buffers or reagents from the freezer/fridge and allow them to reach room temperature.

3.5.1 Purification of Poly-adenylated mRNA

The first component of generating a library for RNA sequencing of fungal mRNA is the isolation of poly-adenylated transcripts from total RNA, and the cleavage of these mRNA transcripts into fragments suitable for sequencing, typically 100–200 bp. Depending on the number of samples being processed you can use individual 0.3 mL PCR tubes, 0.3 mL 8-tube PCR strips or 0.3 mL 96-well PCR plates for setting up these reactions. Ensure that the sample tubes are compatible with your magnetic block.

1. Dilute RNA sample so that there is a total of 0.1–4 μg of total RNA in a final volume of 50 μL using RNAse free distilled water (*see* **Note 13**).

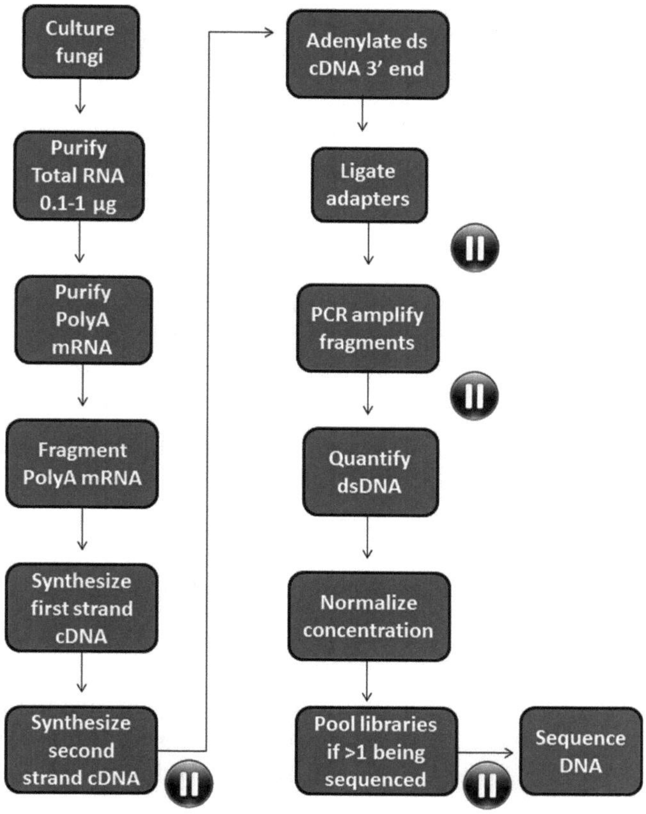

Fig. 1 Flowchart outlining steps involved in generating a strand-specific library for RNA-seq. *Pause symbols* indicate steps where the protocol can be safely paused and the products of that step stored at −20 °C for up to 1 week

2. Vortex RNA purification Oligo-dT beads (included in kit) to fully resuspend them and add 50 μL of beads to the diluted RNA sample. Gently pipette the whole solution up and down to completely mix beads and RNA, and then seal the tube(s)/plate.
3. Place sealed sample in a thermal cycler set to 65 °C and incubate for 5 min to denature RNA and facilitate binding to beads
4. Remove sample from thermal cycler and place in magnetic stand. Allow sample to sit in the stand at room temperature for 5 min to fully capture the magnetic Oligo-dT beads that are now bound to poly-adenylated mRNA. (*see* **Note 14**).
5. After 5 min, while the sample tube(s) is still in the magnetic block, carefully remove the supernatant.
6. Remove sample tube(s) from the magnetic block and resuspend beads in 200 μL of Bead Washing Solution (included in kit). Gently but thoroughly mix the beads and washing solution.
7. Place tube(s) into magnetic block and incubate at room temperature for 5 min to pellet beads. Carefully remove the supernatant leaving a bead pellet in the tube(s). The supernatant

can be discarded as it contains the unwanted ribosomal RNA and noncoding RNA.

8. Remove tube(s) from the magnetic block and resuspend the beads in 50 μL of Elution Buffer (included in kit).
9. Elute the polyadenylated mRNA by incubating the sealed tube(s) at 80 °C for 2 minutes in a thermal cycler. Cool the samples to 25 °C prior to removing them from the thermal cycler.

3.5.2 Fragmentation of Purified mRNA

1. To rebind the purified polyA mRNA, add 50 μL of Bead Binding Buffer (included in kit) to each sample tube and gently resuspend beads. Incubate the beads at room temperature for 5 min to allow for complete binding of mRNA.
2. Place tube(s) in magnetic block for 5 min to capture beads and then carefully remove supernatant.
3. Remove tube(s) from magnetic block and then gently but thoroughly resuspend beads in 200 μL of Bead Washing Solution (included in kit).
4. Capture beads in magnetic block for 5 min and remove supernatant.
5. Remove tube(s) from magnetic block and gently resuspend beads in each sample tube with 19.5 μL of Elute, Prime, Fragment mix (included in kit).
6. Seal tubes and fragment RNA by incubating in a thermal cycler at 94 °C for 8 minutes. Cool samples to 4 °C and remove from thermal cycler.
7. Proceed to synthesis of first strand cDNA. The Elute, Prime, Fragment mix contains random hexamer primers that will be used by reverse transcriptase in first strand synthesis to synthesize cDNA.

3.5.3 Synthesis of First Strand cDNA (See Note 15)

1. Place tube containing purified, fragmented RNA into magnetic block to capture beads. Remove 17 μL of supernatant containing the fragmented RNA and transfer it into a clean 0.3 mL PCR tube/strip/plate.
2. Remove First Strand Master Mix (included in kit) from −20 °C freezer and allow it to reach room temperature.
3. Mix the First Strand Master Mix (included in kit) and Superscript II Reverse Transcriptase (Invitrogen) in a ratio of 9:1
4. Add 8 μL of First Strand Master Mix with added reverse transcriptase to each sample tube and mix thoroughly but gently. Seal tube(s)/plate (*see* **Note 16**).
5. Place sealed tube(s)/plate into thermal cycler with lid preheated to 100 °C. Run the following program to synthesize first strand cDNA: 10 min, 25 °C; 50 min, 42 °C; 15 min, 70 °C; hold at 4 °C

3.5.4 Synthesis of Second Strand cDNA (See Note 17)

The next steps require fresh 80% DNAse/RNAse free ethanol and AMPure XP beads (Beckman Coulter). To make the ethanol solution mix 84 mL 95% ethanol with 16 mL of nuclease free water. AMPure XP beads must be thoroughly resuspend with a vortex and equilibrated to room temperature prior to use.

1. Add 20 μL of Second Strand Marking Master Mix (included in kit) and 5 μL of Resuspenion Buffer (included in kit) to each sample tube/well (these are the same tubes that were used in the first strand synthesis reaction). Mix thoroughly with a pipette and seal tube(s)/plate.
2. Place sealed tube(s)/plate in a thermal cycler set to 16 °C and incubate for 1 h. to synthesize second strand DNA.
3. Add 90 μL of AMPure XP beads to each sample tube/well and mix thoroughly by pipetting up and down 10×.
4. Incubate AMPure XP beads with samples for 15 min at room temperature and then capture beads using magnetic block for a minimum of 5 min.
5. Without disturbing pelleted AMPure XP beads, remove 135 μL of supernatant from each sample tube/well.
6. With samples still in magnetic block, add 200 μL of freshly prepared 80% ethanol to each tube/well without disturbing beads.
7. With sample tube(s)/plate still on the magnetic block, incubate beads in 80% ethanol for 30 s and then carefully remove supernatant without disturbing beads.
8. Repeat **steps 6** and 7 one more time so that the beads have been washed twice with 80% ethanol.
9. After removing the final ethanol wash allow the sample tube(s)/plate to air dry at room temperature for 15 min. Leave samples in the magnetic block while drying.
10. Add 17 μL of Resuspension Buffer (included in kit) to each tube/well and gently mix with beads by pipetting up and down 10×. Incubate mixture for 2 min at room temperature.
11. Place sample tube(s)/plate on magnetic block and capture beads for 5 min. Carefully transfer 15 μL of supernatant containing double stranded DNA into a new 0.3 mL PCR tube/strip/plate (*see* **Note 18**).

3.5.5 Adenylation of 3′ Ends (See Note 19)

1. Add 2.5 μL of Resuspension Buffer (included in kit) and 12.5 μL of A-Tailing Mix (included in kit) to each sample tube/well and mix by gently pipetting up and down 10×.
2. Seal tubes/plate and place into thermal cycler with lid preheated to 100 °C. Run the following program: 30 min, 37 °C; 5 min, 70 °C; hold at 4 °C.
3. Immediately proceed to adapter ligation when samples have reached 4 °C.

3.5.6 Ligation of Sample Indexing Adapters (See Note 20)

1. Add 2.5 μL of Resuspension Buffer, 2.5 μL of Ligation Mix (included in kit) and 2.5 μL of each Adapter (included in kit) to each sample tube/well. Mix well with pipette, seal samples and place in thermal cycler preheated to 30 °C for 10 min.
2. Remove sample from thermal cycler and add 5 μL of Stop Ligation Buffer (included in kit) to each sample tube/well.

3.5.7 Adapter Ligation Reaction Cleanup

1. Add 42 μL AMPure XP beads to each sample tube/well and mix by pipetting up and down 10×. Incubate at room temperature for 15 min and then capture AMPure XP beads by placing samples on a magnetic stand for 5 min.
2. Without removing tube(s)/plate from the magnetic stand or disturbing beads, carefully remove and discard 79.5 μL of supernatant.
3. Wash the captured beads with 200 μL of freshly prepared 80% ethanol for 30 s. Be careful not to disturb beads during washing.
4. Remove ethanol wash from tubes/wells and repeat wash step for a total of two ethanol washes.
5. After removing supernatant from second ethanol wash, allow the tubes/wells to air dry for 15 min at room temperature while samples are still in magnetic plate.
6. Remove the dried samples from magnetic block and add 52.5 μL of Resuspension Buffer (included in kit) and gently mix by pipetting until beads are fully resuspended. Incubate for 2 min and capture beads using magnetic stand.
7. Remove 50 μL of supernatant without disturbing beads and transfer into a fresh tube/well.
8. Perform a second cleanup step by adding 50 μL of AMPure XP beads to each sample tube/well and mix by pipetting up and down 10×. Incubate at room temperature for 15 min and then capture AMPure XP beads by placing samples on magnetic stand for 5 min.
9. Without removing plate from magnetic stand, carefully remove and discard 95 μL of supernatant without disturbing beads.
10. Wash beads a total of 2× with freshly prepared 80% ethanol and dry washed beads as described in **steps 3–5**.
11. Add 22.5 μL of Resuspension Buffer (included in kit) to each tube/well and gently pipette to resuspend beads. Incubate beads in resuspension buffer for 2 min at room temperature and then capture AMPure XP beads using magnetic stand.
12. Carefully transfer 20 μL of supernatant from each sample tube/well to a new sample tube (*see* **Note 18**).

3.5.8 Enrichment of DNA Fragments (See Note 21)

1. Add 5 μL of PCR primer cocktail (included in kit) to each sample tube/well containing the 20 μL of cleaned cDNA with adapters from the previous step.
2. Add 25 μL of PCR master mix (included in kit) to each sample tube/well and mix by pipetting up and down 10×.
3. Seal tubes/wells and place samples in thermal cycler with lid preheated to 100 °C. Run the following cycle. 30 s, 98 °C; 15× (10 s, 98 °C; 30 s, 60 °C; 30 s, 72 °C); 5 min, 72 °C; Hold at 10 °C.
4. Remove tubes from thermal cycler and add 50 μL of AMPure XP beads to 50 μL PCR amplified library.
5. Incubate for 15 min at room temperature and then capture AMPure XP beads with magnetic stand for 5 min.
6. Without removing tubes/plate from magnetic stand or disturbing beads, carefully remove and discard 95 μL of supernatant.
7. Without disturbing beads, wash with 200 μL of freshly prepared 80% ethanol for 30 s.
8. Remove ethanol wash from tubes/wells and repeat wash step for a total of two ethanol washes.
9. After removing second ethanol wash supernatant, allow tubes/wells to air dry for 15 min at room temperature while samples are still in magnetic plate.
10. Remove dried samples from magnetic block and add 32.5 μL of Resuspension buffer (included in kit) and gently mix by pipetting until beads are fully resuspended. Incubate for 2 min and capture beads using magnetic stand.
11. Remove 30 μL of supernatant without disturbing beads and transfer amplified library into a fresh tube/well (*see* **Note 18**).

3.5.9 Final Quality Control, Normalization, and Pooling of Libraries

1. Run 1 μL of the amplified library on a Bioanalyzer 2100 using the high sensitivity Agilent DNA 1000 chip to check the size and purity of the amplified fragments. The fragments should range between 200 and 400 bp with a median size of approximately 260 bp (*see* **Note 22**).
2. Quantify the concentration of DNA in the generated libraries using either qPCR or Quant-iT PicoGreen dsDNA quantification assay kit according to the manufacturer's instructions (Thermo Fisher Scientific).
3. Normalize the concentration of DNA in each library to 10 nM (*see* **Note 23**).
4. Pool the libraries from each sample by adding an equal volume of each library (*see* **Note 24**).
5. Sequence DNA with appropriate Illumina instrument (*see* **Note 25**).

3.6 De Novo Transcript Assembly and Data Processing

The following sections will provide users with the programs required to process and analyze the sequence data obtained from the sequence provider. Ensure that the software listed in Table 1 is installed and that the executables are on the system search PATH. We provide examples of the commands that are input using a command line interface to run each step. These commands may need to be modified to work with your system and the file names you have selected. Commands are preceded by the > symbol indicating separate command lines. This section assumes that users are familiar with unix-based operating systems and are comfortable working from the command line.

1. Create a working directory and copy the RNA-Seq read FASTQ files there. We will assume that the reads are paired-end and in files named R1.fastq.gz and R2.fastq.gz

```
> mkdir transcriptome.
> cd transcriptome.
> cp R*.fastq.gz
```

2. Correct read sequence errors using Rcorrector [7].

```
> mkdir ./Rcorrected
> perl run_rcorrector.pl -1 R1.fastq.gz -2 R2.fastq.gz -k 25 -ek 150000000 -o./Rcorrected -t 12 1> Rcorrector.log 2>&1
```

3. Discard reads with unfixable errors and their mates. A python script to filter out these reads is available on request.
4. Trim sequencing adapters from reads and remove low quality reads using Skewer [8] (*see* **Note 26**).

```
> mkdir -p ./Rcorrected/trimmed
> skewer -x Adapters/illumina_TruSeq_mRNA_adaptors.fa --mode tail --end-quality 2 --min 80 -t 12 -o ./Rcorrected/trimmed/correct ./Rcorrected/correct.1.fastq ./Rcorrected/correct.2.fastq
> rm ./Rcorrected/correct.?.fastq
```

5. Interleave read pairs.

```
> source activate khmerEnv.
> interleave-reads.py -o ./Rcorrected/trimmed/correct-trimmed-interleaved.fastq /Rcorrected/trimmed/correct-trimmed-pair1.fastq
> ./Rcorrected/trimmed/correct-trimmed-pair2.fastq
```

6. Remove ribosomal RNA reads using SortMeRNA [9] (*see* **Note 27**).

```
> mkdir -p Rcorrected/trimmed/rRNA
> mkdir -p Rcorrected/trimmed/non_rRNA
```

```
> RRNA_REF=$SORTMERNA/rRNA_databases/silva-
euk-18s-id95.fasta,$SORTMERNA/index/silva-
euk-18s-db:$SORTMERNA/rRNA_databases/
silva-euk-28s-id98.fasta,$SORTMERNA/index/
silva-euk-28s:$SORTMERNA/rRNA_databases/
rfam-5.8s-database-id98.fasta,$SORTMERNA/
index/rfam-5.8s-db:$SORTMERNA/rRNA_data-
bases/silva-bac-16s-id90.fasta,$SORTMERNA/
index/silva-bac-16s-db:$SORTMERNA/rRNA_
databases/silva-bac-23s-id98.
fasta,$SORTMERNA/index/silva-bac-23s-
db:$SORTMERNA/rRNA_databases/rfam-5s-data-
base-id98.fasta,$SORTMERNA/index/
rfam-5s-db:$SORTMERNA/rRNA_databases/silva-
arc-16s-id95.fasta,$SORTMERNA/index/silva-
arc-16s-db:$SORTMERNA/rRNA_databases/
silva-arc-23s-id98.fasta,$SORTMERNA/index/
silva-arc-23s-db
> sortmerna --ref. $RRNA_REF --reads. /
Rcorrected/trimmed/correct-trimmed-inter-
leaved.fastq --aligned. /Rcorrected/trimmed/
rRNA/rRNA --other. /Rcorrected/trimmed/non_
rRNA/non_rRNA --fastx --log --paired_in
--num_alignments 1 --blast 1
> rm. /Rcorrected/trimmed/correct-trimmed-
interleaved.fastq
```

7. Save the cleaned reads file using khmer [10] with the script.

```
> split-paired-reads.py --gzip -1 R1.
cleaned.fastq.gz -2 R2.cleaned.fastq.gz ./
Rcorrected/trimmed/non_rRNA/non_rRNA.fastq
```

8. Convert read files to FASTA format and remove excess reads from high-abundance transcripts using khmer [10] (*see* **Note 28**).

```
> fastq-to-fasta.py -n -o - ./Rcorrected/
trimmed/non_rRNA/non_rRNA.fastq
| normalize-by-median.py -k 25 -U 100000000
-M 30000000000 -C 20 -q -p
-o. /Rcorrected/trimmed/non_rRNA/inter-
leaved.C20.fasta -
> rm. /Rcorrected/trimmed/non_rRNA/non_
rRNA.fastq
```

9. Assemble reads with MEGAHIT [11] (*see* **Note 29**).

```
> megahit --min-count 3 --k-min 21 --k-max
91 --k-step 10 --prune-level 3 -t 12 --12
Rcorrected/trimmed/non_rRNA/combined.C20.
fasta -o Rcorrected/trimmed/non_rRNA/
megahit_out
```

10. Copy the assembly result to a new folder called transcripts.fasta

```
> cp. /Rcorrected/trimmed/non_rRNA/megahit_out/final.contigs.fa. /transcripts.fasta
```

11. Make a directory for evaluating transcriptome quality and copy the files in transcripts.fasta, R1.fastq.gz and R2.fastq.gz into it. Use cleaned reads if available.
12. Evaluate transcriptome accuracy using TransRate [12] (*see* **Note 30**)

```
> transrate --assembly=$PWD/transcripts.fasta --left $PWD/R1.fastq.gz --right R2.fastq.gz --output $PWD/transrating
```

13. Evaluate transcriptome completeness with BUSCO [13] (*see* **Note 31**).

```
> usearch -fastx_findorfs transcripts.fasta -orfstyle 7 -mincodons 50 -aaout orfs.faa
> python3 $BUSCO/BUSCO_v1.1b1.py -in orfs.faa -l $BUSCO/fungi -m OGS -o BUSCO-OGS
```

3.7 Identification of Differentially Expressed Transcripts.

Once sequencing data has gone through quality control checks and been assembled into contigs, a statistical analysis program can be used to identify genes that are either up or down regulated in response to changes in carbon source. There are several available programs but we will limit our description to the Bioconductor package DESeq2 [14].

1. Make a working directory and copy reads and transcript files there.

```
> mkdir expression
> cd expression
> cp R*.fastq.gz
> cp transcripts.fasta
```

2. Using a text editor, prepare a metadata.tsv file with a tab-delimited line for each RNA sample specifying the filename and experimental variables such as carbon source (Csource). An example is shown below.

```
Csource
R_G1    Glucose
R_G2    Glucose
R_G3    Glucose
R_A1    Alfalfa
R_A2    Alfalfa
R_A3    Alfalfa
R_B1    Barley Straw
R_B2    Barley Straw
R_B3    Barley Straw
```

3. Index the transcripts.

```
> salmon index -t transcripts.fasta -i transcripts_index --type quasi -k 31
```

4. Count both the paired-end and single-end reads mapping to each transcript in each sample.

(a) Paired-end reads in files named Sample_R1.fastq.gz and `Sample_R2.fastq.gz`:

```
> for f in *_R1.fastq.gz ; do salmon quant -i transcripts_index -l ISR -1 $f -2 ${f%1.fastq.gz}2.fastq.gz -o {f%_R1.fastq.gz}_quant -p 12 --useVBOpt; done
```

(b) Single-end reads in files named Sample.fastq.gz:

```
> for f in *.fastq.gz ; do salmon quant -i transcripts_index -l SR -r $f -o {f%.fastq.gz}_quant -p 12 --useVBOpt; dNext
```

5. Collect the read counts into a tab-delimited text file that can be used as input for DESeq2 in R. A Python script to collect the counts is available on request from the authors.
6. Start R.
7. Make DESeq2 available.

```
> library("DESeq2")
```

8. Load counts and metadata tables.

```
> counts <- read.delim("counts.tsv",header=T)
> metadata <- read.delim("metadata.tsv",header=T)
```

9. Calculate DESeq2 statistics assuming carbon source (Csource in the metadata.tsv file created in **step 14**) is the only experimental variable and Glucose is the control level of Csource.

```
> dds <- DESeqDataSetFromMatrix(countData = counts[,-1], colData = metadata, design = ~ Csource ).
> dds$Csource <- relevel(dds$Csource, "Glucose").
> dds <- DESeq(dds)
```

10. Extract results for transcripts that are differentially expressed with Alfalfa as Csource. This step should be repeated using Barley straw as Csource to identify transcripts differentially expressed with this substrate.

```
> DE.Alfalfa <- results(dds, contrast=c("Csource","Alfalfa","Glucose"))
> resOrdered <- DE.Alfalfa[order(DE.Alfalfa$padj),]
> resSignificant <- subset(resOrdered, padj < 0.1)
```

11. Save these results.

```
> write.table(as.data.frame(resSignificant),file='Alfalfa_vs_Glucose.tsv',sep='\t')
```

12. Exit from R.

```
> q()
```

13. Open Alfalfa_vs_Glucose.tsv in a spreadsheet to browse the transcripts differentially expressed in this condition (*see* **Note 32**).
14. Open Barleystraw_vs_glucose.tsv in a spreadsheet to browse transcripts differentially expressed in this condition.

4 Notes

1. The Hungate system is a specialized apparatus for preparing anaerobic media and can be used to enable researchers to work with anaerobic cultures and media on the bench-top while maintaining strict anaerobic conditions. Readers are referred to the review by Wolfe for details of setting up and working with a Hungate system [15].
2. Care must be taken to ensure that the media remains anaerobic. Resazurin is included in the media to serve as an oxygen indicator. The media will change colour from pale yellow/clear to pink if oxygen has been introduced into the media.
3. Some carbon sources such as glucose will caramelize if autoclaved for too long or at too high of temperature. Be sure to set the autoclave cycle appropriately to prevent this. It is also important to secure the cap on the bottle to prevent it from coming off during the autoclave cycle.
4. This experiment will examine whether the particular microbe being studied utilize unique mechanisms to degrade the cell wall of monocot (alfalfa hay) and dicot plants (barley). The growth of cultures on glucose, or another simple sugar such as fructose, will be used to determine baseline protein production and will serve as a reference to identify proteins associated with the digestion of the plant cell wall.
5. It is essential that cultures are grown under the same conditions and are harvested at the same growth stage.
6. The cell free growth media can be retained and used for proteomic analysis of extracellular proteins expressed to deconstruct the plant cell wall.
7. Retain the cell free culture media for proteomic analysis. See the accompanying Chapter 22. Be careful when removing liquid media so as not to lyse fungal cells. This will interfere with identification of secreted proteins.
8. TRIzol contains phenol and guanidine isothiocyanate. Care should be taken when working with this reagent. Always wear gloves, proper personal protection equipment, and work in a fume hood.

9. The sample will separate into a lower red organic phase and an upper colorless aqueous phase. All of the RNA will be in the aqueous phase. This should constitute roughly 50% of the volume in the tube.
10. Placing tube in a heating block at 50–60 °C will help dissolve pellet.
11. Carrying out both elution steps significantly improves yield. In our experience the concentration of the first and second elutions are similar so skipping this step will result in a significant decrease in overall yield.
12. A key determinant of the success or failure of a transcriptomic study is the quality of the RNA that is used as starting material. A commonly used metric to assess the extent of RNA degradation in a sample is the RNA integrity number, or RIN. Samples with a RIN value of 8–10 (A RIN of ten means that no degradation has occurred) are ideal for RNA-seq experiments. Using the approach for RNA extraction and purification outlined here our lab is consistently able to generate 5–10 μg of total RNA with RIN values >8 from 50 mL fungal cultures.
13. The TruSeq Stranded mRNA sample preparation kit has been optimized for use with 0.1–4 μg of total RNA. Using less than 0.1 μg of starting material may result in inefficient ligation of adapters and low yields.
14. Many of the steps involve the use of a magnetic stand that captures the magnetic beads that are used to purify the nucleic acids. Two potential stands are the Ambion Magnetic stand-96 (ThermoFisher) or the Agencourt SPRIPlate Super Magnetic Plate (Beckman Coulter).
15. Reverse transcriptase is required for this step. This enzyme is not included in the Illumina kit and must be purchased separately. As recommended by Illumina, our lab uses Superscript II Reverse Transcriptase from Invitrogen.
16. The First Strand Master Mix with reverse transcriptase is not sensitive to freeze-thaw cycles however, it is advisable to prepare small aliquots of this mixture to minimize the chances of contamination.
17. The product of the first strand synthesis is an RNA:DNA duplex. Synthesis of second strand cDNA involves digesting the template RNA with RNAse H and replacing it with DNA using DNA polymerase I. Strand specificity is achieved through the incorporation of dUTP in place of dTTP in the second strand. The product of this reaction is double stranded cDNA.
18. At this point in the protocol samples can be safely stored in sealed tubes/plates for a week at −20 °C.

19. This step adds a single adenine nucleotide to the 3′ end of the blunt fragments of end repaired, double stranded cDNA. This prevents the fragments from ligating to one another and facilitates ligation of fragments to adapters during the adapter ligation step.
20. Illumina indexing adapters are ligated to the blunt ends of the double stranded cDNA prepared in the previous steps. These adapters serve multiple purposes including (1) binding to the flow cell during sequencing, (2) allowing for PCR based enrichment of adapter labeled DNA fragments and (3) barcoding of samples when multiplexing several libraries into a single sequencing lane. The TruSeq Stranded mRNA sample preparation kit comes with 24 unique indexes. It is important to use the adapter combinations recommended by Illumina to ensure that colour balance is maintained during sequencing to minimize the risk of base calling errors in the index region of the transcripts.
21. This step uses polymerase chain reaction to selectively amplify cDNA fragments that have Illumina adapters ligated to both ends. Additionally, this step amplifies the amount of DNA in the library. The minimum number of amplification cycles should be used to prevent the introduction of amplification bias into the library.
22. The presence of significant amounts of adapter dimers can interfere with the sequencing reaction. If present, gel-purify the amplified library prior to pooling libraries.
23. The amount of DNA present in the library is a key determinant in obtaining optimal cluster density during sequencing. Use of too much or too little DNA can cause failure of the sequencing reaction.
24. This step can be skipped if only one sample is being run on the sequencer.
25. Most labs use commercial sequencing providers or core sequencing facilities for both library preparation and sequencing. The sequencing can be carried out using several instruments including the Illumina MiSeq, Next-Seq, or Hi-Seq. The exact protocol that is required for each of these instruments differs and users are encouraged to consult the Illumina website or your sequencing provider for details of this process. When choosing which sequencing approach to use for a study, consideration of the number of samples being analyzed, the complexity of the transcriptomes, and the experimental question being asked are important variables. A useful resource to help users design experiments and determine the most appropriate experimental approach is the "Standards, Guidelines and best practices for RNA-seq" prepared by the ENCODE

consortium in 2011 [16]. Users are also encouraged to consult the expertise of your service provider to identify the most appropriate approach.

26. Adapters/Illumina_TruSeq_mRNA_adaptors.fa should contain the sequences of the adapters used in the RNA-Seq library construction, in FASTA format; the -t parameter should be no more than the number of available processor cores.
27. SORTMERNA should be set to the directory where SortMeRNA is installed.
28. This command requires at least 30 GB of available RAM.
29. The --k-max parameter should be an odd number at least 9 less than the read length; the -t parameter should be no more than the number of available processor cores.
30. A summary of the transcriptome quality will be output in evaluation/transrating/assemblies.csv and quality statistics for each transcript will be output in evaluation/transrating/transcripts/contigs.csv.
31. Set BUSCO to the directory where BUSCO is installed. Set the -l parameter to an appropriate taxonomic group for which you have installed BUSCO data files. A summary of the results will be saved in evaluation/run_BUSCO-OGS/short_summary_BUSCO-OGS.
32. The functional significance of these differentially expressed transcripts can be assessed following functional annotation of the sequences.

References

1. Tsang A, Butler G, Powlowski J, Panisko EA, Baker SE (2009) Analytical and computational approaches to define the *Aspergillus niger* secretome. Fungal Genet Biol 46:S153–S160
2. Kolbusz MA, Di Falco M, Ishmael N, Marqueteau S, Moisan MC, da Silva BC, Powlowski J, Tsang A (2014) Transcriptome and exoproteome analysis of utilization of plant-derived biomass by *Myceliophthora thermophila*. Fungal Genet Biol 72:10–20
3. Gruninger RJ, Puniya AK, Callaghan TM, Edwards JE, Youssef N, Dagar SS, Fliegerova K, Griffith GW, Forster R, Tsang A, McAllister T, Elshahed MS (2014) Anaerobic fungi (phylum Neocallimastigomycota): advances in understanding their taxonomy, life cycle, ecology, role and biotechnological potential. FEMS Microbiol Ecol 90(1):1–17
4. Couger MB, Youssef NH, Struchtemeyer CG, Liggenstoffer AS, Elshahed MS (2015) Transcriptomic analysis of lignocellulosic biomass degradation by the anaerobic fungal isolate *Orpinomyces* sp. strain C1A. Biotechnol Biofuels 8:208
5. Solomon KV, Haitjema CH, Henske JK, Gilmore SP, Borges-Rivera D, Lipzen A, Brewer HM, Purvine SO, Wright AT, Theodorou MK, Grigoriev IV, Regev A, Thompson DA, O'Malley MA (2016) Early-branching gut fungi possess a large, comprehensive array of biomass-degrading enzymes. Science 351(6278):1192–1195
6. Pollegioni L, Tonin F, Rosini E (2015) Lignin-degrading enzymes. FEBS J 282(7): 1190–1213
7. Song L, Florea L (2015) Rcorrector: efficient and accurate error correction for Illumina RNA-seq reads. GigaScience 4:48
8. Jiang H, Lei R, Ding SW, Zhu S (2014) Skewer: a fast and accurate adapter trimer for next-generation sequencing paired-end reads. BMC Bioinformatics 15:182

9. Kopylova E, Noe L, Touzet H (2012) SortMeRNA: fast and accurate filtering of ribosomal RNAs in metatranscriptomic data. Bioinformatics 28:3211–3217
10. Crusoe MR, Alameldin HF, Awad S et al (2015) The khmer software package: enabling efficient nucleotide sequence analysis. F1000 Research 4:900
11. Li D, Liu CM, Luo R, Sadakane K, Lam TW (2015) MEGAHIT: an ultra-fast single-node solution for large and complex metagenomics assembly via succinct de Bruijn graph. Bioinformatics 31:1674–1676
12. Smith-Unna RD, Boursnell C, Patro R, Hibberd JM, Kelly S (2015) TransRate: reference free quality assessment of de-novo transcriptome assemblies bioRxiv 021626
13. Simao FA, Waterhouse RM, Ioannidis P, Kriventseva EV, Zdobnov EM (2015) BUSCO: assessing genome assembly and annotation completeness with single-copy orthologs. Bioinformatics 31:3210–3212
14. Love MI, Huber W, Anders S (2014) Moderated estimation of fold change and dispersion for RNA-seq data with DESeq2. Genome Biol 15:550
15. Wolfe R (2011) Techniques for cultivating methanogens. Methods Enzymol 494:1–22
16. Encode consortium. https://genome.ucsc.edu/ENCODE/protocols/dataStandards/ENCODE_RNAseq_Standards_V1.0.pdf

Chapter 22

Isolation and Preparation of Extracellular Proteins from Lignocellulose Degrading Fungi for Comparative Proteomic Studies Using Mass Spectrometry

Robert J. Gruninger, Adrian Tsang, and Tim A. McAllister

Abstract

Fungi utilize a unique mechanism of nutrient acquisition involving extracellular digestion. To understand the biology of these microbes, it is important to identify and characterize the function of proteins that are secreted and involved in this process. Mass spectrometry-based proteomics is a powerful tool to study complex mixtures of proteins and understand how the proteins produced by an organism change in response to different conditions. Many fungi are efficient decomposers of plant cell wall, and anaerobic fungi are well recognized for their ability to digest lignocellulose. Here, we outline a protocol for the enrichment and isolation of proteins secreted by anaerobic fungi after growth on simple (glucose) and complex (straw and alfalfa hay) carbon sources. We provide detailed instruction on generating protein fragments and preparing these for proteomic analysis using reversed phase chromatography and mass spectrometry.

Key words Exo-proteome, Fungi, CAZy, Proteomics, Lignocellulose, Mass spectrometry

1 Introduction

The efficiency with which the anaerobic fungi (phylum Neocallimastigomycota) degrade plant biomass is well recognized and in recent years has received renewed interest [1, 2]. Aerobic fungi utilize powerful oxidative enzymes to break down lignin and expose cellulose and hemicellulose [3]. This mechanism is not available to anaerobic fungi and these microbes have likely evolved a unique approach for breaking lignin-carbohydrate bonds. Extensive genomic and transcriptomic efforts have been undertaken and this work has revealed that these fungi utilize a large repertoire of carbohydrate-active enzymes (CAZY enzymes) to degrade the plant cell wall [4–6]. Many of these enzymes show low levels of sequence identity to proteins that have been characterized to date [4–6]. Fungi obtain nutrients from their environment by secreting a potent mixture of enzymes that digest carbohydrates, proteins, and lipids.

D. Wade Abbott and Alicia Lammerts van Bueren (eds.), *Protein-Carbohydrate Interactions: Methods and Protocols*, Methods in Molecular Biology, vol. 1588, DOI 10.1007/978-1-4939-6899-2_22, © Springer Science+Business Media LLC 2017

Identifying and characterizing secreted proteins is essential to fully understand the biology of these microbes. Proteomics is a powerful tool to study complex mixtures of proteins and understand how the proteins expressed by an organism change in response to different conditions. The proteome is an inherently complex system to analyze due to its dynamic nature. The composition of the proteome changes and is dependent on a number of factors such as cell cycle stage, metabolic state, and environmental conditions. Selecting a method that enables specific fractions of the proteome to be targeted for analysis is crucial to the design of successful proteomics experiments.

A combination of computational and experimental tools can be used to predict and identify secreted proteins. The commonly used bioinformatic program SignalP predicts whether a protein is secreted by identifying the presence of an N-terminal secretory signal peptide [7]. One of the primary drawbacks to this approach is the lack of selectivity in distinguishing extracellular proteins from resident proteins of the Golgi apparatus and endoplasmic reticulum. Consequently, many false positives may be generated when bioinformatically predicting extracellular proteins [8]. Proteomics studies examining the soluble extracellular fraction of microbial cultures by protein mass spectrometry can be used to provide strong direct evidence that detected proteins are indeed extracellular. A key aspect of proteomic identification of proteins is the ability to search peptide fingerprints against a database of known proteins. This requires the organism being studied, or a close homologue, to have a well-annotated genome and/or transcriptomes that can serve as this database.

In this chapter, we will provide a detailed protocol for qualitatively examining what effect varying the carbon source used to grow lignocellulose degrading fungi has on the composition of the exo-proteome. We will specifically provide information on the growth of cultures, isolation and concentration of secreted proteins, in solution trypsin digestion of proteins, and preparation of peptides for mass spectrometry analysis (Fig. 1). We will not go into step-by-step details of results interpretation as this analysis area is study dependent and there are a number of approaches that can be applied. Our laboratories work on the biochemical mechanisms of plant cell wall digestion in the phylum Neocallimastigomycota and these cultures will be used specifically as an example of proteome analysis. However, it should be noted that this approach can be used to examine the exo-proteome of any microbe grown in liquid culture.

2 Materials

2.1 Cell Culture

1. Lowes semidefined anaerobic media with appropriate carbon source.
2. Vacuum filter fitted with Büchner funnel.

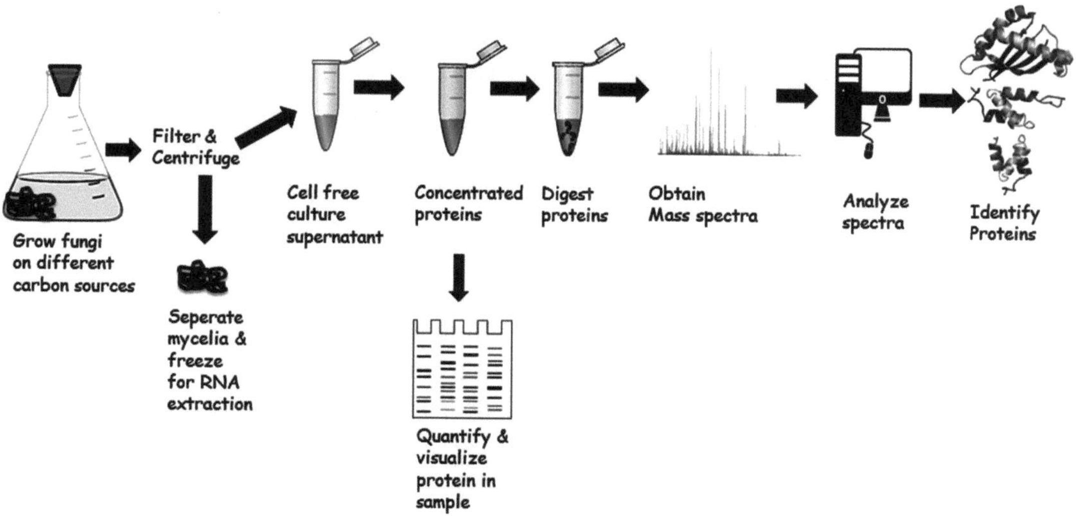

Fig. 1 Representation of the steps involved in the identification of proteins that are secreted during growth of fungi in liquid culture using a mass spectrometry-based proteomics approach

3. Whatmann quantitative 50 fast flow filter paper.
4. Microfuge.
5. 2.0 mL Eppendorf tubes.

2.2 Isolation and Enrichment of Secreted Proteins

1. SDS-PAGE gel and electrophoresis apparatus.
2. Sigma Brilliant Blue G (other Coomassie reagents for staining of total protein from other manufacturers can be substituted).
3. Biorad RC DC Protein assay kit (other methods for quantification of total protein or Bradford reagent from other manufacturers can be substituted).
4. 100% (w/v) trichloroacetic acid (TCA)—prepare solution by adding 340 mL of d_2H_2O to 500 g TCA (*see* **Note 1**).
5. Acetone (Mass spectrometry grade).
6. 100 mM Ammonium bicarbonate (NH_4HCO_3), pH 8.5—to make solution dissolve 7.906 g of NH_4HCO_3 in 950 mL of d_2H_2O, adjust pH to 8.5 and top up volume to 1000 mL.
7. Narrow range pH paper (pH range 8.0–9.0).

2.3 Alkylation and Trypsin Digestion

1. 0.1% (w/v) Anionic Acid Labile Surfactant (AALS) II in 100 mM NH_4HCO_3, pH 8.5—to make solution dissolve 5 mg of AALS II in 5 mL of previously prepared 100 mM NH_4HCO_3, pH 8.5.
2. 100 mM dithiothreitol (DTT) in 100 mM NH_4HCO_3, pH 8.5—to make solution dissolve 1.54 g of DTT in 100 mL of previously prepared 100 mM NH_4HCO_3, pH 8.5.

3. 500 mM Iodoacetamide in 100 mM NH_4HCO_3, pH 8.5—to make solution dissolve 0.925 g of Iodoacetamide in 10 mL of previously prepared 100 mM NH_4HCO_3, pH 8.5.
4. 100 ng/μL trypsin (mass spectrometry grade) in 50 mM acetic acid (add 0.0287 mL of glacial acetic acid to 9.97 mL of d_2H_2O to make 50 mM acetic acid solution) (*see* **Note 2**).
5. 1% (v/v) Trifluoroacetic acid (TFA)—add 1 mL of sequencing grade TFA to 99 mL of d_2H_2O to make 100 mL of 1% TFA.

2.4 Peptide Enrichment and Solvent Exchange

1. C18 ZipTips (EMD Millipore).
2. Acetonitrile (mass spectrometry grade).
3. 0.1% (v/v) formic acid (FA), 5% (v/v) acetonitrile—add 1 mL of 1% TFA, 0.5 mL of acetonitrile to 8.5 mL of d_2H_2O to make 10 mL of solvent.

2.5 Preparation of Anaerobic Media

1. Prepare enough liquid modified Lowes media supplemented with 1% w/v carbon source (1 g carbon/100 mL media) to make 3 × 100 mL media bottles per carbon source (*see* Table 1). Combine media components as indicated in Table 1 and add water to a final volume of 950 mL. Add 50 mL 8% w/v Na_2CO_3 to make final volume of media 1 L.
2. Using a Hungate system (*see* **Note 3**), gently bubble reduced, anaerobic CO_2 into the media and bring media to a boil being careful that the media does not boil over. At 0.1 g of cysteine to boiling media to fully reduce it.
3. Boil the media until it changes color from pink to a pale yellow/clear.
4. Dispense 100 mL of reduced media (*see* **Note 4**) into culture bottle containing the 1 g of appropriate carbon source to make the final concentration 1% (w/v for solid substrates or v/v for liquid substrates).
5. Cap bottles containing reduced media and carbon source and autoclave (*see* **Note 5**).

3 Methods

3.1 Growth of Fungal Cultures

Grow anaerobic fungi using aseptic technique under strict anaerobic conditions. When conducting comparative experiments it is important to have at least two biological replicates to allow statistical analysis to be carried out. If possible, three biological replicates are advised. This experimental design can be used for comparative analysis of the exo-proteome of anaerobic fungi grown using three different carbon sources: (1) glucose, (2) barley straw, (3) alfalfa hay (*see* **Note 6**). To determine basal protein secretion

Table 1
Components for preparing 1 L of modified Lowes semidefined anaerobic media without clarified rumen fluid or carbon source

Lowes semidefined media (per 1 L)	PO_4 solution	Macronutrient solution	Trace minerals solution[a]	Volatile fatty acid solution[b]	Vitamin mix
75 mL PO_4 solution	4.5 g KH_2PO_4	9 g KCl	0.25 g $MnCl_2$-$4H_2O$	6.85 mL Acetic acid	0.25 g 1,4-naphthoquinone
55 mL macronutrient solution		9 g NaCl	0.25 g $NiCl_2$-$6H_2O$	3 mL Propionic acid	0.2 g Calcium-d-panthothenate
10 mL Trace minerals solution		7.5 g $MgSO_4$-$7H_2O$	0.25 g $NaMoO_4$-$2H_2O$	1.85 mL butyric acid	0.2 g nicotinamide
10 mL Volatile fatty acid solution		3 g $CaCl_2$	0.25 g H_3BO_3	0.55 mL 2-methylbutyric acid	0.2 g riboflavin
10 mL 0.1% w/v hemin[c]		8.1 g NH_4Cl	0.2 g $FeSO_4$-$7H_2O$	0.47 mL isovaleric acid	0.2 g thiamin
1 mL 0.1% w/v resazurin[d]			0.05 g $CoCl_2$-$6H_2O$	0.55 mL n-valeric acid	0.2 g pyridoxine-HCl
10 mL vitamin mix			0.07 g $NaSeO_3$		0.025 g biotin
1 g tryptone peptone			0.05 g NH_4VO_3		0.025 g folic acid
0.5 g yeast extract			0.025 g $ZnCl_2$		0.025 g cyanocobalamin
1.5 g PIPES buffer			0.025 g $CuCl_2$-$2H_2O$		0.025 g para-aminobenzoic acid

[a]Trace mineral solution must be made using 0.2 M HCl to dissolve components
[b]Volatile fatty acids are added to 700 mL of 0.2 M NaOH and the pH is adjusted to 7.5 with 1 M NaOH. The solution is then diluted with water to a final volume of 1 L
[c]Hemin is dissolved in 0.005 M NaOH at a concentration of 0.1% w/v
[d]Resazurin is dissolved in water at a concentration of 0.1% w/v

by fungi, grow three cultures using glucose as the sole carbon source. To identify proteins that are secreted by the fungus to break down the plant cell wall, grow three cultures with barley straw and three cultures with alfalfa hay provided as the sole carbon source (*see* **Note 7**).

1. When media has cooled, inoculate cultures with fungal mycelia and incubate at 39 °C under anaerobic conditions for 72 h. Do not shake flasks (*see* **Note 8**).
2. Using a vacuum filter fitted with a Büchner funnel and Whatmann quantitative 50 fast flow filter paper (GE Lifesciences, Mississauga ON), carefully separate the liquid culture from the fungal mycelia and any insoluble material (*see* **Note 9**).
3. Further clarify the supernatant by centrifugation at 12,000–15,000 × *g* for 15 min at 4 °C. This cell-free culture supernatant contains the fungal proteins secreted during growth.
4. Carefully transfer supernatant without disturbing pellet and make several 1.5 mL tubes containing clarified cell-free culture supernatant. Freeze any tubes that will not be used right away for protein isolation which can be used as backups, technical replicates, or for additional experiments.

3.2 Isolation and Enrichment of Secreted Proteins

Contamination of samples with keratin can be minimized by always wearing gloves when handling material and samples, working in a laminar flow hood, using a dedicated set of pipettes and pipette tips, and filtering solutions when possible. Most detergents should be avoided as these can interfere with sample ionization and are difficult to remove from samples. All solutions should be freshly prepared using HPLC grade reagents and regularly checked to ensure they do not contain visible particles or fibers.

1. Quantify the total protein concentration in the supernatant using the RC DC protein assay kit (Biorad, Mississauga, ON) as described by the manufacturer. It is important to know how much protein is in your sample so that the correct amount of protease can be used to fragment the protein into peptides.
2. Run an SDS-PAGE gel to visually inspect the presence of protein bands after Coomassie staining the gel. If no protein bands are visible on the SDS-PAGE gel, concentrate proteins by TCA precipitation.

3.3 TCA Precipitation of Proteins in Solution: Reductive Alkylation of Free Cysteines

1. Mix one volume of 100% (w/v) TCA to four volumes of clarified cell-free culture supernatant and incubate for 10 min at 4 °C. A white precipitate should form.
2. Pellet precipitated protein by centrifugation at 10–15,000 × *g* for 10 min. Decant supernatant.
3. Wash pellet with 500 μL of chilled acetone (−20 °C). Break up the pellet by pipetting up and down.

4. Pellet protein by centrifugation at 10–15,000 × *g* for 10 min. Decant supernatant.
5. Repeat **steps 3** and **4** two more times for a total of three acetone washes.
6. Dry pellet in fume hood for 30–60 min. Residual acetone will reduce the ability to redissolve protein pellet.
7. Dissolve pellet in 0.1% (W/V) AALS II in 100 mM NH_4HCO_3 pH 8.5. Solubilizing TCA precipitated protein can be difficult and conditions may need to be optimized for different samples.

3.4 Reductive Alkylation of Free Cysteines

1. Resuspend an appropriate amount of sample material (5 μg of total protein) in 100 mM NH_4HCO_3, pH 8.5 buffer.
2. Check that protein sample is between pH 8.0 and 8.5 by spotting a small amount on pH paper and adjust if necessary with 1.0 M NH_4HCO_3.
3. Add an appropriate amount of AALS II solution to achieve a final concentration of 0.1%. This detergent facilitates trypsin digestion by denaturing proteins and has the advantage of not interfering with the subsequent LC-MS/MS analysis since it is broken down in acidic conditions.
4. Add the 100 mM DTT stock solution to samples to achieve a final concentration of 10 mM DTT (e.g., 10 μL of 100 mM DTT per 90 μL of sample). Incubate at room temperature for 30 min. This will reduce disulfide bonds in the protein.
5. Add the 500 mM iodoacetamide stock solution to fully reduced samples to achieve a final concentration of 50 mM (e.g., add 10 μL per 90 μL of reduced sample). Incubate in the dark at room temperature for 30 min.
6. Adjust concentration of DTT to 50 mM to quench remaining iodoacetamide.

3.5 Trypsin Digestion of Protein

1. Digest proteins into peptides by adding trypsin at a ratio of 1:50 trypsin:protein. Vortex sample to mix and incubate with gentle agitation at 37 °C for 4–18 h (*see* **Note 10**).
2. Stop protease cleavage by adding 5–10 μL of 1% (v/v) TFA to decrease the pH of sample to 2–3. Leave at room temperature for at least 30 min to allow for the degradation of the AALS II (*see* **Note 11**).

3.6 Sample Desalting and Buffer Exchange Using C18 Zip-Tip

Prior to analysis of samples by mass spectrometry, all salts must be removed and the sample placed in a mass spectrometry-compatible solvent system. Several chromatography based commercial systems are available to quickly desalt and buffer exchange protein samples. We will describe the use of ZipTips (EMD Millipore, Etobicoke, ON) for this purpose. ZipTips are pipette tips containing a small amount of C18 resin in the tip with binding capacity of up to 5 μg of protein. A similar product called Pierce C18 tip is available from Pierce (ThermoFishcer Scientific, Mississauga, ON) and has a binding capacity of 8–80 μg and can be used for samples with larger amounts of protein.

1. Place ZipTip on a standard 10-μL pipettor. Single-channel and multi-channel pipettors are compatible and either can be used to accommodate the required throughput for a particular experiment.
2. To bind sample, aspirate into ZipTip and pipette up and down several times. The protein/peptides will bind to the resin in the tip while the buffer salt components will not.
3. To exchange buffers, aspirate the ZipTip with bound protein/peptides into the mass spectrometry-compatible solvent. Wash the resin with several pipette tip volumes to fully exchange solvent.
4. Elute the resin-bound sample material directly into a fresh tube with a volume of high organic solvent. Typically, 80% Acetonitrile: 0.1% Formic acid (v/v) is used as eluent. The sample will be eluted from the resin in a volume of 1–4 μL.
5. Adjust the volume of the eluent solution such that the final acetonitrile concentration is no more than 5% (v/v).

If LC-MS/MS involves reversed phase separation of peptide on a C18 column, a suitable solvent system is 0.1% (v/v) formic acid, 5% (v/v) acetonitrile (*see* **Note 12**). The sample is now ready for analysis by mass spectrometry (*see* **Note 13**).

4 Notes

1. Always use personal protective equipment when working with TCA as it can cause chemical burns.
2. Pierce Trypsin Protease (ThermoFischer Sci) is recommended. This trypsin is highly purified, free from chymotrypsin, and has been chemically modified to enhance its stability and prevent autolysis during protease digestion.
3. The Hungate system is a specialized apparatus for preparing anaerobic media and can be used to enable researchers to work with anaerobic cultures and media on the bench-top while maintaining strict anaerobic conditions. Readers are referred to the review by Wolfe for details of setting up and working with a Hungate system [9].
4. Care must be taken to ensure that the media remains anaerobic. Resazurin is included in the media to serve as an oxygen indicator. The media will change color from pale yellow/clear to pink if oxygen has been introduced into the media.

5. Some carbon sources such as glucose will caramelize if autoclaved for too long or at too high of temperature. Be sure to set the autoclave cycle appropriately to prevent this. It is also important to secure the cap on bottle to prevent it from coming off during autoclave cycle.
6. Barley is a dicot and alfalfa is a monocot so this experiment will also provide an opportunity to examine whether these fungi utilize unique mechanisms to degrade the cell wall of monocot and dicot plants. The growth of cultures on glucose, or another simple sugar such as fructose, is used to determine baseline protein production and to serve as a reference to identify proteins that are produced to digest the plant cell wall.
7. The complex carbon sources included in the study can be substituted to address the particular goals of different end-users
8. It is essential that cultures be grown under the same conditions, and are harvested at the same growth stage. It is also important to be gentle when growing fungal cultures and preparing cell-free culture supernatant to minimize the chance of cell lysis and leakage of intracellular proteins into the media.
9. Transfer fungal mycelia into a separate falcon tube and freeze in liquid nitrogen. Retain this material for RNA isolation which can subsequently be used for comparative transcriptomic studies.
10. Trypsin is optimally active at pH 8.0 and it is advisable to check that the pH of your sample is in this range by spotting a small volume on pH paper. The ratio of protease to sample and time of digestion may need to be optimized if digestion is incomplete.
11. Trypsin digestion is only one approach for fragmenting proteins prior to mass spectroscopy. Other proteases such as LysC, chymotrypsin, AspN or chemicals such as cyanogen bromide can also be used to generate peptides [10].
12. Prior to injection of peptides into the mass spectrometer, peptides are bound to a C18 reversed phase column in an aqueous buffer and eluted using a linear gradient of increasing acetonitrile. To maximize peptide retention to the column, it is important to minimize the organic solvent content of the sample.
13. The protocol outlined here provides users with the tools to isolate the proteins secreted by a microbe when grown on a particular carbon source. This protocol can be applied to other experimental questions aimed at examining the effect of many growth variables on protein production in microbes. For most users, mass spectrometry analysis of samples will be carried out by commercial service providers, core facilities, or collaborating labs with expertise in proteomics. There are a range of approaches for the analysis of proteomic data and users are

encouraged to consult their service providers or collaborator to determine the most appropriate approach to sample analysis. The interpretation of results and their relevance to a particular biological system is study-dependent and should be validated through additional experimentation using complementary approaches

References

1. Gruninger RJ, Puniya AK, Callaghan TM, Edwards JE, Youssef N, Dagar SS, Fliegerova K, Griffith GW, Forster R, Tsang A, McAllister T, Elshahed MS (2014) Anaerobic fungi (phylum Neocallimastigomycota): advances in understanding their taxonomy, life cycle, ecology, role and biotechnological potential. FEMS Microbiol Ecol 90(1):1–17
2. Haitjema CH, Solomon KV, Henske JK, Theodorou MK, O'Malley MA (2014) Anaerobic gut fungi: advances in isolation, culture, and cellulolytic enzyme discovery for biofuel production. Biotechnol Bioeng 111(8):1471–1482
3. Youssef NH, Couger MB, Struchtemeyer CG, Liggenstoffer AS, Prade RA, Najar FZ, Atiyeh HK, Wilkins MR, Elshahed MS (2013) The genome of the anaerobic fungus *Orpinomyces* sp. strain C1A reveals the unique evolutionary history of a remarkable plant biomass degrader. Appl Environ Microbiol 79:4620–4634
4. Couger MB, Youssef NH, Struchtemeyer CG, Liggenstoffer AS, Elshahed MS (2015) Transcriptomic analysis of lignocellulosic biomass degradation by the anaerobic fungal isolate *Orpinomyces* sp. strain C1A. Biotechnol Biofuels 8:208
5. Solomon KV, Haitjema CH, Henske JK, Gilmore SP, Borges-Rivera D, Lipzen A, Brewer HM, Purvine SO, Wright AT, Theodorou MK, Grigoriev IV, Regev A, Thompson DA, O'Malley MA (2016) Early-branching gut fungi possess a large, comprehensive array of biomass-degrading enzymes. Science 351(6278):1192–1195
6. Pollegioni L, Tonin F, Rosini E (2015) Lignin-degrading enzymes. FEBS J 282(7): 1190–1213
7. Petersen TN, Brunak S, von Heijne G, Nielsen H (2011) SignalP 4.0: discriminating signal peptides from transmembrane regions. Nat Methods 8:785–786
8. Tsang A, Butler G, Powlowski J, Panisko EA, Baker SE (2009) Analytical and computational approaches to define the *Aspergillus niger* secretome. Fungal Genet Biol 46:S153–S160
9. Wolfe R (2011) Techniques for cultivating methanogens. Methods Enzymol 494:1–22
10. Swaney DL, Wenger CD, Coon JJ (2010) Value of using multiple proteases for large-scale mass spectrometry-based proteomics. J Proteome Res 9(3):1323–1329

ERRATUM TO

Chapter 5
Colorimetric Detection of Acetyl Xylan Esterase Activities

Galina Mai-Gisondi and Emma R. Master

D. Wade Abbott and Alicia Lammerts van Bueren (eds.), *Protein-Carbohydrate Interactions: Methods and Protocols*, Methods in Molecular Biology, vol. 1588, DOI 10.1007/978-1-4939-6899-2_5, © Springer Science+Business Media LLC 2017

DOI 10.1007/978-1-4939-6899-2_23

Tables 1 and 2 in Chapter 5 were originally published with incorrect molecular drawings. These have been corrected and the chapter references have been updated.

The online version of the updated original chapter can be found at http://dx.doi.org/10.1007/978-1-4939-6899-2_5

D. Wade Abbott and Alicia Lammerts van Bueren (eds.), *Protein-Carbohydrate Interactions: Methods and Protocols*, Methods in Molecular Biology, vol. 1588, DOI 10.1007/978-1-4939-6899-2_23, © Springer Science+Business Media LLC 2017

Table 1
Summary of reported methods for reaction termination and product detection in assays containing *p*NP-acetate, α-naphthyl acetate, and 4MUA

Substrate	Developing reagent	Product detection	Reference (s)
p-Nitrophenyl acetate (*p*NP-acetate) 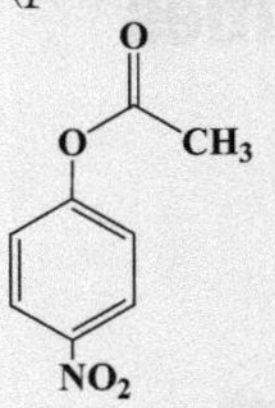	None	*p*-nitrophenol at 410 nm	[5, 9–12]
	None	*p*-nitrophenol at 420 nm	[6, 13–16]
	None	*p*-nitrophenol at 405 nm	[17–22]
	Sodium carbonate (Na_2CO_3)	*p*-nitrophenol at 405 nm	[23]
	none	Acetic acid detection kit from Boehringer Mannheim	[8]
α-naphthyl acetate 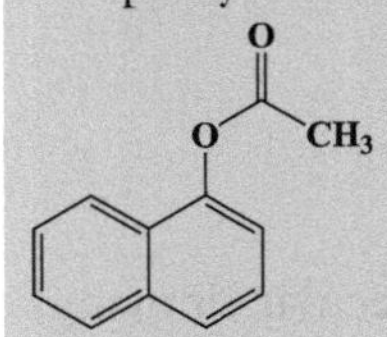	Fast Garnet GBC in sodium dodecyl sulfate; incubate at room temperature for 15 min	α-naphthol in complex with developing reagent at 560 nm	[16–18, 24]
	Fast Corinth V salt in sodium acetate buffer (pH 4.3) containing Tween 20; incubate at room temperature for 10 min	α-naphthol in complex with developing reagent at 535 nm	[25–27]
	none	α-naphthol directly at 321 nm	[28]
4-methylumbelliferyl acetate (4MUA) 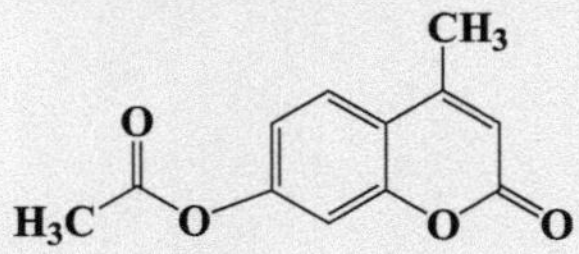	Citric acid to decrease pH to 2–3	4-methylumbelliferone (4-MU) at 354 nm	[7]

Table 2
Summary of noncommercial, synthesized carbohydrate analogs used to detect AcXE activity

Substrate	Reported approach to product detection	Reference(s)
Acetylated methyl β-D-xylopyranosides (e.g., 2,3,4-tri-O-acetylated methyl-β-xylopyranoside) 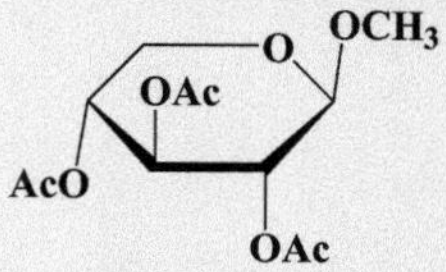 Acetylated xylobiose 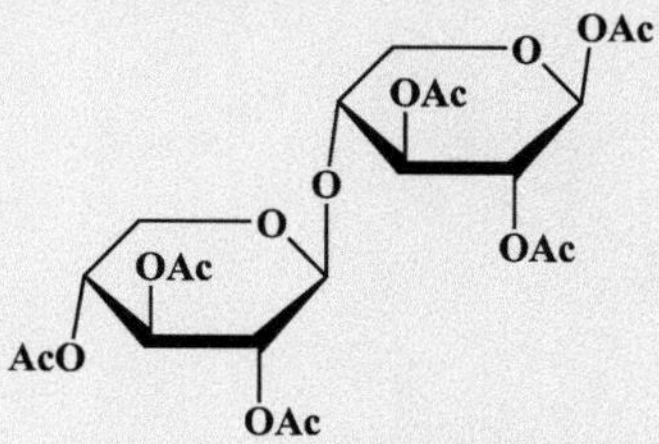 	• Gas-liquid chromatography (GLC-MS) • Acetic acid detection with K-ACETRM acetic acid kit from Megazyme • TLC detection of released sugars • ^{1}H NMR determination of regioselectivity	[2, 9, 29–31]
Deoxy and fluoro derivatives of methyl β-D-xylopyranoside diacetates (e.g., 2-deoxy-2-fluoro-3,4-diacetylated methyl β-D-xylopyranoside) 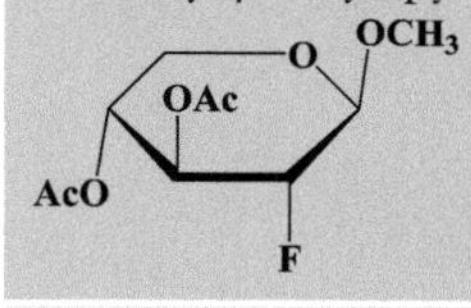 	• TLC detection of released sugars	[32, 33]
Monoacetylated *p*-nitrophenyl β-D-xylopyranosides (e.g., 2-O-acetyl nitrophenyl β-D-xylopyranoside) 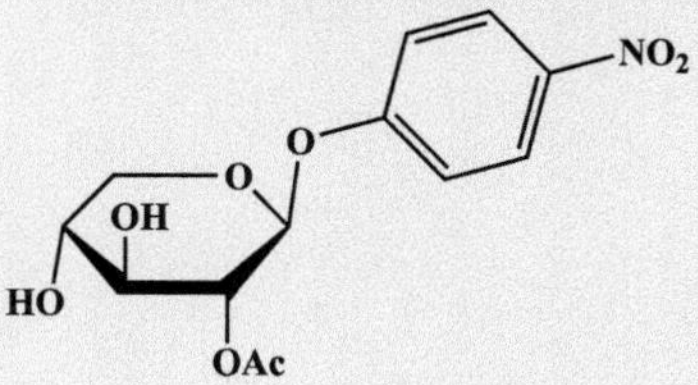 	Addition of $Na_2B_4O_7$ (or N_2CO_3) followed by *p*-nitrophenol detection at 405 nm	[34–36]

INDEX

D. Wade Abbott and Alicia Lammerts van Bueren (eds.), *Protein-Carbohydrate Interactions: Methods and Protocols*, Methods in Molecular Biology, vol. 1588, DOI 10.1007/978-1-4939-6899-2, © Springer Science+Business Media LLC 2017

MIX
Papier aus verantwortungsvollen Quellen
Paper from responsible sources
FSC® C105338

If you have any concerns about our products,
you can contact us on
ProductSafety@springernature.com

In case Publisher is established outside the EU,
the EU authorized representative is:
Springer Nature Customer Service Center GmbH
Europaplatz 3, 69115 Heidelberg, Germany

Printed by Libri Plureos GmbH
in Hamburg, Germany